Satellite Remote Sensing of Polar Regions

Polar Research Series
edited by Bernard Stonehouse

L.W. Brigham, **The Soviet Maritime Arctic**
C. Harris and B. Stonehouse, **Antarctica and Climatic Change**
R.A. Massom, **Satellite Remote Sensing of Polar Regions**

other titles are in preparation

Satellite Remote Sensing of Polar Regions

Applications, Limitations and Data Availability

ROBERT MASSOM

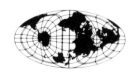

Belhaven Press, London
in association with the
Scott Polar Research Research Institute,
University of Cambridge

Lewis Publishers
2000 Corporate Blvd., N.W.
Boca Raton, FL 33431

First published in Great Britain in 1991 by
Belhaven Press (a division of Pinter Publishers),
25 Floral Street, London WC2E 9DS

British Library Cataloguing in Publication Data
A CIP catalogue record for this book is available from
the British Library

ISBN 1 85293 179 5

First published in the United States in 1991 by Lewis Publishers.

Direct all enquiries to CRC Press, Inc., 2000 Corporate Blvd., N.W.,
Boca Raton, Florida 33431

Library of Congress Cataloging-in-Publication Data
A CIP catalog record for this book is available from
the Library of Congress

ISBN 0 87372 607 8

Typeset by Florencetype Limited, Kewstoke, Avon
Printed and bound by SRP, Exeter

Contents

Acknowledgements

I am deeply indebted to Dr Gareth Rees of SPRI for both helping to knock something rather incongruous into some kind of shape and for his constant encouragement, and to Dr Bernard Stonehouse of SPRI for his excellent editorial skills and the use of his PC. Grateful thanks are due to my colleagues at the Oceans and Ice Branch of the NASA Goddard Space Flight Center, Greenbelt, Maryland, in particular Robert Bindschadler, Jay Zwally, Don Cavalieri, Claire Parkinson, Dave Eslinger, Per Gloersen, Dan Endres, Bob Kirk, Wayne Esaias (for use of his PC, amongst other things), Chuck McClain, Tony Liu and, last but not least, Joey Comiso, whose kindness, patience and stimulating conversation I shall never be able to fully repay.

I would also like to thank the following: Mark Drinkwater of JPL for his willingness to help out at all times and his knowledge of matters active; Jane Ferrigno of USGS; Larry Harding of the University of Maryland; Seymour Laxon of MSSL; Craig Lingle of Geophysical Institute, University of Alaska; Vernon Squire (formerly of SPRI, now of the Department of Mathematics and Statistics, University of Otago, New Zealand); Charles Swithinbank of BAS/SPRI; Ben Holt of JPL; Bob Thomas of NASA HQ; Dave Crane, Norman Davis and Phil Rottier of SPRI; Steve Ackley of CRREL; Jeff Hawkins of NOARL; David Bromwich of the Byrd Polar Center; Nick Digirolamo of SSAI; Mike Darzi of GSC; and my friends of STX: Rico Allegrino, Nick Weiss, Tim Seiss, Mike Martino, Steve Fiegles, Ted Scambos, Patricia Vornberger, and Jamila Saleh.

This work was begun at the Scott Polar Research Institute, University of Cambridge, England with support from the UK Natural Environment Research Council; it was completed while I held a US National Research Council Resident Research Associateship at the Oceans and Ice Branch, NASA GSFC. Grateful thanks are due to these organisations. Further thanks are due to Peter Wadhams, Director of SPRI, and Tony Busalacchi, Head of the Oceans and Ice Branch. Last but not least, I am eternally grateful to Iain Stevenson, Sarah Henderson and Jane Evans of Belhaven Press for all of their efforts in helping the book to materialise from the scraps of paper that I sent them. Special thanks to Kathy Whitefield.

This book is dedicated to my parents.

Acronyms, abbreviations and mathematical symbols

Remote sensing abounds in acronyms (which are pronounced as though they were words, such as RADAR) and abbreviations (such as AVHRR, which is impossible to pronounce), and it is useful to present a list of those which occur in this monograph and others which the reader is likely to encounter. Some of these abbreviations are unique, or almost so, in the sense that (for example) when AGC is written in the English-speaking world it will almost always have the meaning here ascribed to it. Others are local, proprietorial or 'nonce-abbreviations' (an obvious example is FY, meaning very different things to an accountant and to a sea-ice researcher) which may or may not have found universal acceptance.

AB	Antenna Beamwidth	AMRIR	Advanced Medium Resolution Imaging Radiometer (NOAA)
ADEOS	Advanced Earth Observation Satellite (Japan)	AMSR	Advanced Mechanically Scanning Radiometer (EOS)
ADCLS	Advanced Data Collection and Location System	AMSU	Advanced Microwave Sounding Unit (NOAA 'Next' and EOS)
ADRAMS	Air Droppable Random Access Measurement System		
AEM	Applications Explorer Mission (USA, eg HCMM)	APL	Applied Physics Laboratory (Johns Hopkins University, USA)
AES	Atmosphere Environment Service (Canada)	APT	Automatic Picture Transmission (ESSA, ITOS and NOAA)
AGC	Automatic Gain Control		
AIDJEX	Arctic Ice Dynamics Joint Experiment	ARAMP	Arctic Remote Autonomous Measurement Platform
AIRS	Atmospheric Infrared Sounder (EOS)	ASF	Alaska SAR Facility, Fairbanks
ALDCS	Advanced Location and Data Collection System	ASTER	Advanced Spaceborne Thermal Emission and Reflection (EOS)
ALS	Advanced LANDSAT Sensor (LANDSAT 7)	ATN	Advanced TIROS-N (USA)
ALT	Radar Altimeter	ATS	Applications Technology Satellites (NASA)
AMI	Active Microwave Instrument (ERS–1)	ATSR-M	Along-Track Scanning Radiometer-Microwave (ERS–1)
AMIR	Advanced Microwave Imaging Radiometer	AVCS	Advanced Vidicon Camera Subsystem (ESSA and NIMBUS–1 and –2)
AMR	Advanced Microwave Radiometer		

AVHRR	Advanced Very High Resolution Radiometer (TIROS–N)	DAAC	Distributed Active Archive Center (EOSDIS)
AVNIR	Advanced Visible and Near Infrared Radiometer (ADEOS)	DAF	Data Access Facility (NASA)
		DAPP	Data Acquisition and Processing Program (now DMSP)
AWS	Automatic Weather Station	dB	Decibel
		DCP	Data Collection Platform
B&W	Black and White	DCS	Data Collection System
BIL	Band Interleaved by Line	DFVLR	Deutsche Forschungs- und Versuchsanstalt für Luft- und Raumfahrt
BIP-2	Band-Interleaved-by-Pixel-Pairs		
BL	Bilinear (resampling)		
BNSC	British National Space Centre	DMSP	Defense Meteorological Satellite Program (USA)
bpi	bits per inch		
bps	bits per second	DOD	Digital Optical Disc
BRDF	Bidirectional Reflectance-Distribution Function	DOMSAT	Domestic Communications Satellite (USA)
BSQ	Band Sequential	DORIS	Détermination d'Orbite et Radiopositionnement Intégré par Satellite
BUV	Backscatter Ultraviolet Spectrometer (NIMBUS–4)		
		DPT	Delayed Picture Transmission
CBT	Calibrated Brightness Temperature		
CC	Cubic Convolution (resampling)	EDC	EROS Data Center (NASA)
CCD	Charge-Coupled Device	EDIS	Environmental Data and Information Service (USA)
CCRS	Canada Centre for Remote Sensing		
		EEMS	European Environment Monitoring Satellite (ERS–2)
CCT	Computer Compatible Tape		
CDA	Command and Data Acquisition	EGA	Enhanced Graphics Adaptor
		EIFOV	Effective Instantaneous Field of View
CDMS	Cryospheric Data Management System (NSIDC, USA)		
		EMR	Electro-magnetic Radiation
CD-ROM	Compact Disc-Read Only Memory	EMSS	Emulated Multi-Spectral Scanner (LANDSAT)
CEAREX	Coordinated Eastern Arctic Research Experiment	EODC	Earth Observation Data Centre (Farnborough, UK)
CERES	Clouds and Earth's Radiant Energy System (EOS)	EOS	Earth Observing System (USA)
		EOSAT	Earth Observation Satellite Company (USA)
CMB	Composite Minimum Brightness	EOSDIS	Earth Observing System Data and Information System (EOS)
CMT	Composite Maximum Temperature		
		EOSP	Earth Observing Scanner Polarimeter (EOS)
CNES	Centre National d'Études Spatiales (France)		
		EPOP	European Polar Platform (EOS)
CPSST	Cross-Product Sea Surface Temperature (NOAA)	ERB	Earth Radiation Budget (NIMBUS–6 and –7)
CPU	Central Processing Unit		
CRREL	Cold Regions Research and Engineering Laboratory (USA)	ERBE	Earth Radiation Budget Experiment (NIMBUS)
		ERBS	Earth Radiation Budget Satellite (USA)
CZCS	Coastal Zone Color Scanner (NIMBUS–7)		
		ERE	Effective Resolution Element

EREP	Earth Resources Experiment Program (SKYLAB)		GMS	Geostationary Meteorological Satellite (Japan)
ERIM	Environmental Research Institute of Michigan		GMT	Greenwich Mean Time
ERM	Exact Repeat Mission (GEOSAT)		GOES	Geostationary Operational Environmental Satellite (USA)
ERS–1	European Remote Sensing Satellite–1		GOME	Global Ozone Monitoring Experiment (ERS–2)
ERSDAC	Earth Resources Satellite Data Analysis Center (Japan)		GOSSTCOMP	Global Operational Sea Surface Temperature Computation
ERS–DC	ERS–Data Centre (Farnborough, UK)		GPS	Global Positioning System
ERTS	Earth Resources Technology Satellite (LANDSAT, USA)		GR	Ground Resolution
			GRS	Geophysical Research Mission
ESA	European Space Agency		GSFC	Goddard Space Flight Center (NASA)
ESMR	Electrically Scanning Microwave Radiometer (NIMBUS–5 and –6)		H	Horizontal polarisation
			HBR	High Bit Rate
ESSA	Environmental Science Services Administration (USA)		HCMM	Heat Capacity Mapping Mission (USA)
ETM	Enhanced Thematic Mapper (LANDSAT)		HCMR	Heat Capacity Mapping Radiometer (HCMM)
EUMETSAT	European Meteorological Satellite Organization		HDDT	High Density Digital Tape
			HH	Horizontally polarized transmit, Horizontally polarised receive
FD	Fast Delivery		HIMSS	High-Resolution Microwave Spectrometer Sounder (EOS)
FFT	Fast Fourier Transform			
FGGE	First GARP Global Experiment		HIRDLS	High-Resolution Dynamics Limb Sounder (EOS)
Flop	Floating point operation per second		HIRIS	High-Resolution Imaging Spectrometer (EOS–A)
FM	Frequency modulation		HIRS	High-Resolution Infrared Sounder (NIMBUS–6)
FNOC	Fleet Numerical Oceanography Center (USA)		HMMR	High-Resolution Multifrequency Microwave Radiometer (EOS)
FOV	Field of view			
FPR	Flat Plate Radiometer (ESSA)			
FY	First-year sea ice		HR	High-Resolution
			HRIR	High-Resolution Infrared Radiometer (NIMBUS–1, –2 and –3)
GAC	Global Area Coverage			
GARP	Global Atmospheric Research Programme			
GCP	Ground Control Point		HRPT	High-Resolution Picture Transmission (NOAA)
GDR	Geophysical Data Record		HRV	High-Resolution Visual (SPOT)
GEOS–3	Geodetic Earth Observation Satellite (USA)		HRVIR	High-Resolution Visible and Infra-Red sensor (SPOT–4)
GEOSAT	Geodynamic Experimental Ocean Satellite (USA)		HV	Horizontally polarised transmit, Vertically polarised receive
GFD	Geometric Footprint Dimension		Hz	Hertz
GHz	Gigahertz			
GIS	Geographic Information System		ICEC	Ice Centre Environment Canada
GLRS	Geoscience Laser-Ranging System (EOS)		ICSU	International Council of Scientific Unions
GM	Geodetic Mission (GEOSAT)		ID	Identification (number)

IDCS	Image Dissector Camera System (NIMBUS–3 and –4)	LIDAR	Light Detection and Ranging
		LIMEX	Labrador Ice Margin Experiment
IDIAS	Ice Data Integration and Analysis System (Canada)	LIMS	Limb Infrared Monitoring of the Stratosphere (NIMBUS–7)
IDR	Ice Data Record		
IFOV	Instantaneous Field of View		
IGARSS	International Geoscience and Remote Sensing Symposium	LISS	Linear Imaging Self-Scanning sensor (IRS)
IGBP	International Geosphere–Biosphere Programme	LRIR	Limb Radiance and Inversion Radiometer (NIMBUS–6)
INSAT	Indian National Satellite	LRR	Laser Retro-Reflector (ERS–1)
IRIS	Infrared Interferometer Spectrometer (NIMBUS–3 and –4)	LST	Local Standard Time
		Mbps	Megabits per second
IRLS	Interrogation, Recording and Location System (NIMBUS–3 to –6)	MCSST	Multi-Channel Sea Surface Temperature (NOAA)
		MD	Master Directory (NASA)
IRS	Indian Remote-sensing Satellites	MESSR	Multispectral Electronic Self-Scanning Radiometer (MOS–1 and –1b)
ISCCP	International Satellite Cloud Climatology Project		
ISRO	Indian Space Research Organization	METEOR	Meteorological satellites (USSR)
		MHz	Megahertz
ITIR	Intermediate Thermal Infrared Radiometer (EOS)	MIMR	Multi-frequency Imaging Microwave Radiometer (EOS)
ITOS	Improved TIROS Operational Satellites (USA)	MISR	Multi-angle Imaging Spectro-Radiometer (EOS)
ITPR	Infrared Temperature Profiling Radiometer (NIMBUS–5)	MIZ	Marginal Ice Zone
		MIZEX	Marginal Ice Zone Experiment
JERS–1	Japanese Earth Resources Satellite–1	MLA	Multispectral Linear Array
		MLS	Microwave Limb Sounder (EOS)
JIC	Joint Ice Center (Suitland, USA)		
JPL	Jet Propulsion Laboratory (NASA)	μm	micron
		MODIS-N/-T	Moderate-Resolution Imaging Spectrometer-Nadir/-Tilt (EOS)
JPOP–1	Japanese Polar Platform–1		
LAC	Local Area Coverage	MOMS	Modular Opto-electronic Multi-spectral Scanner (RADARSAT)
LAGEOS	Laser Geodynamics Satellite (USA/Italy)		
LASA	Lidar Atmospheric Sounder and Altimeter (EOS)	MOPITT	Measurements of Pollution in The Troposphere (EOS)
LASER	Light Amplification by Stimulated Emission of Radiation	MOS–1	Marine Observation Satellite (Japan)
		MRIR	Medium-Resolution Infrared Radiometer (NIMBUS–3)
LAWS	Laser Atmospheric Wind Sounder (EOS)	MS-DOS	MicroSoft-Disc Operating System
LBR	Low Bit Rate		
LCC	Lambert Conformal Conical (projection)	MSR	Microwave Scanning Radiometer (MOS–1 and –1b)
LEO	Low Earth Orbit		
LFMR	Low Frequency Microwave Radiometer (N–ROSS)	MSS	Multi-Spectral Scanner (ERTS/LANDSAT)

MSU	Microwave Sounder Unit (NOAA)		Remote Ocean Sensing System (USA)
MTF	Modulated Transfer Function	NRSC	National Remote Sensing Centre (UK)
MY	Multi-Year sea ice		
		NSCAT	NASA Radar Scatterometer (ADEOS)
NAS	National Academy of Sciences (USA)	NSF	National Science Foundation (USA)
NASA	National Aeronautics and Space Administration (USA)		
NASDA	National Space Development Agency (Japan)	NSIDC	National Snow and Ice Data Center (USA)
NAVSTAR	Navigation System using Timing and Ranging	NSSDC	National Space Science Data Center (USA)
NCDC	National Climatic Data Center (SDSD, USA)	OCI	Ocean Color Imager
NCDS	NASA Climate Data System	OCTS	Ocean Color and Temperature Scanner (ADEOS and MOS–2)
NEMS	NIMBUS E Microwave Spectrometer (NIMBUS–5)		
NEMSOT	NEMS Output Tape	OLS	Operational Linescan System (DMSP)
NERC	Natural Environment Research Council (UK)		
NESDIS	National Environmental Satellite, Data and Information Service (USA)	PAF	Processing and Archiving Facility (ESA)
		PC	Personal Computer
NESS	National Environmental Satellite Service (Suitland, USA)	PIPOR	Programme for International Polar Oceans Research (ESA)
		Pixel	Picture element
NET	Noise Equivalent Radiance	PLDS	Pilot Land Data System (NASA)
NEΔT	Noise Equivalent Temperature Difference	PMR	Passive Microwave Radiometer (GOES 'Next') or Pressure Modulated Radiometer (NIMBUS–6)
NIMBUS	NASA Meteorological Research and Development Satellite		
nlw	normalised water-leaving radiance	PMW	Passive Microwave
		POAM	Polar Ozone and Aerosol Measurement
nm	nanometre $= 10^{-9}$ metres	PODS	Pilot Ocean Data System (NASA)
NN	Nearest Neighbour		
NOAA	National Oceanic and Atmospheric Administration (USA)	ppt	parts per thousand
		PR	Polarisation Ratio
NOARL	Naval Oceanographic and Atmospheric Research Laboratory (USA)	PRARE	Precise Range and Range-Rate Experiment (ERS–1)
		PRF	Pulse Repetition Frequency
NODS	NASA Ocean Data System	PS	Polar Stereographic (map projection)
NORDA	Naval Ocean Research and Development Activity (USA)	PSF	Point Spread Distribution
NPOC	Naval Polar Oceanography Center (USA)	PTT	Platform Transmitter Terminal (ARGOS)
NRC	National Research Council (USA)	QD	Quick Dissemination
		QL	Quick Look
NrIR	Near Infrared		
NROSS	Naval Research Oceanographic Satellite System or Navy	RAA	Rapid Access Archive
		RADAR	Radio Detection and Ranging

RAE	Royal Aircraft Establishment (UK)	SLAR	Sideways-Looking Airborne Radar
RAMS	Random Access Measurement System (NIMBUS–6)	SMMR	Scanning Multi-channel Microwave Radiometer (NIMBUS–7 and SEASAT)
RBV	Return Beam Vidicon (LANDSAT–1 and –3)	SMS	Synchronous Meteorological Satellite (USA)
RESTEC	Remote Sensing Technology Center (Japan)	SNR	Signal-to-Noise Ratio
RF	Radio Frequency	SOM	Space Oblique Mercator (projection)
rms	root mean square	SPAN	Space Physics Analysis Network (NASA)
s	second	SPOT	Système Probatoire pour l'Observation de la Terre (France)
SAGE	Stratospheric Aerosol and Gas Experiment (NIMBUS–7)		
SAM–II	Stratospheric Aerosol Measurement–II (NIMBUS–7)	SPRI	Scott Polar Research Institute (UK)
SAMS	Stratospheric and Mesopheric Sounder (NIMBUS–7)	SR	Scanning Radiometer (ITOS and NOAA)
SAP	Sensor AVE Package (DMSP)	SSC	Special Sensor C (DMSP)
SAR	Synthetic Aperture Radar	SSE	Special Sensor E (DMSP)
SASS	SEASAT–A Scatterometer System	SSH	Special Sensor H (DMSP)
		SSM/I	Special Sensor Microwave/Imager (DMSP)
SBUV	Solar Backscatter Ultraviolet and Total Ozone Mapping Spectrometer (NIMBUS–7)	SSM/IS	Special Sensor Microwave/Imager Sounder (DMSP)
SCAMS	Scanning Microwave Spectrometer (NIMBUS–6 and –7)	SSM/T	Special Sensor Microwave/Temperature (DMSP)
SCAR	Scientific Committee on Antarctic Research	SST	Sea Surface Temperature
SCATT	Radar Scatterometer (eg SEASAT)	SSU	Stratospheric Sounding Unit (NOAA)
SCMR	Surface Composition Mapping Radiometer (NIMBUS–5 to –7)	STIKSCAT	Stick Scatterometer (EOS)
		SW	Short Wave
		SWE	Snow-Water Equivalent
SCR	Selective Chopper Radiometer (NIMBUS–4 and –5)	SWH	Significant Wave Height
		SYNRAMS	Synoptic RAMS
SDP	Standard Data Product		
SDR	Sensor Data Record	TAT	Antenna Temperature Tape
SDSD	Satellite Data Services Division (USA)	TB	Brightness Temperature
		TBM	Terabit Memory
SEASAT	Sea Satellite (USA)	TCT	Temperature Calibrated Tape
SeaWiFS	Sea Wide Field Sensor (LANDSAT)	TDI	Time Delay and Integration
		TDRS	Tracking and Data Relay Satellite (USA)
SEM	Space Environment Monitor		
SIGRID	Sea Ice Grid System	TDRSS	Tracking and Data Relay Satellite System (USA)
SIO	Scripps Institution of Oceanography (USA)	THIR	Temperature-Humidity Infrared Radiometer (NIMBUS–4 to –7 and SEASAT)
SIR	Shuttle Imaging Radar (USA)		
SIRS	Satellite Infrared Spectrometer (NIMBUS–3)	TIP	TIROS Information Processor

TIR	Thermal Infrared		VHR	Very High Resolution
TIROS	Television Infrared Observation Satellite (USA)		VHRR	Very High Resolution Radiometer (NOAA)
TM	Thematic Mapper (LANDSAT–4 and –5)		VIRR	Visible and Infrared Radiometer (SEASAT)
TOMS	Total Ozone Mapping Spectrometer (NIMBUS–7 and JERS–1)		VIS	Visible
			VISSR	Visible Infrared Spin Scan Radiometer (GOES)
TOPEX	Dynamic Ocean Topography Experiment (USA)		VMS	Virtual Machine System
TOS	TIROS Operational Satellites (USA)		VNIR	Visible and Near-Infrared Radiometer (JERS–1)
TOVS	TIROS Operational Vertical Sounder (NOAA)		VTIR	Visible and Thermal Infrared Radiometer (MOS–1 and –1b)
TRANET	Tracking Network		VTPR	Vertical Temperature Profile Radiometer (NOAA 2–5 and DMSP)
TWERLE	Tropical Wind Energy Conversion and Reference (NIMBUS–6)			
TWT	Travelling Wave Tube		VV	Vertically polarised transmit, -Vertically polarised receive
UARS	Upper Atmosphere Research Satellite (NASA)		WBDCS	Wide-Band Data Collection System (EOS)
USAF	United States Air Force		WCRP	World Climate Research Programme
USGS	United States Geological Survey			
UTM	Universal Transverse Mercator (projection)		WDC–A	World Data Center–A (USA)
			WDR	Waveform Data Record
V	Vertical polarisation		WEFAX	Weather Facsimile
VAS	VISSR Atmospheric Sounder (GOES)		WMO	World Meteorological Organization
VGA	Very high resolution Graphics Adaptor		WOCE	World Ocean Circulation Experiment
VH	Vertically polarised transmit, Horizontally polarised receive		WORM	Write-Once-Read-Many
			WRS	World Reference System

List of mathematical symbols used in the text

ϵ emissivity

L   upward flux of longwave radiation emitted from the surface, (Wm^{-2})

O   downward flux of longwave radiation, (Wm^{-2})

s   Stefan-Boltzmann constant, $(5.67 \times 10^{-8}$ $Wm^{-2}/K^4)$

T_B brightness (apparent) temperature, (K)

T_S physical temperature of the surface, (K)

PART 1

Remote sensing and polar regions

1. Introduction

The need for synoptic study of polar regions

Snow and sea ice are among the most dynamic and ephemeral geophysical features on the Earth's surface. Together with land ice, they exert a profound influence on local and global climatic and oceanic regimes. The vast ice sheets enveloping Greenland (surface area 1,700,000 km^2, maximum elevation 3,000m) and Antarctica (area 13,900,000 km^2, maximum elevation 4,000m) constitute almost 10% of the Earth's surface area; as such, their high albedos and thermal and radiative properties are climatically highly significant. About 90% of the Earth's freshwater is locked up in the Antarctic ice sheet alone. Moreover, polar ice masses, and the ice sheets in particular, are believed to be sensitive indicators of climate change, which it is feared may result from an increase in atmospheric 'greenhouse gases' such as methane and CO_2, the latter caused by burning of fossil fuels (Houghton and others 1990; ICSU/SCAR 1989; Meier 1985) (Figure 1.1). Modelling studies predict that the effects of climate change may be amplified, and thus detectable sooner, at high latitudes (Manabe and Stouffel 1979). The seasonal sea ice covers of both polar regions may react to an increase in atmospheric temperature by decreasing their areal extent; the subsequent reduction of sea ice extent/concentration in global climate models leads to a strong positive feedback to the warming.

Long-term, internally consistent, global datasets are required to detect any climatic change-related trends that may be occurring in ice/snow volume, extent and related parameters (ie surface, tropospheric and stratospheric temperatures, and the radiation fluxes at the top of the atmosphere)(Hall 1988; Parkinson 1989). This approach alone will enable the separation of significant trends on the decadal scale from the large-scale, natural seasonal and interannual trends that occur at high latitudes, eg around Antarctica due to temporary meridional shifts in the circumpolar low pressure belt.

The recent discovery on satellite imagery (from Nimbus-7) of ozone holes over both polar regions (Farman and others 1985) has captured the public imagination, and has underlined their relevance in a global context (Figure 1.2). There is a growing concern that the destruction of the protective layers of the atmosphere may expose polar organisms to increasing amounts and intensities of ultraviolet radiation. The threat of possibly irreversible climate change has forced governments to re-evaluate their positions on polar issues. Warning bells are being rung (ICSU/SCAR 1989); it remains to be seen whether governments will act decisively to encourage further research. Yet in spite of their significance, the polar regions, and Antarctica in particular, remain the least explored and most poorly understood regions of the Earth. Satellite data hold the key to increasing our knowledge of these remote

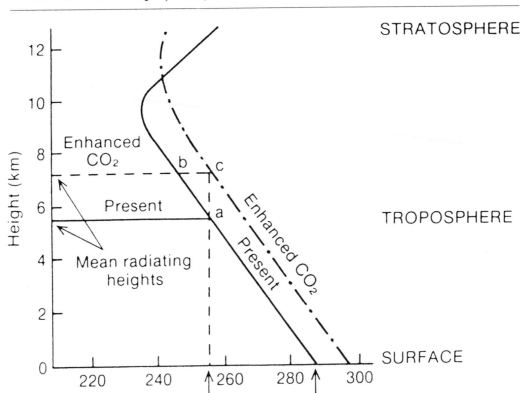

Figure 1.1 The effect of increasing CO_2 on the vertical profile of atmospheric temperature (schematic). Enhancing CO_2 raises the mean radiating height, and the temperature profile warms until the temperature at the new mean radiating height reaches 255K (the current effective radiating temperature of the Earth). From Mitchell 1989).

and infinitely challenging regions at a time when anthropogenic impact threatens as never before.

Land ice

Ice sheets exist as a consequence of the global climate, and are in turn critical in controlling the climate. The dominant effect is a positive feedback mechanism which functions roughly as follows. An increase in mean atmospheric temperature will tend to increase the size of the ice sheets, ie by increasing precipitation. These ice sheets, being highly reflective of visible radiation (ie being white), tend to reduce the amount of the sun's radiation which is absorbed by the Earth, and thus any increase in their size may then lower the mean atmospheric temperature. Other studies suggest that CO_2-induced climatic warming would have the effect of increasing the rate and area of ablation, leading to a net negative mass balance (Meier 1985). Thus, the detailed mechanisms involved are extremely complex, and substantial research is still needed to unravel them. The equation is further complicated by the fact that

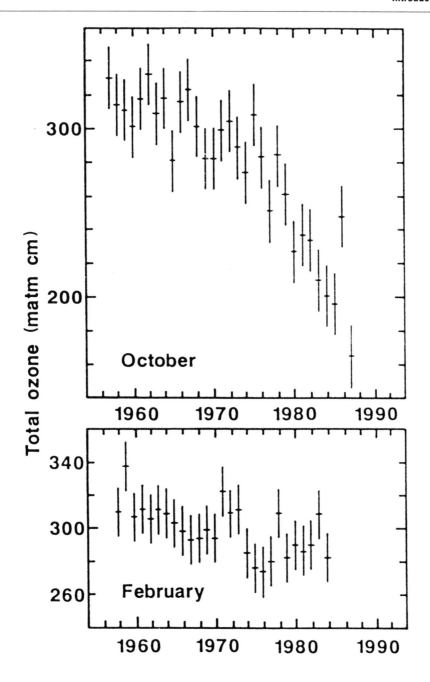

Figure 1.2 The variation in the mean total amounts of ozone in the stratosphere over Halley Bay base, Antarctica, during October 1957—1987 (top) and February 1954—1984 (bottom). From ICSU/SCAR (1989), after Farman and others (1985).

sulphur dioxide emissions may be acting to counteract global warming by blocking the downward path of solar radiation.

In addition to the ice sheets, which can be terrestrial (resting on land which would be above sea level if the ice were removed) or marine (eg the West Antarctic Ice Sheet which rests on rock as much as 2,000m below sea level), and which are several thousand metres thick, we should consider the ice shelves, where the ice flowing from the ice sheets floats onto the sea. The important ice shelves are all found in the Antarctic (Bindschadler 1990) (Figure 1.3). They border 45% of the Antarctic coastline and vary in thickness from about 150 to 400m.

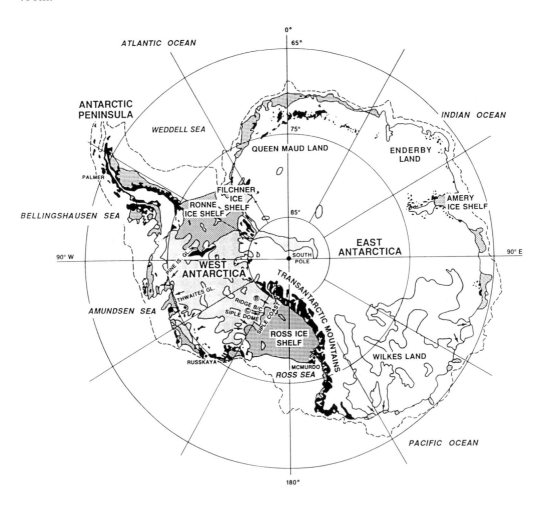

Figure 1.3 Map of Antarctica. Light shaded areas represent portions of the ice sheet grounded on bedrock more than 500m below sea level, and represent a large proportion of the marine ice sheet. Arrows represent major ice streams which drain marine ice sheet regions. Heavily shaded regions are ice shelves that buttress ice flow, and beneath which ocean circulation is an important process of mass exchange. Solid areas represent the approximate distribution of exposed rock. The dashed line represents the edge of the continental shelf, and defines the region between the ice sheet margin and the shelf edge where sediment cores can contribute historical information on ice sheet behaviour. After Bindschadler (1990).

Delineation of the margins and grounding lines of the polar ice sheets and ice shelves, together with measurements of their surface topography, slope and elevation, are essential inputs for the modelling of long-term dynamics and short-term motions or surges as they affect changes in overall volume. For example, freezing or melting processes beneath the Antarctic ice shelves may, through complex coupling with ice streams, affect the mass of the inland ice sheet. However, even though the combined volume of the Greenland and Antarctic ice sheets (approximately $15 \times 10^6 \text{km}^3$) accounts for 99% of the world's ice mass (Drewry 1983), no systematic complete measurements have been made of either. Even their overall shapes are poorly defined.

In spite of years of research, it is still not known whether the large polar ice sheets are growing or shrinking (Thomas and others 1985); the estimated accuracy in mass balance calculations has been put at between 10% (Giovinetto and Bentley 1985) and 50% (Zwally and others 1981). Certain regions (eg West Antarctica) may be inherently unstable, with long periods of slow growth followed by shorter periods of rapid drainage and thinning (Thomas and others 1979). Current research (eg Bindschadler 1990; Lingle 1984) suggests that this retreat, once initiated, may be irreversible, resulting in a serious rise in sea level which may engulf low-lying regions whose inhabitants have never seen ice. Indeed, if the entire Antarctic ice sheet were to melt, the global sea level would rise by about 60m, with enormous social, economic and political consequences; the melting of the West Antarctic portion alone would cause a 6m rise. Similarly, the ablation zone of the west Greenland ice sheet may potentially be a significant contributor to a rising mean sea level due to CO_2-induced climatic warming (Figure 1.4). Global sea level has been rising at an average rate of 1.0 to 1.5mm/yr during the past century (Barnett 1983).

Given that large-scale steady state motion is a function of ice sheet geometry, further detailed information on ice dynamics is necessary to separate those glacial fluctuations that are directly related to climatic change from those that are controlled by internal instabilities, largely unrelated to climatic change. Each active element of the ice sheet system, comprising the inland ice sheet, outlet glaciers, floating ice shelves, etc, has its own distinctive set of boundary conditions and processes governing its flow. In terms of modelling, all must be coupled in a realistic fashion to understand the present day behaviour and to predict the effect of any changes that may be occurring. Improved datasets are thus essential.

Sea ice and oceanography

Southern hemisphere sea ice at maximum extent covers an area of $17-21 \times 10^6 \text{ km}^2$ in August–October, receding to $3-6 \times 10^6 \text{ km}^2$ in January–February (Figure 1.5); these figures represent the average range of ice extents from 1973–82, derived from satellite passive microwave data (Ropelewski 1983). Arctic sea ice extent varies seasonally far less, covering some $15 \times 10^6 \text{ km}^2$ in March and $8 \times 10^6 \text{ km}^2$ in August. The growth of the Antarctic ice cover takes about seven months, significantly longer than its annual retreat; this contrasts sharply with the Arctic sea ice cycle (Ackley 1981b). Superimposed on these seasonal changes are fluctuations on a synoptic scale.

By virtue of its high reflectivity, high latent heat and low thermal conductivity, a sea ice cover greatly modifies i) ocean mixing processes, ii) surface roughness and thus momentum transfer between atmosphere and ocean, iii) the vertical and horizontal transfer of heat and moisture between atmosphere and the relatively warm ocean, and iv) atmospheric boundary layer structure. Long- and medium-range weather forecasts therefore require the inclusion of

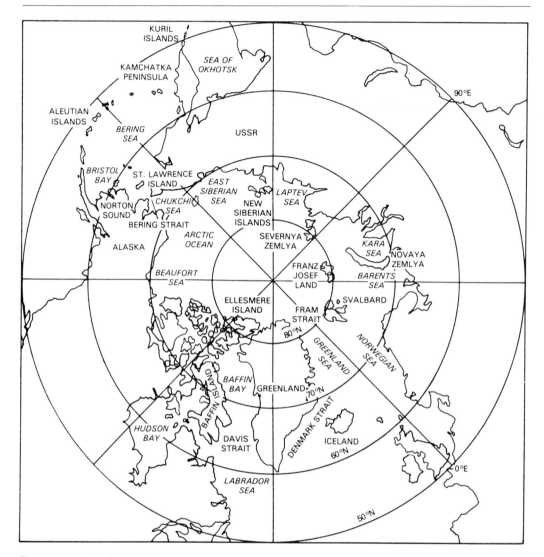

Figure 1.4 Map of the Arctic.

sea ice within both the climate models and the global observational system; variations in ice extent are recognised by the World Climate Research Programme as being a crucial element in the global climate system (WMO 1987a). The exact nature of the ensuing modifications and the complex nature of feedback mechanisms within the air-sea-ice system are, however, poorly understood. For example, a larger-than-average maximum extent of sea ice in one region of Antarctica commonly occurs at the same time as a smaller-than-average maximum in a neighbouring region.

Snow on sea ice (and elsewhere) acts as a highly efficient insulator and selective radiator, with a very high albedo for shortwave radiation and almost zero albedo (hence high emissivity) for infrared radiation. Albedo can be defined as the ratio of the reflected radiant flux to the flux of shortwave radiation incident upon a surface. Accurate measurements of snow and ice albedo from space are important for the following reasons:

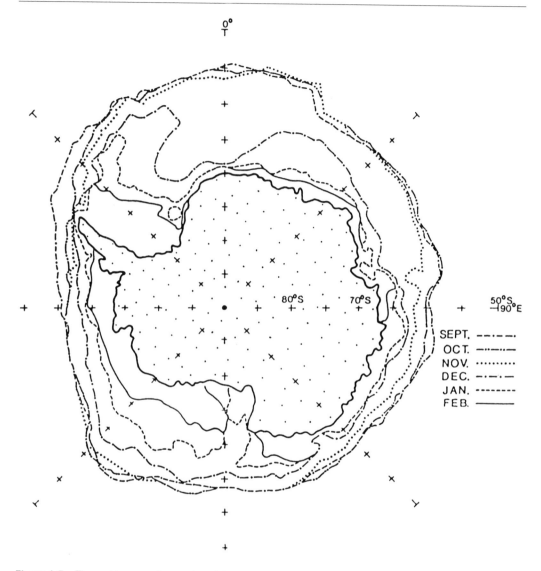

Figure 1.5 The rapid seasonal recession of Antarctic sea ice areal coverage, based on monthly average extents (of concentrations >10%) from Nimbus-5 ESMR data, 1973–6. After Zwally and others (1983a).

i)   As an indicator of snow and ice conditions and their variation over periods of hours to days.
ii) To study the surface radiation budget and its role in surface–air energy exchange.
iii) As a climatic parameter affecting interannual meteorological variations over polar regions.

Thermodynamic changes of state in snow covers occur on shorter time scales than those of sea ice, and are more directly linked to atmospheric circulation processes.

The detection and measurement of open water extent in the form of leads and polynyas within the pack is of crucial importance, as is the distinction between thin and thick ice.

During winter, the rates of turbulent heat and moisture loss to the atmosphere, ice growth, and salt rejection to the underlying ocean are strongly dependent on ice thickness, being up to two orders of magnitude larger over a refreezing lead than over multi-year (MY)($\geqslant$3m thick) ice (Figure 1.6) (Maykut 1985). Constantly-opening leads and polynyas are the major 'ice factories' within the ice pack during freeze-up (initiated when surface air temperatures fall below about 271K) (Figure 1.7). Rapid ice formation alters the local brine content of the water column, changing the density, stratification and mixing of the ocean. The freezing and melting of the ocean surface and the associated fluxes of heat and salt are the major factors in driving the thermohaline circulation of the oceans (Warren 1981). These processes also affect climate by releasing energy in the winter and absorbing energy in the summer, thereby reducing seasonal temperature extremes. Leads and polynyas in spring/summer, having a lower albedo than the surrounding pack, are sites of rapid lateral ablation, enhanced by wave action.

Figure 1.6   Deep within the Weddell Sea pack during the austral winter (August 1986), the relatively warm ocean (surface physical temperature approximately −1.8°C) is exposed to the much colder atmosphere by the formation of leads. This results in vigorous heat and moisture fluxes between ocean and atmosphere. The resultant steaming is called 'frost smoke'.

In addition, the ecology of polar marine biological organisms, from phytoplankton to marine mammals, is profoundly influenced by the distribution and seasonal cycle of the sea ice cover, and by the behaviour of dynamic marginal ice zones (MIZs), recurrent polynyas and shorelead systems in particular (Massom 1988). Changes in sea ice extent will affect primary productivity, carbon fixation and hence the global CO_2 budget (Figure 1.8). To carry out surface measurements in such areas is, however, fraught with difficulty, not to

Figure 1.7 A recurrent coastal latent heat polynya along the Antarctic coastline in the vicinity of 0° longitude (the Weddell Sea), August 1986. Such locations are often the sites of heavy sea ice formation, with the ice being blown seawards by persistent katabatic winds as quickly as it is formed. Note the heavy ice conditions in the foreground.

mention danger. The impact (often detrimental) of human activities, natural variations and pollution on the polar terrestrial and marine environments must be monitored. Pressures on the highly sensitive polar regions are increasing at an alarming rate, with Antarctic mineral exploration and exploitation looming large on the horizon. Oil spillages from transports are a constant threat eg that from the Argentine ship *Bahia Paraiso* close to ecologically rich sites on the Antarctic Peninsula in 1989.

Both sea ice covers are intricately involved in underlying water mass modification processes and thermohaline circulation patterns. Unlike the mediterranean Arctic Ocean, the Southern Ocean is not strongly stratified. Oceanographic processes occurring in the Antarctic alone affect the properties of 60% of the volume of the world's oceans over long time scales (Sarmiento and Tottweiler 1984). Antarctic bottom water produced by brine rejection under heavy ice formation conditions in coastal polynyas and/or under ice shelves in the southern Weddell Sea, for example, flows as far north as 40°N in the Atlantic Ocean on decadal time scales (WMO 1987b). Similarly, Arctic deep water is thought to form in latent heat polynyas in the north. Moreover, freshwater is exported equatorwards in both polar regions, in the form of sea ice, to be inserted into the ocean upon melt at lower latitudes.

Improved estimates of a number of fundamental variables are essential to drive the increasingly sophisticated array of global circulation models (GCMs). Surface wind stress is the principal driving force of large-scale oceanic circulation, and represents the lower boundary condition for models of the atmospheric circulation over the ocean. Ocean

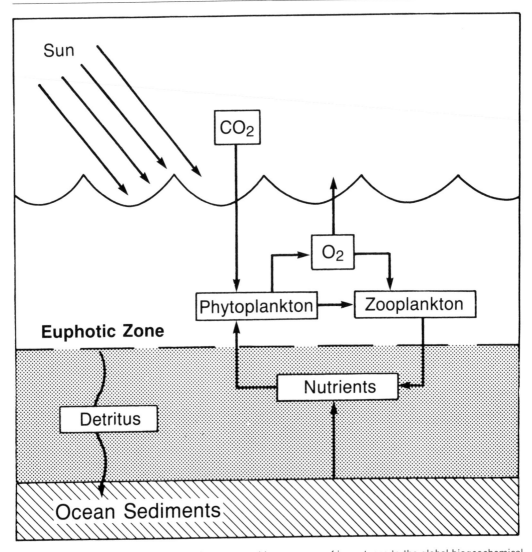

Figure 1.8 Schematic representation of oceanographic processes of importance to the global biogeochemical cycles, particularly the carbon cycle. After NASA (1989).

topography, another important parameter, is closely related to wave and current regimes. Changes have been observed in the latter in the Bering Sea in response to El Niño-Southern Oscillation events. Sea-surface temperature, which is a measure of the upper ocean layer heat content, is a critical factor affecting the exchange of heat energy between the atmosphere and ocean. All three parameters are poorly understood in polar regions. The Southern Ocean is also thought to play a significant role in influencing the interannual variability of the world ocean's capacity to take up CO_2 (Gordon 1988). The presence of a sea ice cover not only greatly modifies the combined effects of these and other parameters, but also hinders their direct measurement.

Research problems in the field of sea ice range in scale from microns, of particular interest in the understanding of electromagnetic, optical and thermal properties of the ice cover, to basin-wide and global (Carsey 1989; Untersteiner 1986). Large-scale problems are centred on the overall effect of ensembles of events and features. Micro-structural properties of the ice assume less importance as the scale of observation increases. Crucial interactions in the air-sea-ice interaction system are normally studied on the basin-wide or macro-scale (100 to 1,000s of km) and the regional/meso-scale (about 10–100km). The former is primarily concerned with seasonable changes in sea ice extent, driven mainly by the annual cycle in the surface heat balance: the latter is important in the study of atmospheric and oceanic circulation on the synoptic temporal scale.

Most sea ice research (related to ice concentration, ice edge delineation, floe size distribution, lead and polynya distribution, and ice dynamics/kinematics) ideally requires a repeat period of coverage of three days at most. Sea ice concentration is defined as the proportion (percentage) of a given area covered by sea ice as opposed to open water. Non-dynamic aspects, including pressure ridge density, height and orientation, ice type and ice thickness distribution studies, do not require such a short repeat period, but rather one of the order of seven days (Figure 1.9). Ice sheet glaciologists and climatologists, on the other hand, generally require data on a less frequent basis than sea ice scientists, oceanographers and meteorologists, but over longer periods (WMO 1987b). Observational requirements for sea ice and snow parameters, in terms of spatial and temporal resolution, are well documented in a number of sources (eg NASA Science and Applications Working Group 1979; Untersteiner 1984).

Figure 1.9 Large pressure ridges, formed by convergent forces in MY ice, Fram Strait, July 1984. Such features are a hazard to shipping, and are a favourite habitat of polar bears outside the breeding season.

Particularly intense lateral heat and mass exchanges, and biological productivity, occur in MIZs, driven by the fronts which often separate ice-free from ice-covered waters. The MIZ is the transition zone between the ice pack and open ocean, and has properties of both. The different thermal and surface roughness characteristics of open ocean and sea ice affect the planetary boundary layer, leading to localised wind speed changes, ocean upwelling and cyclone steering (and possibly formation) at the ice edge. Surface gravity waves penetrating the ice cover determine the floe size distribution in the MIZ, and are in turn attenuated by the ice cover. However, even fundamental parameters, such as the spatial distribution of floes and the location of the ice edge and its variability in response to oceanic and atmospheric forcing, are poorly understood. The complexity of the MIZs, where many ice types tend to occur under a wide variety of conditions over relatively small temporal and spatial scales, tends to defy generalised measurement (Figure 1.10). An excellent review of the nature of these critical, highly dynamic regions is given by Wadhams (1986).

Figure 1.10   Divers from the NOAA ship *Discoverer* recovering buoys from the Bering Sea ice edge during MIZEX West, March 1983. The fragmented FY floes are the result of the passage of a heavy storm; the buoys had been deployed in a continuous ice cover two days earlier.

It has thus become a major requirement of polar oceanographic, meteorological and sea ice programmes to investigate how a sea ice cover modifies vertical and lateral energy exchange, and conversely how the energy fluxes themselves influence the extent and motion of the sea ice cover itself. In an effort to unravel and comprehend the complex feedback processes involved, a number of coupled dynamic and thermodynamic air-ocean-ice (and -land) models (eg the Fine Resolution Antarctic Model, or FRAM; Hibler 1979) have been developed, each requiring different data sampling intervals (ranging from 1 day to 30 days) and different spatial geophysical scales (from 1km to about 100km) for verification and process identification (Untersteiner 1983). All require accurate, improved data on a regular basis in order to describe and ultimately predict observed variations in key variables. Ground-based data gathering techniques are hopelessly inadequate as a means of providing input data to both drive and test these increasingly sophisticated models. Satellite remote sensing again holds the key.

Apart from the strict requirements of data for research purposes, there is a need for routine, near real-time synoptic observations of sea ice and iceberg behaviour on an operational basis for the shipping, fishing and offshore hydrocarbon industries, regulation, search and rescue, and meteorological forecasting (Figure 1.11). The UK Meteorological Office runs weather prediction models twice daily using three-dimensional analyses of the atmosphere to produce forecasts for several days ahead, satellite data from polar regions forming an integral part of this system. The routine availability of satellite imagery has also transformed the ice reconnaissance services offered by organizations like the Atmospheric Environment Service in Canada, where ice forecasting is a major component of national environmental services. Moreover, with the advent of compact, lightweight automatic picture transmission (APT) receivers, the meteorologist in the field can now process the raw imagery.

Figure 1.11
A large pinnacle iceberg, draft 50 feet, heading northwards from the Weddell Sea, June 1986.

Meteorology

Arctic and Antarctic boundary conditions affect global atmospheric circulation through pole-to-equator temperature gradients. Polar regions are the heat sinks that help to drive global atmospheric circulation. However, a number of important questions remain largely unanswered. For example, the extent to which sea ice and cloud distribution are related is unknown (ICSU/SCAR 1989). Clouds over sea ice generally affect the outgoing longwave radiation at the top of the atmosphere. Conversely, a low cloud cover over the Antarctic ice sheet may increase longwave radiation loss due to strong temperature inversion effects. Cloud amount, type and height also relate to precipitation rates, which are equally poorly understood. An improved database on high latitude radiation (Charalambides and others 1986) and atmospheric temperature budgets will be fundamental to the detection of possible global change (Figure 1.12). The processes affecting ozone distribution in the stratosphere are equally poorly understood.

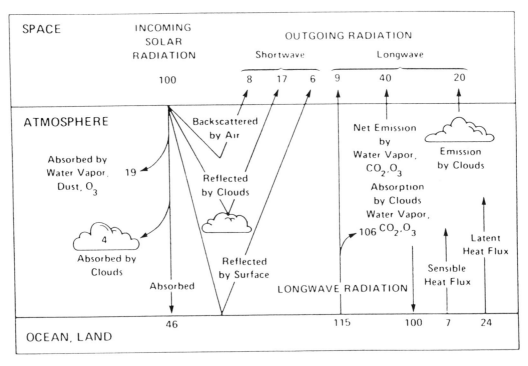

Figure 1.12 Schematic representation of the atmospheric heat balance. Solar fluxes are shown on the left hand side, and longwave (TIR) fluxes on the right. From MacCracken and Luther (1985).

Space-borne observations of polar regions

Polar regions, apart from being vast, remote, inaccessible and notoriously inhospitable, are shrouded in darkness, twilight and cloud for much of the year. It is impractical, hazardous and expensive to carry out conventional surface observations. About 90% of the total cost of a typical Antarctic research operation is consumed by logistics. The majority of Antarctic research stations are clustered around the coast, and it is extremely costly to push field programmes inland and out to sea. Before the advent of remote sensing from space, estimates of sea ice extent and associated meteorological and oceanic variables came from a combination of ship, island station and aircraft observations, invariably conducted on an opportunity basis and often concentrated in areas of logistic convenience during the summer navigation season. Although useful in the operational sense, these datasets are inherently biased and therefore of limited scientific value. In the last 25 years, the great upsurge of interest in high latitudes has served to underline our lack of understanding of even the most fundamental polar parameters. Fortunately, however, it has coincided with great advances in satellite platform, sensor and ground segment technology.

The remoteness and scale of the features and processes involved are uniquely well suited to regular surveillance from space. The difficulty of carrying out *in situ* experiments to help interpret the satellite data will inevitably remain; at present, these tend to be statistically sparse in terms of both temporal and spatial coverage (although recent international programmes have made significant contributions). Satellite remote sensing, however, offers the only realistic method of obtaining the systematic, repetitive and reliable coverage required by many studies. In all fields of polar study, a large number of pre-conceived theories need to be re-examined, and remote sensing from space provides one opportunity to carry out this re-evaluation. Conversely, it must be emphasised that there is no point in generating vast all-encompassing mountains of satellite data just for their own sake; the data should be used to address specific problems, preferably the ones that need to be addressed.

Satellites have a practical role to play in snow and ice studies outside polar oceans and ice sheets, eg in monitoring the freezing and breakup of lake ice. For example, after monitoring the ice on Lake Erie for several years on National Oceanic and Atmospheric Administration (NOAA) meteorological satellite and Landsat data, Wiesnet (1979) was able to ascertain that characteristic patterns of breakup exist. Such findings are of obvious importance to ship navigational planning. Satellite imagery can reveal ice growth, ice jams/dams, decay and movement on large, navigable rivers. Dey and others (1977), for example, used NOAA-4 and -5 data to monitor the breakup on the Mackenzie River in Canada's Northwest Territories from March to July in 1975–7. Other important studies include those of McGinnis and Schneider (1978) and Schneider (1977).

Due to the sparsity of conventional meteorological data in polar regions, remotely-sensed data make a substantial contribution to the meteorological database. Vertical profiles and horizontal data from sensors such as the NOAA TIROS Operational Vertical Sounder (TOVS, providing global temperature and moisture profiles) and Advanced Very High Resolution Radiometer (AVHRR, cloud pattern monitoring) are used on an operational basis by meteorological organisations throughout the world to determine critical weather patterns both within and skirting polar regions (Wiesnet and Matson 1983). The problems identified above are best addressed using satellite data, preferably by integrating data from different radiometric sounders and scanners. The suite of improved sensors to be launched in the next decade will facilitate the development of suitable algorithms for mapping cloud fraction and thickness. Current algorithms are invariably unsuccessful in polar regions due to both the complexity of the vertical temperature profile and the problems of cloud–snow discrimination.

Passive systems measure the natural radiation emitted by a target material and the energy from other sources reflected from it; reflected solar radiance (present only during daylight hours) and the Earth's thermal emission radiance (which can be used during the day or night) are the major sources of energy detected by passive space sensors. The radiometer has so far been the work-horse of space-borne remote sensing, usually measuring the reflected or emitted radiance of a target medium within a given spectral range. Active systems, on the other hand, transmit their own coherent signal and measure the energy reflected or scattered back from the target material. The only active system to have been used for remote sensing from space are microwave systems, although lasers may well be employed on future spacecraft.

Significantly, both active and passive microwave systems are less sensitive to atmospheric conditions and darkness than non-microwave systems, although this depends upon the frequency of the sensor. The spatial resolution achieved by passive systems is inherently poorer than that offered by active systems such as synthetic aperture radar (SAR), but the former can cover a much wider area over a shorter period of time. The common basis of all remote sensing from space is the fact that different kinds of interactions between radiation and a medium participate in the radiative transfer process. Consequently, different types and amounts of information on the observed medium can be gained by measuring different radiation processes, ie by using different wavelengths. The satellite sensor systems discussed can be broadly subdivided into the following categories: photographic systems, eg the Hasselblad cameras onboard the manned space mission Apollo 9 (Mairs 1970) and Skylab-4; television cameras; scanning radiometers; imaging radars; non-imaging radars; and lasers. The geographical coverage offered and the timeliness of data availability are determined by whether the satellite is operational, research and development, military or commercial, or whether it possesses on board recording facilities to enable data collection when out of line of site of suitable receiving stations.

It will become apparent that no one sensor alone has been launched, or is ever likely to be, which can fulfil every requirement. Rather, a careful integration of data from different sensors, frequencies or wavelengths will invariably yield more information than can be extracted from the individual sensor datasets alone. Technological trade-offs exist, no matter what the system. Satellite data alone are not a panacea, and should be combined wherever possible with related data (eg surface measurements) to gain optimum results. Considerable effort is often necessary to convert raw satellite data into useful geophysical or biological parameters; in other cases, useful data products have already been generated by principal investigators or archiving agencies.

Satellite measurements will never completely replace surface and near-surface measurements. Certain parameters, such as sea ice and land ice thickness, the vertical energy fluxes over ice/open water within the ice, winds over ice and the vertical stratification of the ocean cannot yet be measured directly from space, and it is not envisaged that this situation will change in the foreseeable future. Sea ice thickness, for example, although detectable indirectly by submarine sonar (Wadhams 1980), cannot be measured from space due to the high losses in saline ice and the huge power requirements.

Remote sensing from aircraft will continue to be a key tool for snow and ice research (NOAA/NASA/NSF 1989). In ice sheet and glacier studies, for example, much use is made of radio frequency instruments for subsurface probing of the ice and its bedrock, ie radio echo-sounding (Bogorodsky and others 1985); such studies are presently impossible from space. An excellent bibliography of radio echo-sounding techniques in glaciology is provided by Macqueen (1988). Moreover, the detection and monitoring of icebergs by International Ice Patrol aircraft will continue to play an important role in reducing hazards to shipping and oil/gas platforms in Canadian and Greenland waters.

Recent advances in digital recording techniques have opened up the possibility of identifying very faint echoes by automated data stacking and signal processing. Moreover, NASA has recently carried out a successful coupling of air-borne LIDAR (Light Detection and Ranging) systems with navigation by a global positioning system, providing absolute aircraft positioning and elevation measurements to a precision of tens of centimetres (*Ad Hoc* Panel on Remote Sensing of Snow and Ice 1989). This breakthrough has obvious applications, ie making routine estimates of local ice sheet elevation and volume and their change with time. Aerial photography will continue to be an important technique in monitoring glacial fluctuations in regions of dramatic topography, and where broad spatial (synoptic) coverage is not essential. Aerial photography of sea ice is an integral part of field experiments, providing high resolution base-line data on floe size distribution, etc that are useful to a number of disciplines, for example wave-ice interaction and acoustics studies in the MIZ (Rottier 1989).

Localised problems in meteorology, including atmospheric boundary layer studies and micro-meteorology over ice and leads/polynyas still have to be studied from the surface. Moreover, *in situ* measurements are essential in order to validate what the satellite is seeing (ie surface truthing) (Figure 1.13). The general technology available to carry out such measurements is reviewed by Dozier and Strahler (1983); current technological advances in surface snow/ice measurement instruments are highlighted in scientific reports published regularly by the US Army Cold Regions Research and Engineering Laboratory (CRREL), Hanover, New Hampshire. Some kind of reference data and a priori knowledge of the medium being observed, sensor and interference are essential for a correct inversion, eg interpretation of the measured quantities into the natural parameters of the object.

Figure 1.13 Dr Joey Comiso of NASA GSFC next to a passive microwave radiometer on the deck of F/S *Polarstern* (19 July 1986, c.61°S, 1°W) during the Winter Weddell Sea Project. This instrument collected multi-frequency data continuously as the ship traversed the pack. Such surface validation experiments are essential to the improved interpretation of satellite data.

Cross-comparisons of data from different sources can provide valuable insights into interpretations (eg Carsey 1985; Burns and others 1987). Indeed, intensive programmes of surface and aircraft measurements (eg MIZEX) have greatly improved our understanding of snow and ice as remote sensing targets (Luther 1988). Moreover, the huge tanks at CRREL have been used to study the microwave signatures of artificially grown sea ice under controlled conditions (Grenfell and Comiso 1986). Surface and near-surface validation experiments, and the use of a hierarchy of satellite data of increasing spatial resolution, are critical in advancing us towards improved interpretations of the satellite data.

Few of the sensors flown in space so far have been designed specifically to collect data over snow and ice, yet they have made and will continue to make invaluable contributions to the difficult study of ice and snow masses. They also provide measurements of critical related variables such as sea surface temperature (SST) and surface wind and wave conditions, as well as providing a means of relaying data from remote *in situ* sensor packages, eg automatic weather stations (Figure 1.14), ice-strengthened buoys (Figure 1.15), and migrating (and non-migrating) wildlife.

The role that satellites play in polar navigation, rescue and communications should not be overlooked (Rosenberg and Jezek 1987). It has been estimated that continuous communication coverage of both polar regions can be obtained by four satellites (eg the proposed UK CERS or T SAT) in highly eccentric orbits inclined at 63° (Molniya orbits). Meteorological satellite data received routinely at Antarctic research bases play a critical role in determining suitable conditions for ship and aircraft resupply sorties. The part played by remote sensing in Antarctic operations and logistics is covered by Thomson (1989).

For a more detailed technical description of the different satellite sensor systems, the reader is encouraged to consult the two mighty volumes of the *Manual of Remote Sensing* (Colwell 1983), and also Chetty (1988). Indispensible and detailed reference books describing the nature of electromagnetic radiation and the general principles of remote sensing include Asrar (1989), Curran (1985), Elachi (1987), Rees (1990), Sabins (1987), Schanda (1986), Slater (1980), Stewart (1985), Tsang and others (1985) and the three volumes of Ulaby, Moore and Fung (1981; 1982 and 1985). The complex realms of data preprocessing, processing and reprocessing are covered in detail by Bernstein (1983), Billingsley (1983), Jensen (1986), Mather (1987), Muller (1988) and Richards (1986). Map projections commonly used to display satellite imagery are described by Bernstein (1983). Digital hardware technology, an ever-changing field, is reviewed by Nichols (1983). Beckman (1983) and Douglas (1988) discuss satellite communications and data transmission technology. The rapidly evolving field of Geographical Information Systems (GIS) is covered by Star and Estes (1990) and the *1990 GIS Sourcebook* (the latter is available from GISWORLD, PO Box 8090, Fort Collins, CO 80526, USA). Sloggett (1989) covers satellite data processing, archiving and dissemination infrastructure and design. The sea ice terminology used is based on upon WMO (1970).

Structure of the present work

This monograph outlines the evolution of satellite-borne remote sensing of polar regions, reviews satellites currently in operation (in 1991), and looks at probable (and definite) future developments. It follows on from the reference manual by Cornillon (1982), an important work that describes sensors, data products and distribution centres for US satellites, though not exclusively related to polar regions. The present monograph, also designed to be a

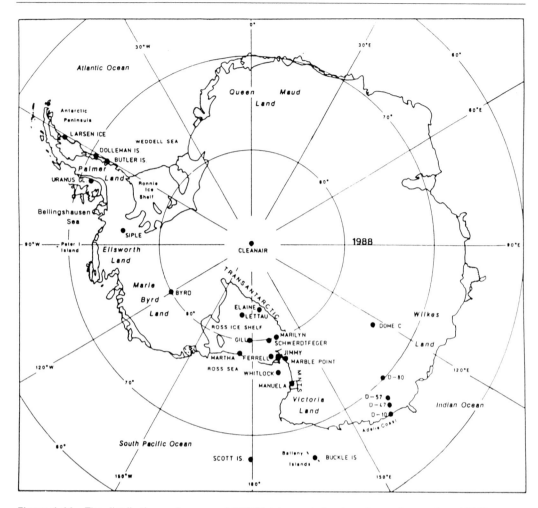

Figure 1.14 The distribution and names of ARGOS-interrogated automatic weather stations (AWS) around Antarctica, 1988, operated by the University of Wisconsin. More than 100 values at 10-minute intervals are recorded every 24 hours. Five AWS have been installed on the Greenland Ice Sheet in support of ice coring efforts. From Keller and others (1988).

manual, concentrates on sensors that detect and measure not only polar snow and ice masses (both sea and land ice) but also those atmospheric and oceanic variables affecting their growth and decay (including waves, wind stress and SST). The application of remote sensing to the biological sciences is covered in a section on ocean colour and its detection from space. Another aim is to provide insight into the inevitable trade-offs that occur with every sensor system.

Such important fields as solar terrestrial physics, monitoring of ozone, the radiation budget, and the seasonal terrestrial snow cover of the northern hemisphere are covered, although in less detail. The maximum areal extent of the latter varies between 37 and 45 $\times$ $10^6 km^2$ (Matson and Wiesnet 1981), and is an important factor in the global heat/radiation balance. The albedo of snow is 0.8 to 0.9, whereas that of bare ground is about 0.2 to 0.3. Snow monitoring from space is reviewed by Andersen and Odegaard (1980), Foster and

Figure 1.15 The deployment of an ARGOS buoy (with suspended thermistor chain) in Weddell Sea FY ice, August 1986 at *c*.68°S, 4°W. The floes consist of individual pancakes cemented together (see Figure 3.8).

others (1984) and Meier (1980). The importance of satellite remote sensing to snow hydrologists has increased dramatically, as it provides the only practical means of obtaining data on snow-covered area, surface albedo and snow-water equivalence. These data are necessary inputs for several runoff forecasting and management systems. Moreover, variations in albedo with snow cover changes have an important effect on global climate. Satellite imagery has revealed a previously unseen large interannual variability (Barry 1985).

Satellite multi-spectral remote sensing, collected mainly at visible, near-infrared and microwavelengths (particularly radar), is also proving to be extremely useful in allowing remote estimates of the biomass of high latitude regions (eg Viereck and others 1990). Arctic ecosystems contain huge amounts of carbon in the form of soil organic matter, and have the potential to act as major sinks or sources for atmospheric CO_2 (depending on future climatic conditions). Tundra and boreal forest soils contain approximately 30% of the global terrestrial soil carbon reservoir; a large proportion of this is immobilised in peat and permafrost, a situation which would change in response to a wetter, warmer climate (Reeburgh and Whalen 1990). High latitude terrestrial ecosystem studies are largely outside the scope of this book.

Part 1 consists of a review of the physical basis and general principles of satellite-borne remote sensing as they relate to snow and ice, along with the history and development of its application in polar regions. In order to determine which particular dataset is best suited to one's needs, it is necessary to have some understanding of the satellite sensor parameters, the frequencies and wavelengths of the electromagnetic radiation that they measure (and/or transmit), and how the radiation interacts with the properties of the medium being studied to yield information of interest.

Part 2 is structured as a reference manual, devoted to the detailed description and evaluation of individual satellites and sensors; limitations as well as applications are described. Another primary aim is to aid potential data users by describing primary data sources, archiving extent, availability and format of data (both raw and with specific data products extracted) for each individual sensor. The user community invariably has great difficulty in determining what data are available and what information exists concerning their quality and possible limitations (Barry and others 1984). The widely dispersed and rapidly evolving literature on remote sensing, combined with the swamp of jargon and acronym-mania, renders the choice of optimum data source difficult if not impossible. Sources tend to be poorly documented, especially for older data where the networks set up for processing and distribution have tended to be ephemeral. Worse still, the data may not even have been retained at all. Moreover, the polar scientist may be totally unaware of the existence of relevant data sets that might make his/her life much easier.

The reader is encouraged to follow up the extensive reference list at the end of Part 1, and after each individual satellite in Part 2, for more detailed discussion of the fields of interest.

New ideas and satellite sensors are being launched at a rapid rate. In order to keep up to date with the exciting developments that are taking place in the field, it is desirable to subscribe to a number of remote sensing journals, not devoted specifically to polar regions, but which often include important articles of polar interest. These include *Remote Sensing Reviews*, *Geo Abstracts – G. Remote Sensing*, *Photogrammetry and Cartography*, *Canadian Journal of Remote Sensing*, *Photogrammetric Engineering and Remote Sensing* (also a valuable source of information on current trends in GIS), *Space*, *COSPAR Information Bulletin*, *Remote Sensing of Environment*, *IEEE Transactions of Geoscience and Remote Sensing*, *The International Journal of Remote Sensing*, and *The Journal of Geophysical Research (Oceans)*. Moreover, many of the important developments are published in annual conference proceedings, including those of the *International Symposium on Remote Sensing of Environment* and the *International Geoscience and Remote Sensing Symposium (IGARSS)*.

For readers in the United Kingdom, the National Remote Sensing Centre (Library) publishes a regular *Current Awareness Bulletin*, listing articles, references, teaching aids and conference proceedings which can be borrowed upon request. For further details contact The Librarian, NRSC, Space Department, Royal Aircraft Establishment, Farnborough, Hampshire GU14 6TD, UK. Similarly, the NASA Scientific and Technical Information Division in Washington DC publishes a very useful list of abstracts in the form of their *Scientific and Technical Aerospace Reports*.

2. Satellite system parameters

In this chapter I present a brief overview of the most important parameters which describe the observing system. This will help to place the following sections in context. It is convenient at the outset to distinguish between instrumental and viewing parameters. The principal instrumental parameters, namely frequency (or equivalently wavelength), polarisation and sensitivity (radiometric resolution), are determined by the design of the transmitter (if there is one), receiver, antenna, and data handling system. The principal viewing parameters (determined both by the instrument design and by the orbital parameters of the satellite) are revisit interval, resolution, swath width, illumination (and/or observation) angle, and mission lifetime. Properties of polar remote sensing targets with respect to electromagnetic radiation, and the consequent choice of instrumental parameters, are dealt with in the next section.

Instrumental parameters

Frequency

All of the remote sensing systems described in this monograph make use of information carried by electromagnetic radiation, of which there is an (in principle) infinite range of possible frequencies. In practice, however, the transparency or otherwise of the Earth's atmosphere limits the possible wavelength ranges to about 0.4–15 μm (the visible and infrared regions) and 1mm to 1m (the microwave region, corresponding to a frequency range of 300–0.3 GHz) (Figure 2.1). The visible (VIS) and infrared (IR) region is conveniently subdivided into, on the one hand, the VIS/near-infrared (NrIR) region (approx. 0.4–2 μm) and on the other hand the thermal infrared or TIR region (approx. 2–15 μm) (Figure 2.2). Sensors designed to detect atmospheric constituents utilise spectral bands between the atmospheric 'windows'.

Naturally-occurring radiation from the Earth's surface is found in all of these ranges of wavelength. In the VIS/NrIR band it is predominantly reflected sunlight, and the most important parameters of the target material is thus its reflectance (see Chapter 3). In the TIR band, on the other hand, the main source of radiation is, as the name suggests, the blackbody thermal mechanism by which all objects above absolute zero emit radiation. For an object at a typical terrestrial temperature, most of this radiation is emitted at wavelengths around 10 μm (Figure 2.3). Detected radiation essentially contains information on two parameters, namely the temperature of the target material and its effectiveness at emitting radiation in this waveband, called the *emissivity*. Emissivity is a unique characteristic of the target

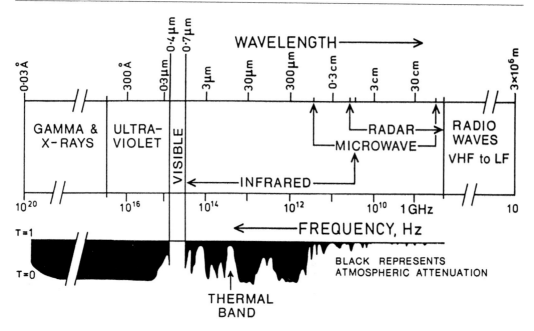

Figure 2.1 Regions of the electromagnetic spectrum exploited by satellite remote sensing, and showing atmospheric transmission windows in the microwave, IR and VIS sectors. After Suits (1983).

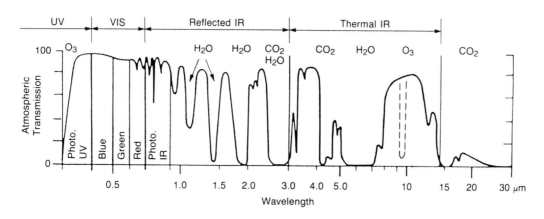

Figure 2.2 An expanded diagram of the VIS and IR region of the electromagnetic spectrum, showing atmospheric windows and intervening regions of atmospheric attenuation.

material and its state or condition. Wolfe and Zissis (1989) provide an excellent handbook of infrared theory and techniques.

Small but significant and measurable amounts of blackbody thermal radiation can also be detected in the microwave band, and this kind of remote sensing is known as passive microwave radiometry. Again, the detected signal is governed by both the target temperature and its type and condition. The other main use of the microwave region is for *active* remote sensing, by which radiation is emitted by the remote sensing instrument and detected after its reflection from the target material. Active remote sensing in the microwave region is generally called *radar*, and the main observable parameters are the range to the target (from the time delay of the returned signal) and the reflectivity of the material (which in turn is determined by many of its physical properties).

Polarisation

In general, the reflective and emissive properties of a material are different for different polarisations (orientations of the electrical field vector in the electromagnetic radiation, where H is horizontal and V is vertical), so that further information on the physical properties of the target material may be obtained by observing different polarisations. In practice, this has so far only found significant application in passive and active microwave remote sensing, not in the VIS and IR bands.

Sensitivity

The sensitivity of a remote sensing system measures the response produced by radiation of a given intensity and wavelength. Other things being equal, it should be as large as possible, but because the output data are usually digitised, they can only span a finite range of values so that a high sensitivity (low value of the minimum detectable signal) implies a low value for the maximum signal which can be detected. This then requires some kind of optimisation, and what is optimal for one kind of target material may not be optimal for another. This has been particularly problematical for VIS remote sensing of snow, which has a very high reflectivity and can often cause saturation of the detecting system.

Spectral resolution refers to the width of the discrete spectral bands employed by a given sensor. An increase in spectral resolution for mechanical scanners is generally accompanied by a reduction in the signal-to-noise ratio. *Radiometric resolution*, on the other hand, refers to the number of digital levels used to record the data. It is commonly expressed as the number of binary digits (bits) required to store the maximum level value. Thus, the number of bits required for 2, 4, 16, 64 and 256 levels is 1, 2, 4, 6 and 8 respectively. The use of binary notation and the storage of digital data is covered by Mather (1987).

Viewing parameters

Revisit interval

Geostationary satellites maintain a fixed location with respect to the Earth's surface, hovering approximately 36,000km above a point on the equator. Conversely, a satellite in a low polar orbit traces out a curving path over the Earth's surface, as a consequence of the satellite's orbital motion and precession and of the Earth's rotation about its axis. This path (the subsatellite track) wraps itself round the Earth from east to west like a ball of string,

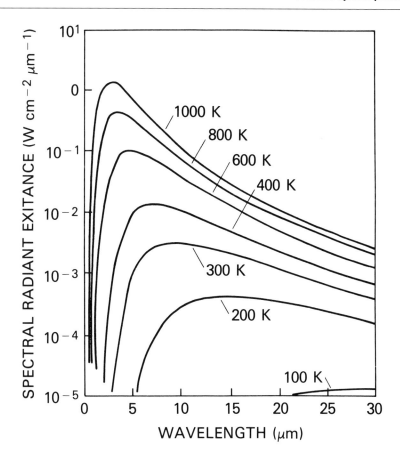

Figure 2.3 The spectral distribution of energy radiated from a blackbody at temperatures from 100 to 1,000K. The average temperature of the Earth's surface is 300K; the peak of emitted radiation at this temperature occurs in the thermal infrared region.

oscillating between equal north and south latitudes in a pattern set by the *inclination* of the orbit. The path may close up on itself if the orbital parameters (inclination, height and eccentricity) are suitably chosen, in which case the satellite will revisit a given location at regular intervals. This interval may in general be any integral number of days, though other constraints on the orbital parameters may limit the choice. The point where the satellite, travelling northwards, passes directly over the equator is called the ascending node; the descending node describes the southward crossing. A detailed description of the orbital dynamics involved is not within the scope of this monograph, and the interested reader is referred to Stewart (1985) and Rees (1990).

If a further restriction is placed on the orbital parameters, the orbit may be chosen to be *sun-synchronous*, in which case the time at which the satellite revisits a given location will be the same on each occasion. This is very useful for VIS and IR observations, since the level of solar illumination can be chosen. Biological studies, for example, require very specific timings for the collection of data.

Although most satellites have a fixed repeat period, others are able to vary their repeat period after launch, although this is limited by fuel availability. For example, the European

Space Agency satellite ERS-1, to be launched in 1991, will be able to change its orbital configuration, and thus its revisit interval, up to six times during its lifetime to satisfy all interested scientific parties. Quite small changes in orbital altitude offer a wide range of repeat periods.

Spatial resolution

Spatial resolution is a complex concept which can, for the purposes of remote sensing of polar regions, be defined as the smallest object that can be detected and distinguished from a point. The most frequently used measure, based upon the geometric properties of an imaging system, is the instantaneous field of view (IFOV) of a sensor. The IFOV is the area on the surface that is theoretically viewed by the instrument from a given altitude at a given time.

The spatial resolution is usually determined by instrumental parameters and by the height of the satellite above the ground. With the exception of active microwave systems, the resolution of a system cannot be better than approximately $H\lambda/D$ (the diffraction limit), where H is the height, λ is the wavelength and D is the diameter of the objective lens, objective mirror or antenna. This limit is typically of the order of 10 to 100m for VIS and IR systems operating from satellites in low orbits, and typically 1 to 10km when the satellite is geostationary. For passive microwave observations, the resolution limit is much coarser (of the order of tens of km) because of the larger wavelength measured.

It was stated that the best achievable spatial resolution is of the order of $H\lambda/D$ (except for some types of radar system), although some non-radar systems may not reach this resolution because of other instrumental effects. Two important examples are sensors in which the incoming radiation is focused onto an array of discrete detecting elements, and photographic systems. The detecting element or film imposes its own maximum resolution, again proportional to the height H and, if this is poorer than the diffraction-limited resolution, it will dominate.

The spatial resolution achievable by radar systems is very dependent on the way in which data from the system are processed, and can be better than the simple diffraction-limit formula would suggest. Such systems are often pulsed, and one important factor is the length of the emitted pulse. Synthetic aperture radars (SARs) also integrate the return signal for a period of time while the radar is carried forward on its platform, and the integration time also influences the resolution. It is not possible to give here a statement of the general principles determining radar spatial resolution, and the interested reader is referred to treatments given, for example by Ulaby, Moore and Fung (1981 and 1982), Elachi (1987) and Rees (1990).

The situation is actually more complicated than the preceding paragraphs would suggest. The achievable spatial resolution, in addition to being strongly affected by instrumental parameters and orbital height, is also affected by atmospheric conditions and by the nature of the target material itself. For example, linear features such as leads (Figure 2.4), although narrower than the specified spatial resolution of a sensor, may be detectable on a VIS/IR image due to the high brightness contrast between ice and open water.

Swath width

Swath width is defined as the width of the strip, parallel to the satellite's track, from which radiation is received. It is generally desirable that this should be as large as possible, whereas

Figure 2.4   Leads in the largely divergent FY ice cover of the eastern Weddell Sea, July 1986.

the IFOV should be small, and these two conflicting requirements are often reconciled by arranging that the IFOV should be *scanned* from side to side as the satellite moves forward. If the array of detecting elements is larger in the across-track direction, such scanning is not necessary, and with some microwave techniques scanning is not feasible because of the large and massive antenna. In the latter case, the swath width and the IFOV will be the same.

The maximum swath width attainable by a sensor is effectively limited by its data bandwidth, the maximum value for which is about 100 megabits per second for present transmission systems. High resolution systems are thus limited as to the swath width that they can cover. Typically, radar and visible scanners achieving a 15 to 20m resolution are confined to about a 100 to 200km swath. In contrast, systems with a 1km spatial resolution, such as the NOAA Advanced Very High Resolution Radiometer (AVHRR), can scan about 3,000km across the subsatellite track. However, trade-offs always occur in satellite remote sensing, and this advantage of wide coverage is offset to some extent by image forshortening and an increase in the atmospheric slant towards the edge of the swath.

It is clear that if one wishes to obtain spatially complete coverage of a given area which is wider than the swath width of a sensor, the satellite should be in an orbit such that spatially-adjacent suborbital tracks are closer together than the swath width. The spacing between adjacent suborbital tracks decreases towards the Earth's poles, so this is less of a problem for polar remote sensing than for the study of equatorial regions. Interorbital spacing is strongly dependent on the revisit interval, in the sense that small spacings require long revisit intervals, so that the choice of orbital parameters for a satellite mission usually represents a compromise between conflicting requirements.

Illumination and observation angle

Many satellite systems observe at the nadir (directly below the satellite, viewing angle = 0°), and this is especially valuable for imaging systems since distortion is reduced if the image and object planes are parallel. Some systems, however, observe at non-zero viewing angles. This is usually to exploit the dependence on incidence angle of the emissivity or reflectivity of the surface, for which a physical model may exist. Observation away from nadir can thus assist in the discrimination of one target material from another, or the measurement of some physical property of the target material.

Mission lifetime

Satellite remote sensing missions are of much greater duration than airborne operations, and the continuity of data thus provided is one of the substantial advantages offered by satellites. The factors which limit the operational lifetime of a satellite remote sensing mission are the robustness of the equipment carried by the satellite, and decay of the orbit through atmospheric friction. It is not yet routinely possible to 'service' satellites in orbit, so once a piece of equipment fails the corresponding data are for evermore degraded or unavailable. The space environment is fairly harsh, and lifetimes of a few years are typical. For a satellite in an orbit at a height of 500km or more, the decay of the orbit itself is negligibly slow, but at a height of say 200km, the orbit lifetime can be as short as one month or so, and this then becomes the dominant factor. Very low altitude orbits such as this are essentially only used for high-resolution military reconnaissance satellites.

3. Remote sensing of snow and ice: physical basis and general principles

Introduction

Remote sensing can be defined as the process of obtaining information from the Earth's surface by measuring electromagnetic radiation, the observations being made from considerably above the target. In the case of satellite remote sensing, we need not worry about a definition of 'considerably', since any system will qualify. The advantages of satellite remote sensing are the facts that large and inaccessible areas can be studied and the data from them can be obtained very rapidly, and that some systems can function at night or through fog, cloud or rain. This latter point is especially valuable in polar remote sensing, since these regions undergo six-month nights and are typically cloud-covered.

It is not appropriate here to provide a general introduction to the physical principles underlying the operation of remote sensing systems: that task is undertaken in a number of books such as Curran (1985), Elachi (1987) and Rees (1990), and it will be assumed that these are familiar to the reader. Instead, this chapter will discuss the electromagnetic properties of 'polar' materials (essentially ice and snow) as they affect remote sensing observations. The first of these properties is the *reflectivity* (or *albedo*), which measures the radiation reflected from the surface and below it (*volume scattering*) (Figure 3.1). If volume scattering is significant, we also need to consider the *penetration length* (or depth), which defines the maximum distance below the surface from which significant amounts of radiation are scattered. For active microwave remote sensing, a concept analogous to the reflectivity is defined. This is the normalised dimensionless backscattering coefficient, usually just called the *backscattering coefficient*. This also measures the reflecting ability of the material, but it is normally expressed with an explicit statement of the incidence angle and state of polarisation of the radiation to which it applies. Finally, for thermal remote sensing in both the TIR and passive microwave bands, the important quantity is the *emissivity*, which relates the power emitted by a real material to the theoretical maximum emitted by a perfect emitter (a 'blackbody' under identical conditions and at the same physical temperature). Naturally occurring objects all have an emissivity of less than 1. All of these electromagnetic properties are in general defined as functions of wavelength or frequency, since they are not in general constant across a waveband.

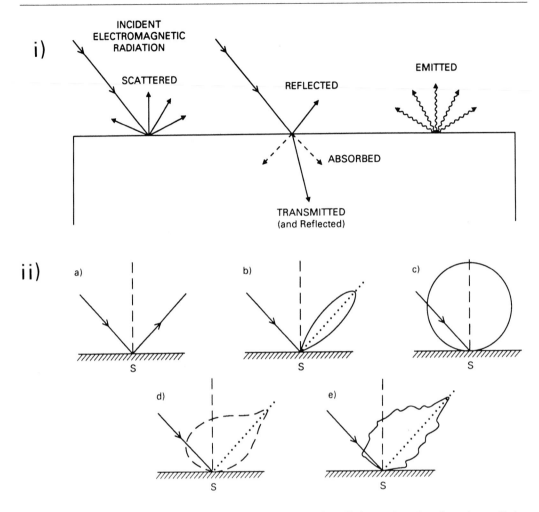

Figure 3.1 (i) Interaction mechanisms between electromagnetic radiation and matter. Specular scattering predominates from surfaces which are smooth relative to the wavelength of the incident radiation; diffuse scattering predominates from rough surfaces. Both are dependent on incidence angle. (ii) Schematic diagram of various types of surface scattering: (a) specular; (b) quasi-specular; (c) Lambertian; (d) quasi-Lambertian; and (e) complex. The lobes are polar diagrams of the scattered radiation. The length of an imaginary line joining the point S (where the radiation is incident on the surface) to the lobe is proportional to the amount of radiation scattered in the direction of the line. Note the increased scattering in the specular direction from most real surfaces. After Rees (1990).

Electromagnetic properties of ice and snow

Visible and near-infrared observations

Remote sensing in the VIS and NrIR parts of the electromagnetic spectrum relies on the sun as a source of illumination. Illumination angles change as a function of time, season and latitude. In the VIS, the nature and intensity of the reflected radiation as measured by the sensor are a product of the incident radiation and the absorption, scattering and reflectance properties of the target medium. The penetration depth is negligible at such short wavelengths, and absorption and scattering processes are concentrated within the uppermost few millimetres of the surface layer (although contributions emanate from greater depths in open water regions, ie of the order of metres). The scattering is a function of geometric properties of the surface and the absorption coefficient, which is a function of wavelength.

Albedo (reflectance) is an important snow and ice bulk parameter that can be measured at short (ie VIS and NrIR) wavelengths (Figure 3.2). The albedo of snow is much greater than that of any other common natural substance at these wavelengths, which has important implications for the global radiation budget and climate. It is mainly a function of the surface roughness and slope, and the electrical (dielectric) properties, density, wetness, illumination angle, grain size, depth (when the snow/ice is thin) and volumetric inhomogeneities of the near-surface layer (Dozier 1989). The sensitivity of albedo to grain size is greatest at 1.0–1.3 µm.

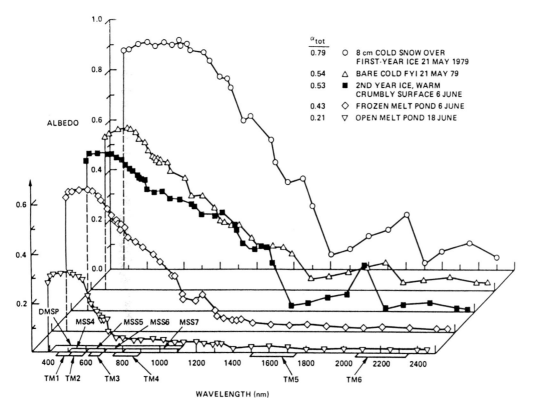

Figure 3.2 Sea ice albedos collected by Dr Tom Grenfell (University of Washington) near Barrow, Beaufort Sea. These values represent samples as no effort has been made to generalise by season or region. Some satellite sensor bands are shown (ie LANDSAT MSS and TM). From Carsey and Zwally (1986).

Spectral variations in the reflectance of snow/ice at VIS and NrIR wavelengths are mainly due to the fact that the absorption coefficient (ie the imaginary part of the refractive index) of these surfaces varies by up to seven orders of magnitude at wavelengths between 0.4µm and 2.5µm (Dozier 1989). Ice and water are optically very similar, except for the region from 1.55–1.75µm where ice is slightly more absorptive (NASA 1987a). In the NrIR wavelengths, ice is moderately absorptive, and the absorption increases with wavelength. Because snow is strongly forward-scattering in the NrIR, albedo increases will illumination angle at these wavelengths.

The complex angular relationships between the satellite, the sun (the solar zenith angle) and the target point (the point on the surface viewed by the sensor) govern the directional sensor response from that target surface. The anisotropically distributed radiance incident on the target is directionally scattered according to the target's bidirectional reflectance distribution function (BRDF), which in turn determines the response of a sensor looking in a particular direction. These factors should ideally be accounted for when using satellite VIS and NrIR data obtained from a fixed look angle to retrieve realistic surface albedos. Certain simplified schemes, such as band ratios, are often implemented to counteract these complex effects, although this particular method is successful only if the two bands chosen have similar atmospheric scattering properties. This arm of VIS and NrIR remote sensing is the subject of intense research (eg Dozier 1989).

The presence of absorbing impurities and increases in grain size reduce reflectance, and reflectance tends to decrease as the snow ages; fresh fallen snow has an albedo of 0.75–0.90, whereas that of old snow is 0.45–0.70 at VIS wavelengths. Similarly, the albedo of growing sea ice exhibits a great variability, as does that of melting snow (Ross and Walsh 1987). The presence of liquid water in the snow does not in itself significantly affect the albedo. It does, however, cause individual grains to cluster (Colbeck 1979) and behave optically as single, enlarged grains, causing decreased reflectance at NrIR wavelengths. It may be possible, with data of a sufficiently high spatial resolution, to use reflectance measurements at 1.03, 1.26 and 1.37 μm to detect the presence of liquid water and improve snow runoff prediction models (Hyvärinen and Lammasniemi 1987). Although grain size and the amount of absorbing impurities can be roughly estimated with present-day sensors, more accurate measurements of snow reflectance will not be possible until the advent of higher spectral resolution sensors in the 1990s (NASA 1987a). The optical properties of snow are extensively reviewed by Benson (1989) and Warren (1982), seasonal snow metamorphism by Colbeck (1982), and freeze-thaw snow layering effects by Colbeck (1991).

Spectral reflectance of sea ice is more difficult to model than that of snow and firn (snow that has survived a summer melt season and has begun the transformation to glacier ice) because columnar ice crystals, the size of which are related to growth rate, cannot be approximated by 'equivalent' spheres. Grenfell (1983) has developed a radiative transfer model for sea ice albedo which is applicable where the ice layer is homogeneous. The albedo is higher at VIS wavelengths for thicker sea ice, whereas it is insensitive to ice thickness in the NrIR. Albedo increases with a decrease in ice density, due to multiple scattering by the air bubbles. The faster sea ice grows, the higher its albedo. Moreover, sea ice is typically covered with snow covers of differing ages and states. The spectral albedo of sea ice is covered by Grenfell (1979), Grenfeil and Maykut (1977) and Perovich and Grenfell (1982).

Thermal infrared observations

As discussed earlier, the important parameters in TIR (wavelengths 2–15 μm) remote sensing are the emissivity of a material and the penetration length. The former determines the efficiency of the material at emitting thermal radiation, the latter the depth below the surface from which this radiation emanates. Penetration depth in all the important geophysical materials in polar remote sensing (ice, water, snow and firn) are very small at TIR wavelengths (of the order of microns), so that the signals detected from space are derived only from the surface and near-surface layer. Emissivities are typically high (0.97–1.00), so that most of the variation in the detected signal is caused by real temperature differences rather than by differences in the target material. Further research is required, however, to refine relationships between emissivity and important physical parameters such as grain size and wetness and instrument parameters such as viewing angle.

All matter radiates energy in proportion to the fourth power of its absolute temperature. Neglecting atmospheric and directional effects, the surface temperature (T_S) may be estimated from the upward flux of longwave radiation (Wm^{-2}) emitted from the surface measured by the radiometer (L) by Stefan's Law

$$L = \epsilon\sigma T_S{}^4 + (1 - \epsilon)\, O$$

where ϵ is the longwave emissivity of the surface (typically 0.99 for snow and 0.97 for bare ice, leads and melt ponds), σ is the Stefan-Boltzmann constant ($5.67 \times 10^{-8}\ Wm^{-2}/K^4$), and O is the downward longwave flux (Wm^{-2}). Because the absorptance of the surface approaches unity in the TIR wavelength interval, almost all of O is absorbed.

In the $3\mu m$–$5\mu m$ region, the energy detected by a satellite sensor represents a mixture of solar reflected and thermally emitted energy; for this reason, this window is normally used for environmental purposes only when measurements are made at night. Neither this nor the 10.5–$12.5\mu m$ atmospheric window is perfectly transparent. Weak absorption by water vapour and CO_2 occurs in both, and absorption by ozone occurs in the 10.5–$12.5\mu m$ range. However, for objects with near-ambient temperatures (approximately 273–300K), the thermal radiation is near maximum for wavelengths in the 8–$13\mu m$ window, and is therefore measured by sensors sensitive to this region (see Figure 2.3).

Presently available TIR detectors are characterised by a lower spatial resolution and longer dwell time than VIS and NIR sensors, and the detectors need to be cooled to about 115K in order to reduce thermal noise. This in itself presents a considerable technological challenge. Dozier and Warren (1982) have shown that snow temperature should be detectable from space in the TIR region. In practice, the interpretation of thermal data and images of areal temperature distribution over snow and ice is far from simple, as the measured radiance depends on the emissivity of the surface (a function of viewing angle) and atmospheric effects. The contribution of the latter can be great in polar regions. Thus, some knowledge of the physical and temporal conditions under which the surface is heated is essential in analysing such data. Differences in reflectance, emissivity and thermal inertia of the target media and variable atmospheric radiance are important factors affecting the observed surface temperatures.

Unfortunately, the usefulness of VIS and IR remote sensing of polar regions is severely limited by its inability to penetrate cloud cover and darkness, and that of TIR sensing by its inability to penetrate clouds. The ability of the latter to penetrate darkness is, however, a considerable advantage.

Passive microwave observations

Passive microwave sensors detect radiation originating by the same thermal mechanism as TIR radiation, but at much longer wavelengths. The important electromagnetic parameters are again the emissivity and the penetration length, but in this case the variability is much greater, and strongly influenced by the nature and condition of the target material.

The intensity of radiation at microwavelengths thermally emitted by a surface is generally expressed as a brightness temperature, or T_B. Neglecting atmospheric and outer space contributions, the T_B measured by a passive microwave radiometer at satellite height follows the relationship $T_B = \epsilon T_S$, where ϵ is the emissivity of the surface and T_S is its physical temperature in degrees Kelvin. The proportionality of T_B to T_S is a consequence of the Rayleigh-Jeans approximation to Planck's Law of Thermal Radiative Emission. In reality, most objects emit only a fraction of the radiation that would be emitted by a blackbody at the same physical temperature. The emissivity of a surface medium is strongly dependent on its physical properties and state; important information on the latter can thus be derived from measurements of T_B from space. Due to the inherently poor spatial resolution of passive microwave systems, the measured brightness temperature represents an integrated value for all the constituent brightness temperatures within the IFOV or pixel (picture element).

The microwave emissivity of snow is very dependent on the rate at which it has accumulated, being as low as 0.65 (at about 10GHz) in regions of low accumulation and as high as 0.9 where the accumulation rate is high. However, Rott (1985) suggests that the dependence on accumulation rate is not simple and it is clear that much more validation work (with simultaneous determination of temperature both at the surface and as a function of depth) is necessary. What is clear, though, is that the emissivity increases with wetness, and the beginning of surface melting is easy to detect. Penetration depth in dry snow and firn vary from of the order of 20m at 7GHz to about 1m at 37GHz (Figure 3.3).

Passive microwave observations of the polar oceans have proved themselves enormously useful in *delineating* ice edges and in estimating the *concentrations* of sea ice by virtue of the large contrast between the microwave emissivities of open water (typically 0.4 near 10GHz) and of sea ice (0.7–0.9). Microwave remote sensing is also invaluable as a tool for *classifying* sea ice, exploiting the fact that brine distribution and state within a given floe depend on the rate at which the ice grew, its age, crystalline structure, temperature profile and its thermo-dynamic history (Weeks and Ackley 1982). Due to its unique dielectric properties, the presence of brine in the freeboard layer affects the microwave emissivity and thus the measured brightness temperature (Figure 3.4). The highly variable and inhomogeneous nature of sea ice renders it a far more complex microwave medium than snow and terrestrial ice; the interpretation of these data is accordingly more complicated.

The determination, both analytically and experimentally, of the bulk dielectric properties of snow and ice is a prerequisite to the development of scattering and emission models to aid in the interpretation of satellite microwave data (Weeks and Ackley 1982). This challenging field is occupying a number of workers in both experimental and theoretical research, and in studies of both the passive (and active) microwave behaviour of ice (eg Onstott and Gogineni 1985; Grenfell and Lohanick 1985).

To further complicate the problem, Arctic and Antarctic sea ice show significant differences in microwave properties. In the central Arctic, the IFOV of a radiometer is generally filled with a mixture of first year (FY) and MY ice (together with many younger and intermediate ice types). First year ice typically has a thickness of $\geqslant$30cm, and, critically in terms of microwave remote sensing, is relatively saline in the freeboard layer and has a small penetration depth (typically $<$ 1 wavelength of the observed radiation at a temperature of about $-10°C$) and thus a high emissivity (assuming that snow cover effects are negligible) (Figure 3.5). Multi-year ice (ie ice that has survived at least one melt season) is generally thicker and more deformed, and the brine in its surface layers has been largely flushed away during periods of melt to be replaced to a large extent by air/gas pockets and ice lenses (see Figure 3.3). Being of the sub-millimetre scale, and on sub-centimetre spacing scales similar to the wavelengths of the measured radiation, these enlarged inhomogeneities act as volume scatterers and suppress the effective emissivity of the radiating portion of the ice. Multi-year

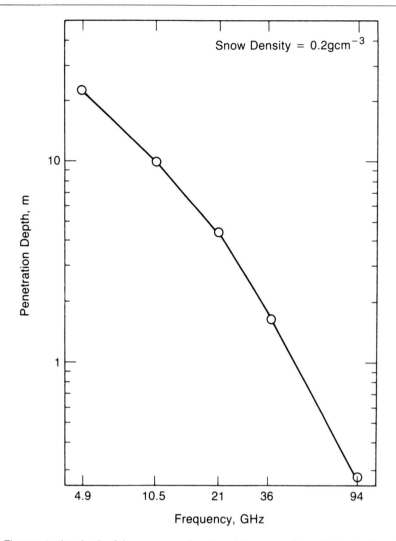

Figure 3.3   The penetration depth of dry snow as a function of frequency. From Rott and others (1985).

ice is thus characterised by a relatively low absorption coefficient, a larger penetration depth and a correspondingly lower emissivity and brightness temperature. The latter tends to be strongly frequency-dependent, the difference in brightness temperature between FY and MY ice increasing with decreasing wavelength in winter (Figure 3.6). As a general rule of thumb, the emissivity of sea ice (with a dry snow cover that is 'transparent' to microwaves) decreases with increasing density, decreasing salinity, decreasing temperature and brine channel scattering (Carsey and Zwally 1986). An excellent review of the growth, structure and electromagnetic properties of sea ice is provided by Weeks and Ackley (1982).

During the spring/summer melt period, weathering and ablation processes act on angular surface features to produce a more rounded topography (Perovich 1983). Snow melt on FY ice in early summer produces a patchy melt water distribution, the low albedo of which causes further locally preferential melting. This process continues until autumn freeze-up, and results in an undulating surface relief with patches of refrozen, relatively non-saline

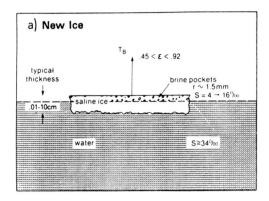

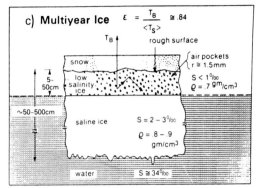

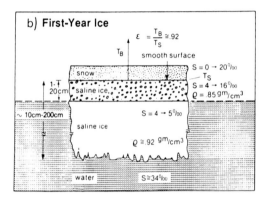

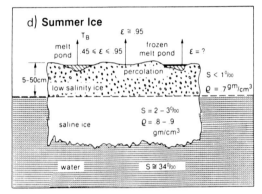

Figure 3.4 Schematic diagram of the physical properties of principal Arctic sea ice types as they affect their radiometric properties at 19.35GHz (wavelength 1.55cm). Differences in observed emissivity are caused mainly by (i) differences in ice thickness (for very thin ice); (ii) variations in dielectric properties (dependent largely on salinity, temperature and wetness); and (iii) the degree of volume scattering. From Zwally and others (1983a).

meltponds (Figure 3.7). Each succeeding melt season enhances the relief on the ice surface, with hummocks becoming higher and meltponds fewer, until it reaches equilibrium after a number of years.

The synopsis presented above essentially describes the evolution of a central Arctic sea ice cover; the situation in Antarctica is sufficiently different to merit the use of different techniques (or the adaption of existing ones) to extract sea ice parameters from satellite passive microwave data. These fundamental differences in sea ice microwave properties emanate from dissimilarities in the meteorological and oceanic variables driving the respective surface energy budgets (Andreas and Ackley 1982). On average, surface winds blowing over Antarctic sea ice are significantly stronger than those over Arctic seas, and the relative humidity of the air tends to be less over the former. As a result, higher air temperatures are required to initiate ice melt, and melt ponds are rare in the south.

As a result of the net effect of these factors, the seasonal brine flushing so characteristic of Arctic sea ice occurs at a much lower level in the south. This combines with a higher proportion of frazil ice (as opposed to columnar ice [Clarke and Ackley 1984; Lange 1988]) to ensure that Antarctic ice surviving the melt season remains relatively saline (commonly

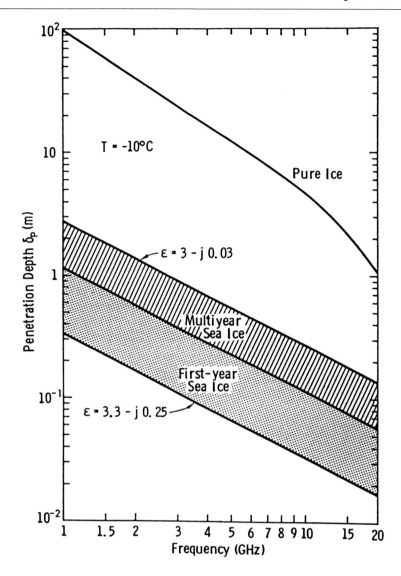

Figure 3.5 Penetration depths in pure ice, FY and MY sea ice. Shaded areas correspond to the range of values of the dielectric constant reported in the literature. Scattering losses due to air inclusions and brine channels were ignored in the computation. From Ulaby and others (1985).

attaining values of 4.6ppt for FY ice and 3.7ppt for second year/MY ice [Ackley and others 1982], as opposed to 3.0ppt and 2.0ppt respectively for the Arctic). The radiometric signature of Antarctic perennial ice therefore remains very similar to that of FY ice at frequencies of less than 40GHz (Gow and others 1982). In terms of remote sensing in the microwave region, Antarctic sea ice therefore represents a far more homogeneous target than its Arctic counterpart (Figure 3.8). The Antarctic sea ice cover is also largely seasonal; perennial ice occurs in significant quantities in the western Weddell, eastern Ross, and central Amundsen Seas. The recognition of Antarctic second/MY ice during the subsequent freeze-up by its microwave signature alone remains a great challenge. Having the ability to monitor

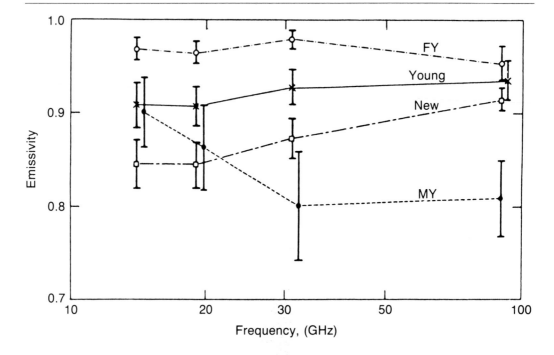

Figure 3.6 The emissivity of four Arctic sea ice types at nadir in the 14–90GHz range. Each vertical bar represents 1 standard deviation in the data, with symbols displaced to avoid overlapping. From Troy and others (1981).

the behaviour of these reservoirs of older ice would greatly enhance our knowledge of the atmospheric and oceanic processes responsible for their existence.

Experiments have shown that a snow cover has little effect on the microwave signature of the underlying ice so long as it is dry, non-saline, fine and no thicker than roughly 35cm. Such a snow cover is an almost loss-free dielectric at microwave frequencies, and is optically thin. However, a thicker snow cover can begin to affect the ice signature in a number of ways, even under freezing conditions. For example, the physical temperature of the ice emitting layer can be changed by the insulating effect of the snow. Towards the snow-ice interface, individual ice grains and water inclusions may act as significant scattering centres, thereby suppressing the amount of radiation emitted (Schanda 1987). Their geometric configurations, dimensions and distributions, measured in wavelengths, are thus decisive factors in determining the scattering behaviour of the snow layer. Wicking of salt from the ice and the accumulation of salt by wind-blown snow can further complicate the equation.

Polarisation effects in passive microwave radiometry of sea ice are poorly understood, yet their analysis may yield additional information on the nature of the surface (Mätzler and others 1984). In general, the brightness temperatures of smooth surfaces display a polarisation effect. The observation of both polarisations enhances the distinction of open water from sea ice, since the real part of the index of refraction is an order of magnitude higher for liquid water than for ice; thus, the polarisation is considerably greater. Vertically polarised

Figure 3.7   Refrozen melt pond on a large FY floe in the Fram Strait, August 1984.

(V) brightness temperatures are invariably greater than those measured at horizontal polarisation (H). The contrast between ice and water is greater with H polarisation, and increases with increasing wavelength, although this polarisation is more sensitive to surface roughness effects. Moreover, the difference between the V and H brightness temperatures is consistently greater for open water than for all ice types at all microwave frequencies (Untersteiner 1984).

Atmospheric contributions are functions of the emission of radiation from the atmosphere itself, and of absorption of radiation emitted from the surface, largely by water vapour (Figure 3.9). These effects are typically small at low frequencies (<10GHz) and at high latitudes due to the low humidity levels characteristic of these regions, particularly during winter. At lower latitudes and in the vicinity of MIZs at maximum ice extent, however, atmospheric attenuation is greater, and its contribution limits the sensitivity of uncorrected brightness temperature measurements to a larger degree. Moreover, atmospheric effects become significant at higher frequencies (>85GHz), as the wavelength of the radiation (approximately 0.33cm) approaches the dimensions of atmospheric particulates (Ulaby and others 1981). Spatial and temporal variations in atmospheric contributions means that their effects cannot be adequately accounted for by a single correction factor applied universally. It is for this reason that surface-sensing MW radiometer systems generally include a channel (often sensing close to 22GHz) that is sensitive to atmospheric water vapour and liquid water (to correct for these effects).

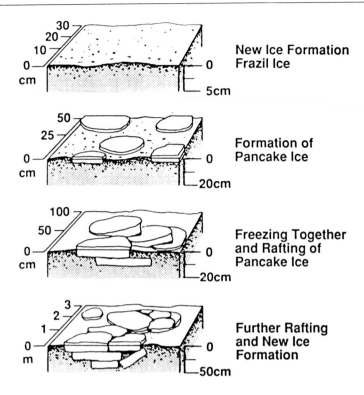

Figure 3.8 Schematic diagram of pancake ice formation in the marginal ice zone of the Weddell Sea, Antarctica. The 'pancake cycle' continues until (i) ocean waves can no longer penetrate into the pack, or (ii) wave energy is not sufficient to raft the pancakes or generate new areas of open water. From Lange and others (1989).

Active microwave observations

Active microwave (radar) techniques, whether imaging (real and synthetic aperture radars) or non-imaging (scatterometers and radar altimeters), measure the power backscattered from a surface; a high return results in a bright target on an imaging radar. The important parameters are the backscattering coefficient (expressed in dB), which is related to the emissivity and the penetration depth (both of which have already been discussed). Thus, the electromagnetic properties which are important in active microwave observations are similar to those in passive microwave radiometry. Because of significant remaining uncertainties in the dependence of backscattering coefficient on physical parameters, the main use of imaging radar to date has been in establishing the two-dimensional geometry of ice masses, by making use of radiometric and tonal/textual contrast.

The interpretation of the SAR image is not as straightforward as it seems, as volume scattering and/or surface roughness effects may contribute significantly to the detected signal (Figure 3.10). Surface roughness effects are a function of wavelength, polarisation and viewing angle. Attempts to classify sea ice and to infer surface characteristics of land ice have

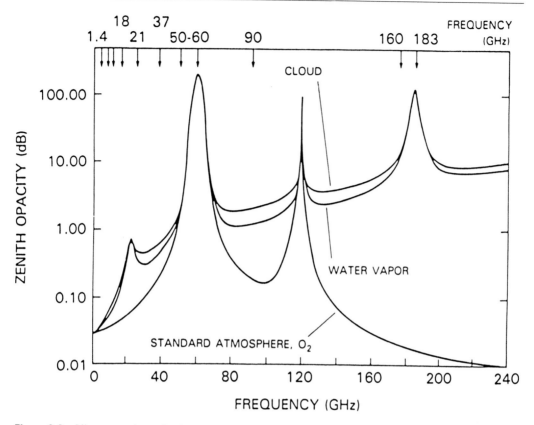

Figure 3.9   Microwave absorption in the atmosphere. The lower curve represents atmospheric opacity due to oxygen; middle curve – opacity with $20 kgm^{-2}$ water vapour added to the oxygen; upper curve – opacity with $0.2 kgm^{-2}$ stratus cloud added to the oxygen and water vapour. From NASA (1987d).

met with limited success (see below), and will require substantially more research before they can be routinely applied. Thanks to the launch of a suite of important SARs in the 1990s, however, the research effort has intensified. The theory of backscattering within sea ice is treated by Winebrenner and Tsang (1988), and within snow and land ice by Rott and others (1985).

SAR imagery is often characterised by random tonal variations from one pixel to the next, a phenomenon known as speckle. This grainy appearance is due to interferences of coherent radiation reflected by neighbouring regions within the same scene. SAR generates images by coherent processing of scattered signals, thereby making it particularly susceptible to speckle noise. Brighter areas are much noisier than darker areas on the image. This factor, alone or in combination with topographic and geometrical effects such as shadows, layover and highlights, can hinder unambiguous interpretation.

Fortunately, speckle effects can be reduced by applying processing technique whereby several independent images of the same area, produced by using different portions of the synthetic aperture and effectively dividing the along-track beam into four or eight sub-beams, are averaged together incoherently (ie with no phase adjustments) to produce a

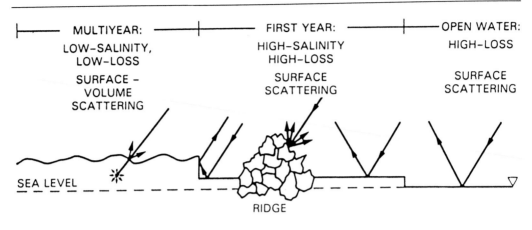

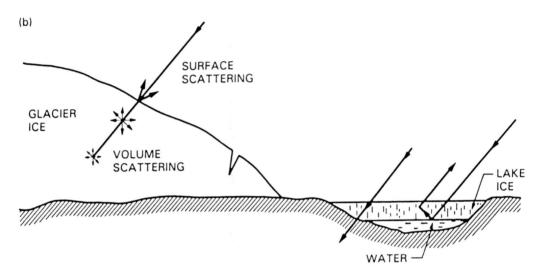

Figure 3.10   Schematic diagram of the interactions between radar and (a) sea ice, and (b) glaciers, ice sheets and lake ice. Interactions will differ at times of melt. From NASA (1987b).

smoother image. This is achieved at the expense of degraded resolution. The number of independent images is called the number of looks; the size of the resolution cell is directly proportional to the latter. Both the digital and optical Seasat images processed by the NASA Jet Propulsion Laboratory are four-look images; the resolution cell is four times larger than, and the speckle intensity one-half that of, a one-look image. Useful guidelines to the analysis of radar data are given by Elachi (1988), Moore (1983a and 1983b) and Simonett and Davis (1983). Radar altimetry and scatterometry are covered in more detail later.

4. History and development of satellite-borne polar remote sensing

Visible and infrared applications

The formation of the US National Aeronautics and Space Administration (NASA) in 1958, and its subsequent planetary exploration programme, was undoubtedly responsible for the development of modern optical space-borne remote sensing systems. For example, the earliest reconnaissance of the Moon, made by the Ranger series, was carried out using a vidicon electronic imaging system. In the race for the conquest of the moon, NASA developed further lunar imaging instruments which required testing on Earth, a factor which led to the establishment of the NASA Earth Resources Survey Program. It was subsequently discovered that mechanical scanning systems turned towards the Earth from space could provide a great deal of information about our planet.

The TIROS series of satellites

The launch of the first Television Infrared Observation Satellite (TIROS) by NASA in April 1960 was an important landmark in the history of remote sensing from space in that it highlighted, for the first time, the huge potential of large-scale overviews of the Earth, albeit cloud- and darkness-limited in the VIS part of the EM spectrum. Atmospheric particulates play a significant role in the VIS, NrIR, IR and TIR regions of the spectrum because their dimensions are similar to the dimensions of the wavelengths of radiation being sensed. Nevertheless, it immediately became clear that one great advantage of satellites is their unique ability to obtain synoptic views on a repetitive basis of large remote areas, even though the orbits of these early spacecraft barely encroached upon the polar regions.

Nine additional TIROS satellites, each carrying a pair of miniature TV cameras and on some missions a TIR Scanning Radiometer (SR) and an Earth Radiation Budget Experiment (ERBE), were launched in the period 1960–5 as research and development testbeds for future operational meteorological sensors. TIROS II, using fairly modest resolution TV cameras, monitored the gross breakup of a sea ice cover for the first time in the spring of 1961 in the Gulf of St Lawrence (Wark and Popham 1962). TIROS II is notable for being the first satellite to carry an IR scanning device. In March 1961, a US Navy transport ship was routed into Stephenville, Newfoundland, largely on the basis of TIROS II photographs of sea ice.

In February and April of 1962, TIROS IV was employed by the US and Canada in a carefully planned reconnaissance programme (Project TIREC) involving the qualitative comparison of satellite and aircraft photographs for ice condition evaluation purposes around the Gulf of St Lawrence. Although it was not always easy precisely to locate surface features, large-scale phenomena such as leads/polynyas and overall ice extent could be distinguished.

In 1963, a follow-up programme was carried out using data from TIROS V and VI in an attempt to develop an operational ice surveillance procedure to use later polar orbiting satellites. US Navy ice observers analysed satellite photographs of ice in the Gulf of St Lawrence, Hudson Bay and along the coasts of Greenland and Labrador. They then transmitted the resulting ice charts to the Canadian Ice Forecasting Central Office in Halifax, Nova Scotia. Consequently, sea ice drifting westwards from Cape Whittle in Labrador, sensed by a satellite on 28 April 1963, became the first reconnaissance data from a satellite source to be used in an official advisory publication to shipping.

TIROS VIII, launched in 1963, provided the first real-time, direct readout of observations to simple ground stations. These initial sunlit images were crude, with an oblique viewing angle causing geometric distortion, and were characterised by a low radiometric sensitivity. In August and September 1964, reconnaissance techniques developed during Project TIREC were used at the Canadian automatic picture transmission (APT) receiving station at Frobisher Bay, Baffin Island, to provide ice and weather information to ships operating between the Labrador Sea and northern Hudson Bay. In the spring of 1964, TIROS VII and VIII images were used to monitor the breakup of ice in the Great Lakes.

The early satellite data were of limited operational value due to the uncertainty of obtaining an observation over a specified area, and few were retained in digital form. Moreover, the orbital inclinations (in the range of 48–58°) mainly precluded extensive coverage of the polar regions. The very short life expectancy of each satellite also upset continuity of data flow.

TIROS IX and X, however, were placed into quasi-polar orbits, and were the forerunners of the present operational meteorological satellite systems. They were the first satellites, along with Nimbus-1, to provide an almost complete global view of the Earth's surface and cloud cover (cyclone distribution) each day. TIROS IX was the first satellite to be placed into a sun-synchronous orbit (on 22 January 1965).

TOS/ESSA series of satellites

The commitment to provide routine daily global coverage without interruption was not met, however, until the launch of the TIROS Observational Satellite (TOS) series in February 1966; it was later renamed the Environmental Science Services Administration (ESSA) series. This marked the inauguration of operational vertical viewing and, with growing assistance from the contemporary revolution in computer technology in the ground segment, the data product line began to diversify. Moreover, the circular sun-synchronous orbits at an inclination of 102° and an altitude in the range of 1,520–1,716km, together with the use of on-board tape recorders for use when out of line-of-site of receiving stations, provided more favourable coverage of polar regions (to within about 10° latitude of the poles).

The nine-satellite TOS/ESSA series used 800-line vidicons to 'photograph' an area of 3,700km² at a ground resolution of 3.7km. By sensing the strong contrast in the albedos of water, snow and ice, these early VIS observations revealed for the first time that the sea ice canopies of both the Arctic and Antarctic undergo large-scale spatial variations over relatively small time scales.

Composite Minimum Brightness (CMB) image mosaics were generated from Advanced Vidicon Camera Subsystem (AVCS) data at the US National Environmental Satellite Data and Information Service (NESDIS), as an aid to research into the large-scale behaviour of snow and ice masses in both polar regions (Figure 4.1) (Booth and Taylor 1969). The even-numbered ESSA satellites provided direct transmission of their imagery to low cost receiving stations by Automatic Picture Transmission (APT), whereas the odd-numbered craft stored their data on onboard tape recorders for subsequent downloading. With ESSA-3, the availability of daily orbital strip coverage represented a great advance in the surveillance of snow and ice from space. Indeed, early satellite images were used for shipping route identification in ice-infested waters. Moreover, Wendler (1973) was able to distinguish five categories of sea ice cover from the imagery, ranging from open water to very compact ice with a dry snow cover.

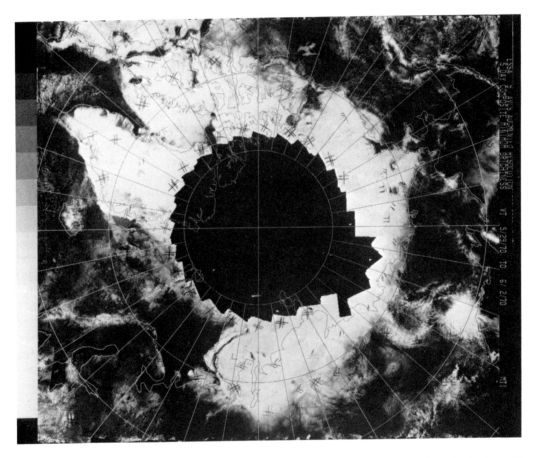

Figure 4.1   A five-day ESSA 9 AVCS composite minimum brightness image of the northern hemisphere, 29 May – 2 June 1970. Courtesy of Charles Swithinbank, SPRI/BAS.

However, many important parameters (eg ice sheet flow lines, leads, pressure ridges, sea ice concentration) were not resolved, the data were difficult to geolocate and they were generally not coupled with detailed (or any) surface truth observations. Other limitations included restricted duty cycles and short mission lifetimes. These factors combined to limit effectively the scientific usefulness of early satellite data. Moreover, they underlined the major drawback in applying VIS and IR sensors alone to the study of polar regions, namely their inability to penetrate cloud cover and darkness.

Cloud cover over Arctic MIZs can be as high as 80% in January. In addition, the presence of thin, low altitude stratus cloud, haze and fog can make the interpretation of surface features difficult if not impossible. Only spring and autumn tend to be acceptably clear, and even then, cloud cover conditions can be unpredictable. It is possible to construct composite image montages from different satellite passes to minimise cloud effects in the VIS and IR when observing ice sheet and glacier margins (McClain and Baker 1969). This approach, however, is less satisfactory when applied to the study of the more rapidly changing sea ice processes. Even under perfect lighting conditions, many different types of ice can yield similar grey signatures when sensed in the VIS or IR, rendering quantitative interpretation difficult unless digital data are available.

Conversely, cloud-contaminated images are themselves important, as reliable cloud cover data from high latitudes are sparse; the study of clouds can give valuable information on the development and passage of weather systems, including intense polar lows (McGuffie and others 1988). Moreover, cloud cover data are critical inputs to coupled dynamic-thermodynamic sea ice growth and decay models, and are important in the study of albedo-climate feedback relationships and radiation budgets (Schiffer and Rossow 1983; Shine and others 1984). Clouds affect the radiance from sea and land ice by changing the radiative balance at the air-ice interface, thereby altering the surface temperature.

Improved TIROS Operational Satellites

The second generation operational polar orbiting meteorological satellite series was launched with the Improved TIROS Operational Satellite-I (ITOS-I) in January 1970 into a sun-synchronous orbit with an inclination of 101.7° and a nominal altitude of 1,742km. It provided the first day (VIS and TIR) and nightime (TIR only) radiometric data in real-time, as well as data stored on-board for subsequent playback, albeit at a poor spatial resolution (2.2km at best) (Figure 4.2).

The evolution of the ITOS system from the proven earlier TIROS and TOS systems permitted significant technological improvements to be made, including the replacement of spin-stabilised spacecraft with three-axis-stabilised platforms. Subsequently, the ITOS system evolved to become the ITOS-D system, the first satellite of which, NOAA-2, was launched on 15 October 1972; this was the first meteorological satellite not to carry cameras, relying instead on scanning radiometers. ESSA was absorbed into the National Oceanic and Atmospheric Administration (NOAA) in October 1970.

The increasing flexibility in platform design, combined with a longer life expectancy, permitted a broader and more sophisticated array of sensors to be carried with only minor alterations to the spacecraft. NOAA-2 and the following three spacecraft in the series each carried a 2-band Very High Resolution Radiometer (VHRR), which measured in both the VIS (0.6–0.7μm) and TIR (10.5–12.5μm) ranges. This sensor, together with the Scanning Radiometer (SR), not only offered superior resolution (about 1km at nadir) but also marked a significant advance in the semi-quantitative use of satellite data to study the large-scale behaviour of snow and ice masses for both operational and research purposes, particularly during polar darkness.

The contribution that these early satellites made to polar research was not limited solely to the domain of sea ice; they also provided, for the first time, an overview of the vast ice sheets of Greenland and Antarctica. They proved to be particularly useful in the large-scale

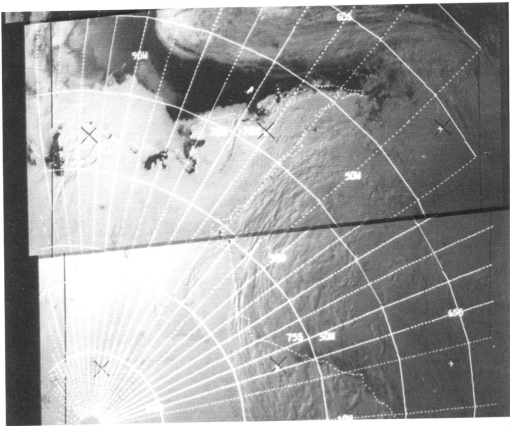

Figure 4.2   An ITOS composite of Antarctica, 22 December 1970 at 22:00 LST. Note the Antarctic Peninsula in the upper centre of the image. Courtesy of Charles Swithinbank, SPRI/BAS.

mapping of previously uncharted mountain ranges, ice sheets/ice shelf margins and large glaciers (Swithinbank 1973). Indeed, they opened up a vast new realm of fascinating and largely unexpected discoveries, and whetted the appetites of glaciologists for the higher spatial resolution sensors that were soon to follow.

TIROS-N and Improved TIROS-N

These sensors were succeeded by Advanced VHRRs (AVHRRs) on the third operational satellite system, launched into a near-polar, sun-synchronous orbit in 1978 and designated TIROS-N. In the present operational configuration, two satellites are positioned with a nominal orbit plane separation of 90°, with one operating in a morning descending orbit and the other operating in an afternoon ascending orbit.

The AVHRR is a significant improvement in terms of spectral resolution (sensitivity), as the broad VIS and NrIR bands of previous sensors are subdivided into separate VIS and NrIR bands. It is also characterised by a fine to medium spatial resolution (1.1km at nadir), a wide swath (approximately 3,000km), good temporal coverage towards the poles and extensive archiving of the data, although its effectiveness as a tool for mapping surface features is limited by cloud cover.

Poleward of 60° latitude, the distance between the ground tracks is less than half the AVHRR's swath width, and each point on the surface is visible in two consecutive passes twice a day. Poleward of 70°, the overlap is large enough to guarantee coverage of a particular point on at least three consecutive passes, rendering the imagery useful in the study of relatively short-lived features. Automatic techniques have also been developed to extract broad-scale pack ice motion from sequential AVHRR imagery by a maximum cross-correlation method (Figure 4.3) (Ninnis and others 1986). Finer scale measurements of motion require finer resolution imagery; this comes at the expense of narrower coverage. Details of floe, lead and fracture patterns stand out more clearly in VHRR/AVHRR data than in the earlier VIS/TIR imagery, and the data have been extensively used to describe sea ice covers on basin-wide to synoptic scales for both operational and research purposes (Figure 4.4).

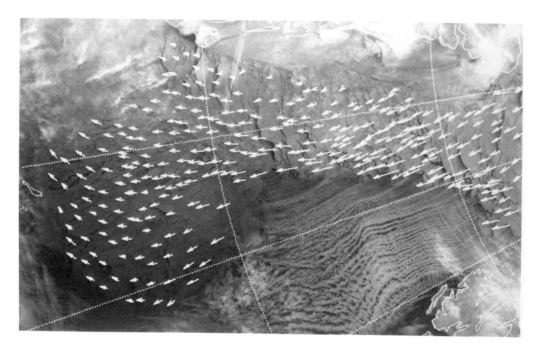

Figure 4.3 The tracking of sea ice floes off East Greenland using sequential AVHRR imagery (25–26 February 1987); average velocities are 30cm s^{-1}. Courtesy of Preben Gudmandsen, Technical University of Denmark.

Over highly reflective snow and land ice masses, the AVHRR sensors are significantly less affected by saturation (over-exposure) problems than those of the Landsat Multi-Spectral Scanner (MSS) and Return Beam Vidicon (RBV), to be discussed subsequently, and also cover regions further poleward. The ten-bit radiometric resolution allows the detection of subtle variations in the topography and emissivity of ice sheet surfaces (Cassasa and others, in press). Berg and others (1982) describe the construction of a full resolution satellite mosaic, using about 70 scenes of NOAA-6 imagery, of the entire continent of Antarctica at a scale of 1:2,000,000. Copies of this map are available from the National Remote Sensing Centre, Royal Aircraft Establishment, Farnborough, Hampshire, UK. Similarly, Wiesnet (1980) has prepared a mosaic of Greenland using NOAA VIS and IR data.

Figure 4.4 An AVHRR band 1 (VIS) image of sea ice in the central Arctic and the N Bering Sea (to the top right), 26 March 1986. A is St Lawrence Island, B Alaska, C the USSR, D the Brooks Range and E Wrangell Island. The Bering Strait can be seen between B and C. Note the polynya in the wake of St Lawrence Island. Courtesy of NOAA/NESDIS.

Due to the high albedo of snow, AVHRR (and Defense Meteorological Satellite Program, or DMSP) imagery have been used extensively to detect and monitor Arctic sea ice snow cover and its evolution (melt, etc) (Crane and Anderson 1984; Dozier and others 1981). Digital archives have been set up of the Northern Hemisphere snow cover extent derived from satellite data (Dewey and Helm 1982).

One major drawback to using VIS, NrIR, IR and TIR imagery in polar regions has been the difficulty in differentiating cloud from snow and ice. Snow and clouds have similar reflected radiances in the VIS and NrIR (up to 1.1 μm). Semi-automatic cloud-snow discrimination is feasible, however, using the wavelength band of 1.55–1.75 μm, where snow appears dark and clouds bright. The reflectance of cirrus clouds is greater than that of snow due to the smaller mean crystal size. Stratus and cumulus clouds have an even higher reflectance because of their greater thickness and the lower absorption coefficient of water, compared to ice, in this spectral region. Persistent cloud cover over the Arctic at times of snow melt is, however, a restrictive problem. Cloud-surface discrimination algorithms are discussed by Arking and Childs (1985), Bolle (1985) and Ebert (1987).

In response to the sparsity of meteorological data from surface-based observation networks in polar regions, Warren and Turner (1988) have applied a technique, used operationally for processing Meteosat geostationary satellite imagery, to AVHRR data to obtain wind speed and direction over Antarctic sea ice. The method operates by tracking clouds on sequential imagery. The relative displacements of the tracers between images are measured by an automatic search procedure using a cross-correlation technique. Heights of the cloud tracers are estimated by comparing AVHRR TIR band brightness temperature data with radiosonde temperature profiles from a neighbouring upper-air station. Both cloud and ice tracking on sequential AVHRR imagery are hampered to a certain extent because the pixel size increases from 1km at the swath centre to 6km towards its edge; consequently, features which may be prominent on one image may be completely obscured on another image from the next pass. In spite of this limitation, these techniques have yielded promising results in the analysis of high latitude mesoscale atmospheric and oceanic circulations.

Sea surface temperature changes demarcate oceanic fronts, which are features of ocean circulation associated with boundary currents and upwelling, ie regions of probable high biological productivity. AVHRR TIR data (and those from earlier sensors, including the VHRR) provided input into the operational Global Operational Sea Surface Temperature Computation (GOSSTCOMP) (Njoku and McClain 1985), now improved through the Multi-Channel Sea Surface Temperature (MCSST) product to the Cross Product Sea Surface Temperature (CPSST) product. Cloud cover and absorption/emission by water vapour and CO_2 are the primary sources of error in these observations. Even in the absence of clouds (ie under clear-sky conditions), the effects of the intervening atmosphere on surface radiances measured in the VIS, NrIR and TIR at satellite height may be significant. In order to better account for these effects (especially water vapour), odd-numbered satellites (from NOAA-7 onwards) are equipped with the improved AVHRR/2 sensor (with 'split window' channels in the TIR, ie 4 and 5).

Atmospheric aerosols are also an important modifier of short- and longwave radiation reaching the satellite sensor from the Earth's surface. NOAA operationally derive the optical thickness of aerosol particles in the total atmospheric column (ie aerosol content) by matching measured radiances (from channel 1 of the AVHRR) with computations in a model of a cloud-free, plane-parallel turbid atmosphere (Nagaraja and others 1989).

Computations to retrieve snow and ice surface physical temperatures from TIR data, a critical variable in the global climate system, must account not only for intervening atmospheric effects but also for spatial and temporal variability of surface emissivity as a function of satellite and solar zenith angle (Dozier and Warren 1982; Price 1984). Typical snow/ice emissivities are in the range of 0.97 to 0.99, whereas that of the sea surface is closer to unity. Algorithms to compute these quantities must also apply corrections to correct for the non-linearity of the TIR channels (Weinreb and others 1990). Unfortunately, AVHRR channel 1 and 2 data are not calibrated on board, and sensor drift becomes a problem, a few years into each mission. Advice on how to compensate for these problems can be gained from NOAA.

The first space-borne TIR sensors were useful only at night and had a low resolution, while recent versions operate equally successfully during the day and night and at finer spatial resolutions. Initially, satellite IR data were also only available in photographic form as a by-product of meteorological cloud imaging, but have recently become more widely available in digital form. Digital data can be radiometrically enhanced more readily to intensify small differences, combined, and geometrically rectified to allow data to be compared for different days and from different data sources. This is particularly important because Earth rotation causes the surface to move beneath the satellite, in general resulting in a geometric distortion of the image formed by a scanning sensor. Moreover, wide swath sensors such as the AVHRR suffer from an increasing amount of atmospheric contamination and geometric distortion towards the swath edge.

The Nimbus series

Concurrent with the NOAA meteorological satellites, NASA initiated the Nimbus pro-gramme in the early 1960s as a research and development testbed for future sensors. Nimbus was a far more advanced and expensive satellite than TIROS, due to the higher development cost of its improved sensors and the larger platform bus. Indeed, this series laid the groundwork for the Landsat satellites. Nimbus-1, for example, carried the first high spatial resolution radiometer into a sun-synchronous orbit with an inclination of 81° (in August 1964). Significantly, it used a new control system, enabling it to orbit in a stable manner; TIROS satellites were stabilised by spinning about their axes.

Seven Nimbus spacecraft have been launched, each representing a significant advance in sophistication, capability and performance over its predecessor. The earliest (Nimbus-1 to -4) carried sensors such as the High Resolution Infrared Radiometer (HRIR) and the Thermal Infrared Radiometer (THIR), which measured cloud patterns and the Earth radi-ation budget and gave another early demonstration of the unique ability of satellites to obtain near-instantaneous synoptic views over vast remote ice masses. An Advanced Vidicon Camera Subsystem (AVCS) image from Nimbus-2 on 8 September 1964 (of the Weddell Sea) represents the first known satellite image of an iceberg (Sissala 1969). Observations from the Nimbus-3 HRIR, operating in the 0.7–1.3μm spectral band, revealed areas of melt (by detecting a decrease in albedo). The Medium Resolution Infrared Radiometer (MRIR) on Nimbus-3 enabled large-scale coverage (albeit at a poor ground resolution, ie 50km) of the albedo of the snow and ice fields of both polar regions (Vonder Haar 1974).

More attention will be paid in the next section to the very important passive microwave radiometers flown on Nimbus-5 to -7 from the early 1970s until September 1987.

NASA Heat Capacity Mapping Mission

Interest in the mapping of the heat capacity of materials on the Earth's surface (particularly from a geological perspective) led to the launch of the NASA experimental Heat Capacity Mapping Mission (HCMM) satellite in 1978. Polar coverage was limited to parts of the Beaufort, Greenland, Chukchi and Bering Seas; these data have been under-used in polar research problems, although the few polar images that do exist are of a high quality (Figure 4.5).

First generation Landsat

Twice-daily coverage from pole to pole makes imagery from the operational meteorological satellites particularly useful for the routine monitoring of large-scale phenomena, including ice sheet margins and sea ice extent; as such, these data complement passive microwave data.

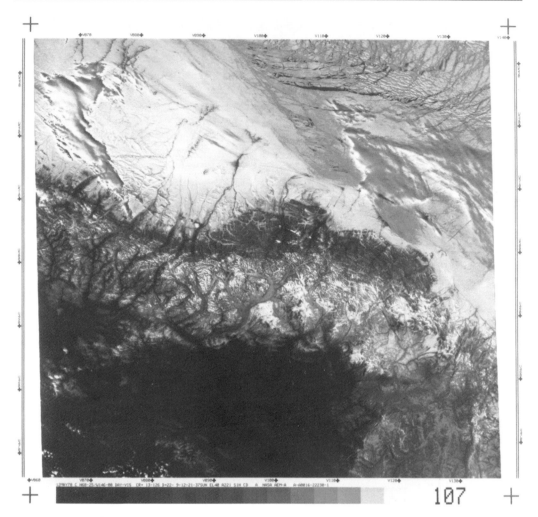

Figure 4.5 An HCMM daytime VIS image (I.D. 016–22230–1) of sea ice off the North Slope of Alaska, 12 May 1978. Courtesy of NSSDC.

Following the proven success of these systems, the US launched the first satellite designed specifically to collect data from the Earth's surface, the Earth Resources Technology Satellite-1 (ERTS-1, later renamed Landsat 1), in July 1972. The payload was designed specifically to meet the requirements of two US Government agencies, the Department of Agriculture and the Geological Survey.

This series has to date included four further satellites. Landsats 1, 2 and 3 carried two types of imaging sensor, namely the Return Beam Vidicon (RBV) and the Multi-Spectral Scanner (MSS), although few polar data were collected by the former due to technical problems (particularly on Landsats 1 and 2). The MSS, which has also been included in the payloads of Landsats 4 and 5, collects data at a resolution of 79m over a 185km wide swath by angular motion of an object plane scan mirror. The oscillating scan mirror provides the cross-track scan, with the along-track component coming from the orbital progress of the satellite. It observes in the VIS and NrIR, and as such is severely limited by cloud and darkness; under clear conditions, however, images of surface features contain superb detail. These data have been used to validate sea ice retrievals from coarser resolution passive microwave imagery (Steffen and others 1989).

The MSS provided the first readily available multi-spectral synoptic images from space in digital form. However, the saturation thresholds of MSS sensors operating over highly reflective snow and ice masses are significantly lower than those of the AVHRR sensors, bands 4, 5, 6 being the worst affected. The problem is greatest at times of higher solar elevation angles (November–January in the Antarctic). Saturated data cannot be computer-rectified. Fortunately, this problem can be predicted to some extent both areally and temporally (Dowdeswell and McIntyre 1986). When the sun's angle of elevation is low, topographic features of both sea ice and land ice show up in great detail (Rees and Dowdeswell 1988).

Of the 3,771 Landsat 3 RBV (30m ground resolution) subscenes acquired over Antarctica (four overlapping subscenes form one standard MSS image), only 100 can be classified as good and 300 as marginal (Ferrigno and others 1983); the rest are saturated. The very few perfectly exposed and processed RBV images are superb, however, in terms of the glaciological feature detail that they reveal (Figure 4.6). Had there been a Landsat receiving station in Antarctica, the RBV problem of sensor saturation over high albedo surfaces would undoubtedly have been diagnosed far sooner (Swithinbank 1985).

Care must be taken when inferring sea ice concentrations from any VIS or IR sensor due to the presence of unresolved narrow leads, small floes and thin ice within the IFOV, ie features must be resolved to be identified. Concentration estimates using these data tend to be too low, the problem growing with decreasing (poorer) spatial resolutions. The MSS can, however, theoretically detect objects smaller than the pixel size; it has been reported that linear features (eg leads) and highly radiative objects, such as icebergs with their low surface temperatures, contrast sharply against a water background in the NrIR spectral region. Iceberg detection within an icefield is more difficult in the VIS.

The narrow swath width (185km) and repeat coverage interval of Landsat are not ideal for many polar applications. Coverage of the same site during a 3–5 day period (at latitudes higher than 70°) is followed by a 13–15 day 'holiday' before the satellite returns to the same area, although coverage improves with increasing latitude due to the convergence of orbits. With two Landsats simultaneously operational, the length of the holiday is reduced by more than half. Even then, however, no coverage is possible poleward of latitude 82°, and cloud-free conditions cannot be guaranteed.

Nevertheless, Landsat imagery played a significant role in the multi-disciplinary Arctic Ice Dynamics Joint Experiment (AIDJEX) in the Beaufort Sea in the early 1970s, whereby sea ice drift, deformation and lead patterns were determined from sequential imagery (Campbell and others 1974). Landsat data provide a detailed 'snapshot' of floe size and shape distribution in the MIZ, and have been valuable in the study of polynyas. The long gaps in coverage, however, make identification of the same sea ice features difficult except during periods of good weather. It is for this reason that Landsat imagery is not suitable for operational use.

Consequently, MSS data have probably contributed more to the study of the great ice sheets, eg providing ice surface velocities from sequential imagery, carefully geolocated (Figure 4.7). Landsat imagery has been particularly valuable as a baseline for planimetric mapping of Vatnajökull in Iceland (Williams 1976) and Antarctica, the latter at a scale of 1:250,000 (Swithinbank 1988). Images were used by the Scott Polar Research Institute to prepare an accurate 1:6,000,000 scale base map of Antarctica (Drewry 1983); more detailed

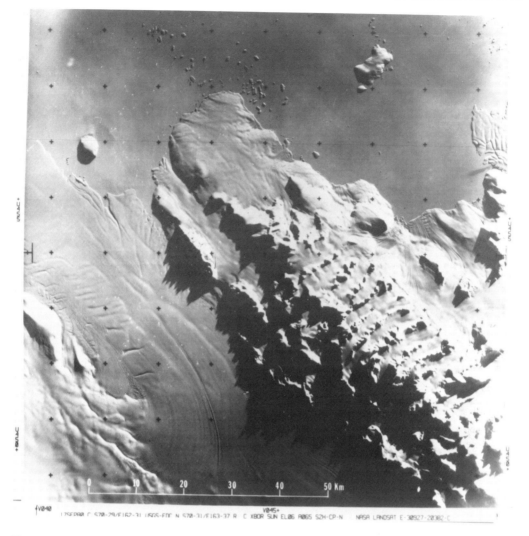

Figure 4.6 Landsat 3 RBV image (I.D. 30927–20382–C) of part of Rennick Glacier, the Bowers Mountains and the Stuhlinger Ice Piedmont, Oates Land, East Antarctica, 11 June 1981. Courtesy of Jane Ferrigno, USGS.

coastal mapping is currently taking place using available data (C. Swithinbank, personal communication). Satellite images have revealed many diverse features never seen before (Swithinbank 1973), including substantial changes in the Antarctic coastline (Ferrigno and Gould 1987). About 55% of the continent now has excellent or good coverage provided by Landsat imagery (Williams and Ferrigno 1988).

Landsat (and NOAA) imagery was used during the late 1970s to locate areas of surface melt on the Greenland Ice Sheet in the vicinity of Sondrestrøm, with a view to the possible development of hydro-power from the runoff (Wiesnet and Matson 1983). This imagery has also proved to be suitable for the study of snow/ice in mountainous regions; Meier (1973) detected the surging of glaciers on early MSS imagery. Moreover, the data have been used to

Figure 4.7   Landsat 2 false colour MSS image (I.D. 2281-07474) of the Jutulstraumen Glacier, Queen Maud Land, Antarctica, 30 October 1975. Flow lines, blue-ice, rock outcrops and crevasses are discernible. The latter have been used on succeeding images to indicate an average glacier velocity of 0.75km/yr. Courtesy of Jane Ferrigno, USGS.

obtain snow area inputs to snowmelt runoff models in lower latitudes (Rango and Martinec 1979).

After 1974, Landsat onboard tape recorders were either inoperative or assigned to higher priority needs, which I shall term the 'Russian Harvest Syndrome'. Consequently, only a handful of MSS images of Antarctica were acquired between 1974 and 1981. Williams and Ferrigno (1988) have prepared invaluable tables of optimal Landsat 1, 2 and 3 images of Antarctica, which will save users much time and effort.

The MSS produces a variety of data products that are map-correct and can be digitally computer-enhanced/combined to highlight features of interest. Digital data should be used wherever possible in preference to hard-copy images, as statistical and mathematical functions can be applied to the former to aid analysis and interpretation. Moreover, the MSS recognises 128 grey tones (0–127) in bands 4–6 and 64 (0–63) in band 7, far more than can be distinguished by the human eye. As a result, much finer spectral distinctions can be retrieved from digital data. An important attribute of Landsat images is that they include the precise time and date of acquisition (Figure 4.8).

Second generation Landsat

Landsats 4 and 5 (launched on 16 July 1982 and 1 March 1984 respectively) carry the Thematic Mapper (TM) in addition to the MSS. An extension of the MSS design, it includes three additional bands, yielding greater detail at an improved spatial resolution (30m for VIS and IR, 120m for TIR), nightime viewing, and offering better calibration and more separation between spectral bands. Another feature is the improved degree of precision/detail offered, as the digitisation was increased from 6 to 8 bits (256 levels).

As satellite-borne remote sensing has advanced technologically, and as the objectives of missions have been more precisely defined, sensor systems have been designed with narrower wavebands. For example, the first three bands of the MSS each have a bandwidth of 0.1μm, whereas the TM is able to employ bandwidths of 0.07, 0.08 and 0.06μm, making it a significantly more sensitive instrument radiometrically. Moreover, the considerable improvement in spatial resolution has led to a great increase in the amount of detail that can be gleaned from TM imagery. For example, TM data are useful in alpine regions, where the measurement of snow and ice properties require data on a spatial scale similar to that of the topographic relief, ie tens of metres.

Landsats 4 and 5 were launched into lower orbits (705km as opposed to 919km for the first three Landsats) with an inclination of 98.2° (as opposed to 99.09°), thereby providing coverage slightly further poleward (82.5°) and changing the repeat cycle from 18 to 16 days. The simultaneous multi-spectral images can be combined or computer enhanced to highlight subtle differences and detail in the low contrast snow surface. Enhanced data from band 4 have even been used to detect Adélie penguin rookeries on the Antarctic coast as a result of the unique spectral characteristics of their guano deposits (Schwaller and others 1986). High quality 1:100,000-scale image maps can be constructed from these data.

Information on bedrock topography and ice sheet dynamics can be inferred from observations of ice surface topography and crevasse patterns (Figure 4.9). Scientists at NASA (Bindschadler and Scambos, 1991) have developed a new method permitting the measurement of ice surface velocity from sequential imagery without the need for nearby rock outcrops to co-register the data. Their study also further highlights the fact that subtle changes in ice sheet elevation, although largely undetectable on the ground, are readily detectable in carefully enhanced multi-spectral imagery.

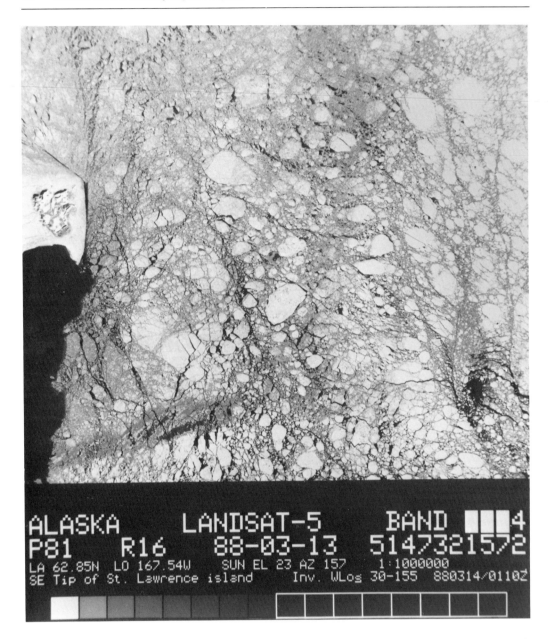

Figure 4.8 Landsat MSS image (band 4) of sea ice in the Bering Sea, 13 March 1988; the SE tip of St Lawrence Island, and the polynya, can be seen to the left. The image was obtained from the University of Alaska. Courtesy of Konrad Steffen, University of Colorado.

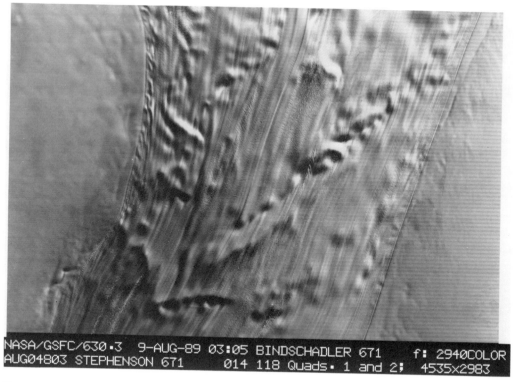

NASA/GSFC/630.3   9-AUG-89 03:05 BINDSCHADLER 671   f: 2940COLOR
AUG04803 STEPHENSON 671   014 118 Quads. 1 and 2;   4535x2983

Figure 4.9 Enhanced subscene of Landsat TM image (Y5105215335XO) of the central section of Ice Stream E, showing patterns of undulations, streamlines and crevasses at the crest. Flow is from top to bottom. From Stephenson and Bindschadler (1990).

Both MSS and TM data have been used to detect and map floating ice fronts, grounded ice walls, ice divides, ice streams and glacier termini behaviour (Orheim and Luchitta 1987). Although the identification of thickly debris-covered ice can present problems, the location of the annual equilibrium line separating areas of ablation and accumulation can be determined. Another interesting application has been the detection of blue ice, which often marks the sites of meteorites embedded in the ice sheet.

In 1987, a consortium of Scientific Committee on Antarctic Research (SCAR) nations initiated a project to acquire cloud-free Landsat imagery of the entire Antarctic coastline. Comparisons with earlier data by scientists at the US Geological Survey have revealed significant changes in the coastal margins in a number of regions. In 1986, more than 11,225km^2 of the Larsen Ice Shelf and 11,500km^2 of the Filchner Ice Shelf calved into the Weddell Sea (Ferrigno and others 1990). A continuing reduction of the Wordie and Larsen Ice Shelves has also been noted.

MSS data have been available, when not limited by duty cycle constraints, on a more or less continuous basis since 1972, although their collection over polar regions has in reality been somewhat sporadic. All Landsat 4 and 5 TM data outside the line-of-sight of receiving stations are transmitted through the Tracking and Data Relay Satellite System (TDRSS); neither satellite has an onboard tape recorder. The technology, however, has remained rather static, with a heavy reliance on mechanical scanning techniques.

Third generation Landsat

Landsat 6 is scheduled for a May 1992 launch, with Landsat 7 possibly following in 1993. This next generation of Landsats will incorporate design improvements. These will include an Enhanced Thematic mapper (ETM) which will sense in eight bands in the spectral range of 0.45–12.5 µm, including a 15m resolution panchromatic band (0.5–0.86 µm). The spatial resolution of the other bands is 30m for the VIS/NrIR, and 120m for the IR/TIR. Although the swath width (185km) is no wider than that of the TM, the improved spatial and spectral resolutions offer exciting possibilities. The ETM bands will not operate simultaneously because of data rate transmission constraints. A low resolution (500m) sensor may also be included in the payload. Known as OMNISTAR, this improved platform will have a life expectancy of 20 years. Latitudinal coverage will still be between 82.5°N and S.

Landsat 7 will also carry an ETM. It may also carry an advanced linear array push broom instrument (the Advanced Landsat Sensor, or ALS), designed to sense in 32 bands in the 0.4–2.5 µm region at a resolution of 10m (VIS/NrIR) to 20m (IR), but over a very narrow swath (41km). The ALS will offer stereo coverage, and will be sensitive to snow/ice angular reflectance properties. Up to 200 minutes (about 400 scenes) of ETM imagery will be acquired every day. Both Landsats 6 and 7 will also carry Emulated Multi-Spectral Scanners (EMSSs), which will use the ETM bands 1–4 to emulate the earlier MSS data. The EMSS bands will have the same wavebands as TM bands 1–4, but will attain a 60m pixel size by processing the ETM data onboard the satellite.

Sensors carried on future Landsats will thus take full advantage of proven subsystem technologies from preceding satellite programmes such as the Advanced TIROS-N and the US Department of Defense Meteorological Satellite Program (DMSP). High-volume onboard tape recorders will ensure near-global coverage when TDRSS is otherwise occupied, eg during Shuttle missions. At the time of writing, the future of the Landsat system is uncertain for financial reasons.

SPOT

The French Centre National d'Études Spatiales (CNES) launched the first Système Probatoire Pour l'Observation de la Terre satellite (SPOT-1) into a polar orbit on an Ariane rocket on 26 February 1986. Its payload of two identical High Resolution Visual scanners (HRVs) provide improved spatial resolution over the TM (10m in panchromatic, 20m in multi-spectral mode), although they lack the ability of the latter to penetrate darkness, view over a very narrow swath (60km for each sensor when viewing vertically), are again cloud-limited, and cannot collect data poleward of 84° latitude. The SPOT system is run as a commercial venture, which is in direct competition with the Landsat system.

Although polar coverage is sporadic and limited by demand, data obtained show a wealth of glaciological detail, and are suitable for mapping at scales of 1:50,000 to 1:25,000. If the SPOT sensors are programmed only for nadir viewing, the revisit frequency for any given point is 26 days, which is clearly unacceptable for the monitoring of short-lived and dynamic phenomena. Flexibility, however, is inherent in the SPOT system, with its ability to adjust both the sensor inclination (up to 27° from nadir) and orbit altitude. This means that the system can provide stereo pairs of images where required, and that the repeat interval can be reduced to 2.5 days for priority regions. SPOT-2 was launched on 22 January 1990, with an identical payload, and SPOT-1 will subsequently be phased out of operation.

The HRV uses the emerging technology of solid-state array imaging sensors, often referred to as multi-spectral linear arrays (MLAs). By this 'pushbroom' technique of scanning, the forward motion of the satellite is used to sweep a linear array of charge-coupled device (CCD) detectors oriented perpendicular to the ground track across the scene being imaged. The orthogonal scan component is provided by electronic sampling in the cross-track dimension to form an image. This method allows a longer dwell time at each scan element, which in turn gives a better (higher) signal-to-noise ratio (SNR) than the mechanical scanner, which has a single detector observing each scan line element sequentially. Thus, narrower bandwidths and a larger number of quantisation levels are theoretically available without decreasing the SNR to unacceptably low levels.

This technique has the advantages of no moving parts, higher geometric fidelity and a longer life expectancy than mechanical scanners. Under present technology, however, push-broom scanners cannot sense at wavelengths longer than approximately 1.05μm, although it is hoped that the second generation of SPOT sensors will add 1.65–2.2μm and 9–11μm bands in the 1990s. An onboard recording capacity of 2 × 23 minutes of data per orbit ensures that some coverage of Antarctica is possible, but only when ordered in advance (Figure 4.10).

SPOT data are inevitably compared with Landsat TM data. The TM includes the appropriate wavelengths for snow/cloud discrimination and night-time (winter) viewing, whereas the HRV has a finer spatial resolution and a pointing capability (covering 1.5° further poleward than the TM). Both the TM and HRV are characterised by a high data volume, and are very expensive (TM less so). Even so, time-lapse measurements using these data offer a rapid, cost-effective method of obtaining average velocities of many ice streams and outlet glaciers near their termini, and the identification of many subtle important features not visible on the ground.

Marine Observation Satellites

Launched in 1987, the first Japanese Earth observation satellite, Marine Observation Satellite-1 (MOS-1), also carried a pushbroom scanning device, the Multi-spectral Electronic Self-Scanning Radiometer (MESSR), up to 82°. MOS1b was launched with an identical payload on 7 February 1990. The MESSR senses in four bands at VIS to NrIR wavelengths, corresponding to bands 1–4 of the Landsat MSS. It offers an improved spatial resolution (50m) over the MSS, but covers a narrower swath (100km). MESSR data will help cover any data gaps left by the probably break in the flow of Landsat MSS data, and are marginally useful in providing ocean colour data to cover the demise of the Nimbus-7 Coastal Zone Colour Scanner (CZCS) in 1986. The wavebands are not ideal for this purpose (see below).

MOS satellites also carry a Visible and Thermal Infrared Radiometer (VTIR). With four bands operating from 0.5–12.5μm, a ground resolution of 0.9km (VIS) and 2.7km (TIR), and a swath width of 1,500km, this sensor is similar to the AVHRR, and has many of the same applications.

The Russian Meteor series

Satellites of the Russian operational Meteor 2 series, launched in 1975, are equipped with vidicons and scanning radiometers, and orbit at an angle of approximately 81° at altitudes of 858–90km (Diesen and Reinke 1978). Two satellites are in orbit at any one time, offering coverage in the VIS (0.5–0.7μm, at a spatial resolution of 1.5km at perigee over a 2,100km swath) and TIR (8.0–12.0μm, spatial resolution 8–10km, swath width 1,400–2,700km). Some of the series also carry a two- or four-band Earth resources multi-spectral scanner; the

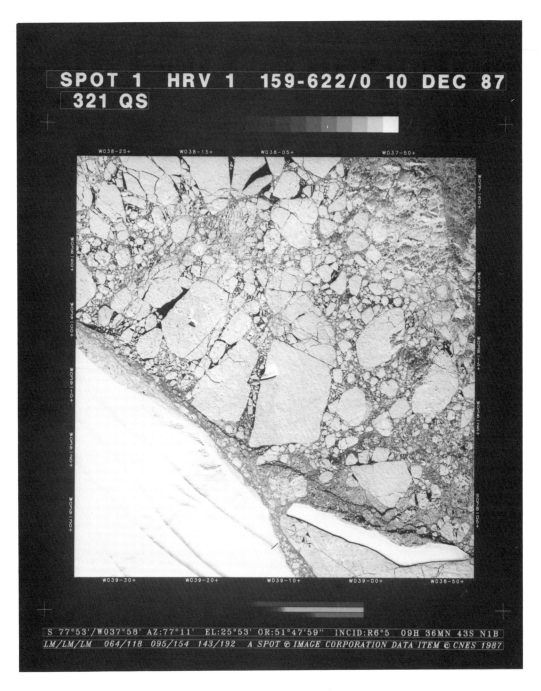

Figure 4.10   SPOT-1 HRV image of sea ice and icebergs in the southern Weddell Sea (77°58′S, 38°21′W), 10 December 1987. The Filchner Ice Shelf can be seen in the bottom left. Copyright 1990 CNES. Provided by SPOT Image Corporation.

two-band sensor, operating in the 0.4–0.7μm and 0.7–1.1μm bands, has a ground resolution of 25m at perigee over a 2,000km wide swath. The four-band scanner, operating in essentially the same spectral bands as the Landsat MSS, provides imagery with a spatial resolution of 1.6km over a 2,800km swath.

The Meteor 2 system delivers data twice daily (at least) on cloud, snow cover and sea ice extent in the VIS and IR; twice daily global data on temperature profiles in the atmosphere, cloud height, SST and the intensity of Earth-emitted radiation; and three times-daily TV images to local receiving stations. Although data are not readily available from the USSR, the automatic picture transmission (APT) data stream is transmitted in the international frequency range of 137–8MHz. Consequently, Meteor data can be obtained using conventional receiving equipment as the satellite passes overhead (although geographical coordinates are not included), and are used for ship routing through ice-infested waters and the monitoring of storms via cloud tracking. Inherent disadvantages of Meteor 2 spacecraft are their inability to process simultaneously VIS and TIR data products, and the low spatial resolution of the TIR product. Information on this and the similar COSMOS series is provided by Caprara (1986), Johnson (1989) and WMO (1989).

The Meteor 3 series, launched in 1989, orbit at the same inclination but at a higher altitude (approximately 1,200km). The standard payload is also similar to that of the Meteor 2 satellites (Johnson, 1989).

A NASA Total Ozone Mapping Spectrometer (TOMS), similar to the Nimbus-7 instrument, will be launched on a Meteor 3 satellite in August 1991. The Hydrometeorological Service of the USSR, Ulitsa Pavlika Morozova, d.12, Moscow 123376, is responsible for the Meteor programme.

The Russian Resurs series

The Soviet Union launched the Resurs series in the late 1980s, the system being divided into three parts: Resurs-F, Resurs-O and Okean-O (Johnson, 1989). The former encapsulates short-duration photographic reconnaissance missions which produce high quality data but suffer from a lengthy processing and distribution turnaround. Resurs-O satellites, the first operational member of which (COSMOS 1939) was launched on 5 July 1988, operate from 82.5°, sun-synchronous orbits at altitudes of 600 to 650km. They are roughly analogous to the US Landsat system, and carry the following standard payload: a multi-band (5) conical scanner (operating at 0.5–1.1μm and 10.4–12.6μm, 600 km swath, 240 × 170km spatial resolution); two multi-band (3) charge-coupled device (CCD) scanners (0.5–0.9μm, 45km swath up to 350km from nadir, 45m resolution); a 4-channel microwave radiometer (0.8–4.5cm wavelength, 1,200km swath, 17–90km resolution); and a sideways-looking synthetic aperture radar (9.2cm wavelength, 100km swath, 200m resolution) with some onboard data processing.

The typical Okean-O payload includes a microwave scanning radiometer (0.8cm wavelength, 650km swath, 6–15km resolution); a sideways-looking airborne radar (VV, 3.15cm, 450km swath, 1–2km resolution); a 4-band scanner (0.5–1.1μm, 1,930km swath, 1–2km resolution); and a high resolution multi-spectral (2-band) scanner (0.5–0.9μm, 1,280km swath, 350m resolution). The satellite also carries a system called CONDOR, which is similar to the ARGOS system of the NOAA satellites in that it collects and relays data transmitted from ground-based instrument packages. The primary role of the Okean-O system is for ship routing in polar regions; ice forecasts are transmitted to ships via EKRAN geostationary satellites. An improved Okean-O satellite is scheduled for launch in 1991; new sensors will operate with more channels and a resolution as great as 50m in the 0.45–1.1μm band (Johnson 1989).

Digital data and data products from these satellites are available from Soyuzkarta, 45 Volgogradsky Prospekt, Moscow 109125, USSR, telephone 95 177–4050, telex 411942 REN SU. The marketing agents in the UK are Sigma Projects. Other national points of contact are being established; contact Soyuzkarta for further details.

The Chinese Feng Yun (FY) series

The People's Republic of China launched its second experimental polar orbiting meteorological satellite FY-1B on 3 September 1990 (FY-1 was launched on 6 September 1988). The payload includes an AVHRR, similar to that flown on the NOAA series, but operating at different wavelengths (0.58–0.68, 0.725–1.1, 0.48–0.53, 0.53–0.58 and 10.5–12.5μm) and with a sub-point ground resolution of 1.1km. The sun-synchronous orbit has an altitude of 900km, an inclination of 99° and a period of 102.86 minutes. Data are collected in three modes, namely HRPT, APT and Delayed Picture Transmission (DPT). The data formats of HRPT and APT are similar to those of the NOAA satellites. The Chinese Satellite Meteorological Center will disseminate the FY-1B orbital prediction via the Global Telecommunications System, enabling worldwide users to receive HRPT and APT data in real-time. The DPT data can be received in China only. Further details are available from NOAA/NESDIS.

The Indian Remote Sensing satellite programme (IRS)

The first dedicated Indian Earth resources satellite IRS-1A was launched in March 1988 into a sun-synchronous, 99.03° inclination, 22-day repeat orbit. The three Linear Imaging Self-Scanning (LISS) pushbroom charge-coupled device (CCD) sensors on board operate in four spectral bands compatible with Landsat TM and SPOT HRV outputs.

US Military Meteorological Satellites

Details were revealed in 1973 of a US Department of Defense Meteorological Satellite Program (DMSP) which has developed in parallel with the civilian programme, and which is strongly influenced by NOAA satellite technology. First launched in 1966 as Block IVA, and formerly known as the Data Acquisition and Processing Program (DAPP), this series routinely transmits data to the US Air Force and Navy for operational purposes.

DMSP polar coverage is similar to that of the NOAA system, but data have not been widely available outside the military community. Generally, two DMSP satellites are in orbit at any one time, providing coverage of any spot on the Earth's surface at least four times daily. Imagery from the Operational Linescan System (OLS) provides an overview of polar regions in the VIS and TIR on a scale (swath width approximately 3,000km, ground resolution 0.618km [2.8km for night-time, smoothed and stored data]) that is intermediate between those of the NOAA AVHRR and Landsat MSS. As such, DMSP provides the highest spatial resolution data available from any meteorological satellite. In addition, darkness is not a limitation (cloud cover is), and the map format, although not as convenient as that of Landsat, is simpler to use than that of the NOAA satellite data (Figure 4.11). DMSP imagery can be geometrically corrected and displayed on standard map projections, with coastlines added. This permits accurate geographical registration of ice features and measurement of their movement by overlying frames.

DMSP OLS data are freely available (in hard-copy and digital formats) at receiving facilities in regions to the south of 60°S; the finer resolution data have only recently been de-classified; all other data, including those collected in the Arctic, are encrypted and are therefore not available in real-time outside the military community. Arctic data are, however, available at a later date from the National Snow and Ice Data Center in Boulder, Colorado (in hard-copy form; negotiations are underway to archive digital Arctic data). DMSP data are used to study variations in summer snow melt on Arctic sea ice (Scharfen and others 1987). Other applications are similar to those of the AVHRR.

The Antarctic Research Center at Scripps Institution of Oceanography (SIO), La Jolla, California has received CCTs containing digital DMSP (and AVHRR) data collected at McMurdo Base in Antarctica since 1987 and latterly at Palmer station (both US National Science Foundation facilities). Together these facilities offer invaluable coverage of a large proportion of the Antarctic continent and its surrounding sea ice cover.

The present series of DMSP satellites (Block 5D-2), launched on 18 June 1987, also carry a third-generation passive microwave radiometer, the Special Scanning Microwave/Imager (SSM/I). SSM/I data, although unclassified, are not available in real-time to civilian re-searchers (see page 81).

Future VIS/IR sensors

It is expected that satellite VIS/TIR imagery will continue to play an important complemen-tary role, alongside passive microwave sensors, in the monitoring of large-scale snow and ice conditions. Future TIR sensors, such as the Along-Track Scanning Radiometer and Microwave Sounder (ATSR-M) on ERS-1 (from 1991), may offer improved detection of new and young ice types (which are often unresolved from space). Although ice and snow surface physical temperatures are theoretically retrievable from TIR data, considerable research is necessary given the uncertainties introduced by emissivity variations and cloud cover. Similarly, further work may enable the retrieval of information on accumulation rates and snow/ice crystal orientation and size from surface emissivity data.

JERS-1, to be launched in 1994, will include a high resolution (25m) Visible and Near-Infrared Radiometer (VNIR). Operating at eight bands in the range of 0.52–2.4μm over a swath width of 150km, potential polar applications are similar to the SPOT HRV, Landsat MSS and MOS MESSR.

In the Earth Observing System (EOS) era, plans are afoot to replace the current NOAA AVHRR/2 with an Advanced Medium Resolution Imaging Radiometer (AMRIR), due for launch on the first European polar platform (EPOP-M1) in 1998. With 11 bands operating in the spectral range of 0.45–12.5μm, a spatial resolution of 0.8 to 3.5km, and a swath width 2,940km, the AMRIR will map snow/ice/albedo at improved spectral and spatial resolutions. NOAA data collection will continue both in parallel and interleaved with the EOS programme.

The EOS-A series (from 1998) will include an Advanced Spaceborne Thermal Emission and Reflection sensor (ASTER), a joint Japanese-US system. Three channels in the VIS-NrIR (0.52–0/86μm) will provide data with a spatial resolution of 15m and will offer a stereo-scopic viewing capability. Six other channels cover the 1.6–2.43μm region at 30m resol-ution, with five further channels covering 8.125–11.65μm at 90m resolution. The swath width of this sensor, which has its inheritance in the MOS MESSR and VTIR and the JERS-1

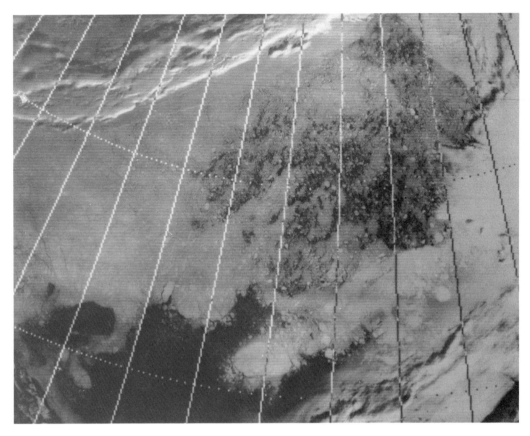

Figure 4.11   Low concentration sea ice in the Canada Basin on DMSP OLS VIS imagery for 2 September 1988. Courtesy of Mark Serreze, University of Colorado.

VNIR, will be 60km. Probable sea ice products will include surface temperature (brightness and kinetic) and meltpond fractional area.

More detailed spectral and spatial sampling will be provided by the High-Resolution Imaging Spectrometer (HIRIS), to be launched on EOS-A in 1998 (please see chapter 5) (NASA 1987a). This instrument will acquire simultaneous images in 192 spectral bands in the range of 0.4–2.5 µm, at a special sampling interval of 10 nanometres (ground resolution 30m, swath width 30km). A pointing capability will allow image acquisition up to +60°/–30° down-track and ±24° across-track. This will enable changes in snow and ice surface spectral bidirectional reflectance-distribution functions (BRDF) and variations in atmospheric attenuation with viewing angle to be studied. At the time of going to press, the future of HIRIS is uncertain. Please see chapter 5 for descriptions of other EOS sensors.

Ocean colour measurements

Sensing in six spectral bands from 0.433–12.5 µm, the Coastal Zone Colour Scanner (CZCS) was launched on Nimbus-7 in 1978 and operated until April 1986. With a ground resolution of 0.825km, the CZCS operated over a wide swath (1,566km), and its narrow band widths allowed enhanced detection of very subtle differences in surface reflectance. It was primarily designed to measure the concentration of chlorophyll-*a* and suspended matter in the upper 10–20m of the open ocean water column. Waters containing a low chlorophyll-*a* concentration have a higher reflectance in the blue region of the VIS spectrum than the green, whereas the reverse is true for waters with a high concentration (Figure 4.12). The detection of particulate matter in the surface layers also allows demarcation of ocean current and circulation patterns.

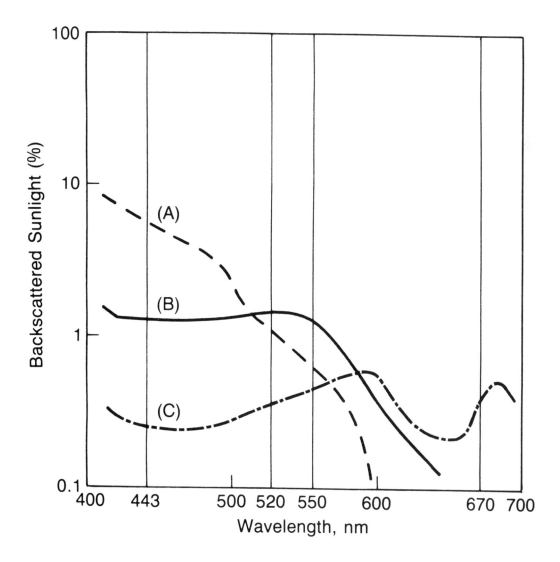

Figure 4.12 The percentage of sunlight backscattered from the upper ocean layers as a function of wavelength (in nanometres), under three conditions: (a) clear open ocean, low phytoplankton concentration; (b) moderate phytoplankton bloom, open ocean; and (c) turbid coastal waters containing sediment and phytoplankton. CZCS observing wavelengths are indicated by vertical lines. After NASA 1989.

The amount of radiation reaching the CZCS under cloud-free conditions depends on factors other than pigment concentration; 80–90% of the total radiance received by the sensor is contributed by atmospheric backscatter, mainly by aerosols and gas/water vapour molecules. This contamination must be removed before any accurate retrievals can be made of the near-surface constituents of the ocean, and improved algorithms have been developed for the purpose (eg Gordon and Castaño 1987). Present CZCS algorithms estimate the atmospheric radiance fraction of the total radiance to accuracies of better than 1% for scan angles within 25° of nadir.

Sun glint is often an important limitation for optical radiometry, becoming insignificant in the TIR at wavelengths longer than approximately 6.0μm. The CZCS scanner could be tilted by up to 20° along-track to avoid this.

CZCS coverage was limited by power restrictions, cloud, darkness and sensor saturation over ice or snow masses (for bands 1–4 under certain lighting conditions). Product retrieval validation from CZCS data is limited by a lack of knowledge of the role of the subpixel cloudiness, various instrumental (electronic problems) and wide pigment/sediment variability. In spite of these inherent limitations, the overall quality of CZCS data is excellent. This factor, coupled with an absence of immediate follow-on sensors, has spurred a concerted effort to extract maximum information from the existing data. Data processing and distribution to the scientific community has been greatly improved, thanks to a cooperative effort by the NASA Goddard Space Flight Center (GSFC) and the University of Miami.

Comiso and others (1990) have used CZCS to look at biological productivity in and around the MIZ of the Weddell Sea, and have produced a number of very high quality images showing both ocean colour and sea ice in great detail. Valuable information related to gross ice extent, residual flows, frontal mixing and mesoscale eddies (horizontal turbulence structures) close to and at the ice edge can be retrieved from CZCS data with careful processing (Figure 4.13).

MOS-2, to be launched by Japan in 1992, will carry a nine-band Ocean Colour and Temperature Scanner (OCTS) into near-polar orbit. Similarly, a dedicated ocean colour satellite, SeaWiFS will be launched by NASA in the mid-1990s. SeaWiFS will sense in eight spectral bands, including four in the VIS, two in the NrIR, and two in the TIR. It will have a one-day revisit interval, with a 2,800km swath width. Nominal resolution will be at 1.0km (local area coverage data) and 4.5km (global area coverage data).

These sensors are improvements over the CZCS in terms of sensitivity to oceanic particulate matter. SeaWiFS combines the best attributes of the CZCS with those of the AVHRR. Although not designed specifically for the study of ice masses, such an instrument could prove very useful in sensing the early stages of ice formation and ice edge processes. Moreover, it will provide improved estimates of primary productivity in the vicinity of the sea ice edge and possibly within polynyas.

The Moderate-Resolution Imaging Spectrometer (MODIS), to be launched on EOS-A in 1998 (see Chapter 5) comprises two companion imaging spectrometers, namely -T (with an in-track tilt capability) and -N (with no tilt capability). The major goal of MODIS-T is to monitor ocean primary productivity; these measurements will be critical in understanding the role of polar marine phytoplankton as a source or sink for carbon (through their intake and emission of CO_2). MODIS-N will measure atmospheric, terrestrial and oceanic processes. The 0.25–1.0km spatial resolution and wide swath (1,500–2,300km) will complement microwave sensor snow and ice data. MODIS-T (sampling in 32 bands in the range of 0.4–0.88μm) and MODIS-N (36 bands, 0.4–14.2μm) may offer improved monitoring of ice surface melt effects, thereby greatly aiding the interpretation of microwave data in summer

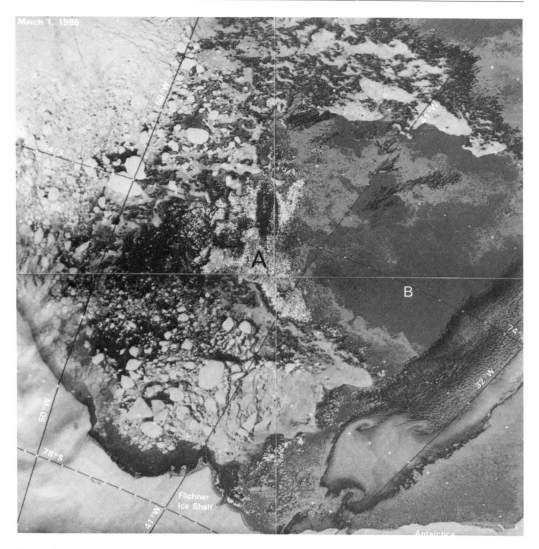

Figure 4.13 CZCS pigment concentrations in the western Weddell Sea, 1 March 1986. The region marked A has a high pigment concentration, whereas B has a low concentration. From Comiso and others (1990).

(NASA 1986). Albedo and reflectance data will be an important input to general circulation models. Moreover, MODIS will produce simultaneous detailed data on ocean colour, SST and sea ice extent in the highly productive MIZs.

Present spatial resolutions have limited the use of CZCS data to the monitoring of oceanic features with dimensions of 5km or more. The fine spatial spectral resolutions of MODIS will allow a more accurate determination of atmospheric effects combined with superior response characteristics. MODIS-T should permit both regional and global studies of the variability of biological and suspended sediment features with scales of 100–200m. It will also offer the flexible option of combining spectral bands or pixels to increase spectral resolution. The large data rate will, however, preclude its use for repetitive mapping of large areas.

Ocean colour sensors, comparable to the CZCS, have also been launched on Russian satellites, eg Intercosmos 21. This satellite was launched on 7 August 1981 into a polar orbit (period 101.7 minutes, altitude 880km and inclination 81.2°).

The Earth's radiation budget and upper atmosphere dynamics

The need for accurate radiation budget measurements over a wide range of time scales is critical to the detection of possible greenhouse and global warming effects. One of the great sources of uncertainty in atmospheric science is whether the presence of a given type of cloud cover heats or cools the planet, ie how it affects the radiation budget. Following on from sensors included in the payloads of NOAA and Nimbus satellites since the early 1960s, sensors of the Earth Radiation Budget Experiment (ERBE) have been collecting very accurate, well calibrated data from two satellites, namely the dedicated Earth Radiation Budget Satellite (ERBS) and the operational NOAA satellites, since February 1985. Although the orbital inclination of ERBS is only 56°, the two NOAA satellites ensure diurnal coverage of polar regions.

The data are archived at, and available from, the US National Space Science Data Center/World Data Center A for Rockets and Satellites, NASA Goddard Space Flight Center, Greenbelt, MD 20771, USA. For more information on the role of satellites in ERBE, please see Barkstrom and others (1989) and Ramanathan (1987). More sophisticated instruments with longer life expectancies will be launched on the polar platforms of EOS in the late 1990s.

Closely related is the monitoring of upper atmospheric dynamics. Scheduled for a 1991 launch, the NASA Upper Atmosphere Research Satellite (UARS) programme will significantly improve our knowledge in the following areas:

i)   The coupled chemistry and dynamics of the stratosphere and mesosphere.
ii) The role of solar radiation in these processes.
iii) The susceptibility of the atmosphere to long-term changes in the concentration and distributions with altitude of key atmospheric constituents, including ozone.

UARS data will be coordinated with results from the Solar Backscatter Ultraviolet Spectrometer/2 (SBUV/2) scheduled for launch on future operational meteorological satellites and the Space Shuttle during the UARS mission. UARS sensor specifications are given in NASA (1990a); the data will also be archived at the US National Space Data Center (see above for address).

Passive microwave applications: (a) sea ice research

Microwave radiometry from space began when a sensor, operating at frequencies of 15.8GHz and 22.2GHz and carried on Mariner 2, made three scans of the planetary disc of Venus in December 1962 (Barath and others 1964). The application of passive microwave radiometry from space to monitoring the Earth's surface was initiated with the launch of the Russian satellite COSMOS 243 on 23 September 1968. This satellite carried a nadir-viewing, non-scanning passive microwave radiometer operating at 8.6, 3.4, 1.35 and 0.81cm wavelengths, corresponding to frequencies of 3.5, 8.8, 22.2 and 37GHz. Basharinov and others (1971) have published what must be one of the first plots showing brightness temperature variations over the Antarctic. Since then, microwave radiometers have been flown on spacecraft in the INTERCOSMOS, COSMOS and Meteor series (Matveev 1976). For example, a 37GHz (0.81cm wavelength) scanner was launched on a Meteor satellite in 1976.

Passive microwave sensors consist of two generic types:

i)   multi-channel spectral radiometers, with spectral line receivers (spectrometers) capable of measuring molecular line profiles from which molecular abundances, temperatures, winds, pressures and other physical parameters can be determined in the vertical atmospheric profile; and

ii) broadband radiometers, used to determine temperature, wind speed, water vapour, precipitation and gross atmospheric profile (the latter four parameters over open ocean only) and, most relevant to this monograph, polar ice extent, concentration and type.

Relevant space-borne passive microwave sensors, past, present and future, are listed in Table 4.1.

First generation passive microwave

NIMBUS-5 Electrically Scanning Microwave Radiometer
Not until the launch (December 1972) of the single channel (frequency 19.35GHz, wavelength 1.55cm) Electrically Scanning Microwave Radiometer (ESMR) on Nimbus-5 did the first synoptic overviews of polar ice masses become available, irrespective of weather, time of day or season. The 3,000km swath width and coarse resolution, with a pixel size of 30 × 30km, allowed the creation of maps of the sea ice covers of both polar regions (poleward of 55°) in their entirety once every 12 hours. The value of passive microwave radiometry on a global scale stems primarily from the large contrast in the emissivities of sea ice and open water (Gloersen and others 1973).

Algorithms to extract ice concentration values from single frequency (and polarisation) ESMR data do so mainly by linear interpolation between the radiance of ice-free open water (with an assumed brightness temperature of 130–5K) and that of a fully consolidated sea ice cover (Carsey 1982; Gloersen and others 1974). Thus, IFOVs containing mixtures of open water and ice will register intermediate brightness temperature values.

An inherent disadvantage of satellite passive microwave observations is that individual ice features can neither be resolved nor tracked. The ice edge and other large-scale features can also only be delineated to the spatial resolution of the instrument. On the other hand, it is not necessary to resolve each open lead/polynya in order to make a quantitative measurement of the open water fraction within a specified area. This technique, by integrating the radiative emission from both open water and ice within the IFOV, avoids the problem encountered by resolving techniques (ie having to resolve open water areas or small ice features).

The marked seasonable variability of snow and ice physical temperature and emissivity, the latter largely as a result of melt and freeze-thaw effects, have a marked effect on the accuracy of sea ice retrievals from passive microwave data (Figure 4.14). The presence of meltponds on Arctic sea ice drastically reduces its emissivity to approximately 0.44 at ESMR wavelengths (Carsey 1985). Although the accuracy of derived sea ice concentrations in areas of mainly first-year (FY) ice has been estimated to be about ±15% (Zwally and others 1983a), it can be as poor as ±25% in areas which have a mixture of FY and multi-year (MY) ice, MIZs and under melt conditions (Parkinson and others 1987). Only if the total ice concentration is known, and in certain seasons and locations, can the relative fractions of FY

Table 4.1 Satellite passive microwave sensors, past, present and future.

Spacecraft and year of launch	Sensor	Frequency (GHz)	Swath width (km)	Spatial resolution at nadir (km)
Mariner 2, 1962	PMW radiometer	15.8, 22.2	Planetary (Venus)	1,300
Cosmos 243, 1968	PMW radiometer	3.5, 8.8, 22.2, 37.0	—	37
Cosmos 384, 1970	PMW radiometer	3.5, 8.8, 22.2, 37.0	—	13
Nimbus-5, 1972	ESMR	19.35 (H)	3,000	25
	NEMS	22.2, 31.4, 53.6, 54.9, 58.8	185	185
Skylab, 1973	S-193	13.9	11–170	16
	S-194	1.4	280	115
Meteor, 1974	PMW radiometer	37.0	—	—
Nimbus-6, 1975	ESMR	37.0 (V & H)	1,270	20 × 43
	SCAMS	22.2, 31.6, 52.8, 53.8, 55.4	2,618	145 to 330
DMSP Block 5D, 1978	SSM/T	50.5, 53.2, 54.35, 54.9, 58.4, 58.825, 59.4	1,600	175
TIROS-N/NOAA, 1978	MSU	50.3, 53.7, 55.0, 57.9	2,300	110
Seasat, 1978	SMMR	6.63, 10.69, 18.0, 21.0, 37.0 (all H & V)	600	149 × 87 to 16 × 27
Nimbus-7, 1978	SMMR	6.63, 10.69, 18.0, 21.0, 37.0 (all H & V)	800	148 × 151 to 27 × 32
MOS-1 (1987), MOS-1b (1990), MOS-2 (1992)	MSR	23.8 (H), 31.4 (V)	317	32, 23
DMSP Block 5D-1, 1985	SSM/T-2	91.5, 150, and 3 channels near 183.0	175–200	50
DMSP Block 5D-2, 1987	SSM/I	19.35, 37.0, 85.5 (all V & H), 22.235 (V only)	1,394	16 × 14 to 70 × 45
ERS-1, 1991	ATSR-M	23.8, 36.5	500	22
NOAA 'Next', 1991 and EOS-A, 1998	AMSU-A	23.9, 31.4, 12 channels in range of 50.3 to 57.29, 89.0	2,300	50
	AMSU-B	89.0, 157.0, and 3 channels near 183.31	2,300	15
GOES 'Next', 1991	PMW radiometer	92, 118, 150, 183, 230	500	35 to 15
TOPEX/Poseidon, 1992	PMW radiometer	18.0, 21.0, 37.0	—	51, 40 and 27
EOS-A, 1998	ESTAR	1.4	1,400	10
	MIMR	6.8, 10.65, 18.7, 23.8, 36.5, 90.0 (all V & H)	1,400	60 to 4.8 (goal)

and MY ice be estimated (in the Arctic) (Gloersen and others 1978; Carsey 1982). During melt periods, the observed microwave signature of Arctic MY ice approaches that of FY ice (Grenfell and Lohanick 1985). Antarctic sea ice represents a far more homogeneous and thus more suitable target for the ESMR than its Arctic counterpart (only 20% of the winter sea ice survives the summer); conversely, the MY fraction is less easily derived in the south.

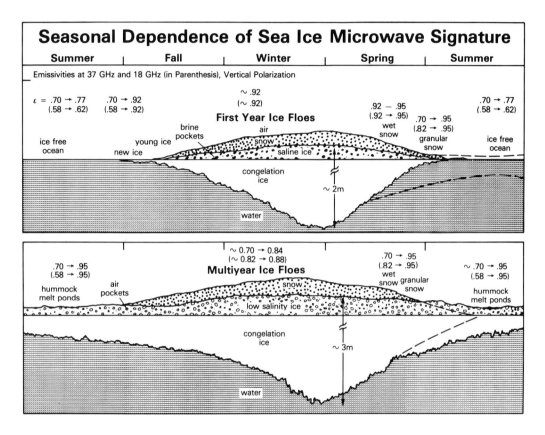

Figure 4.14 Schematic diagram of the seasonal dependence of Arctic sea ice emissivity (at 18 and 37GHz, V polarisation) during its first and subsequent years. Different conditions occur in Antarctica. From Comiso (1985).

Although unambiguous interpretation of its data is not always possible, ESMR-5 ensured virtually continuous polar coverage, and functioned well until 1976. Zwally and others (1983a) derived sea ice concentration and extent maps from ESMR data for the entire Antarctic region from January 1973 to December 1976 by assuming a) a unique emissivity for Antarctic sea ice (ie that of FY ice), and b) that the ice surface temperature is linearly related to the climatological air temperature. Scientists at the Oceans and Ice Branch of NASA GSFC have also produced an Arctic sea ice atlas using ESMR data (Parkinson and others 1987). ESMR data revealed the winter-time existence of a huge polynya in the Weddell Sea (Carsey 1980) (Figure 4.15). This discovery has spawned numerical modelling studies (eg Hibler and Ackley 1983; Martinson and others 1981; Parkinson 1983) which have furthered our knowledge of Southern Ocean air-sea-ice interaction processes.

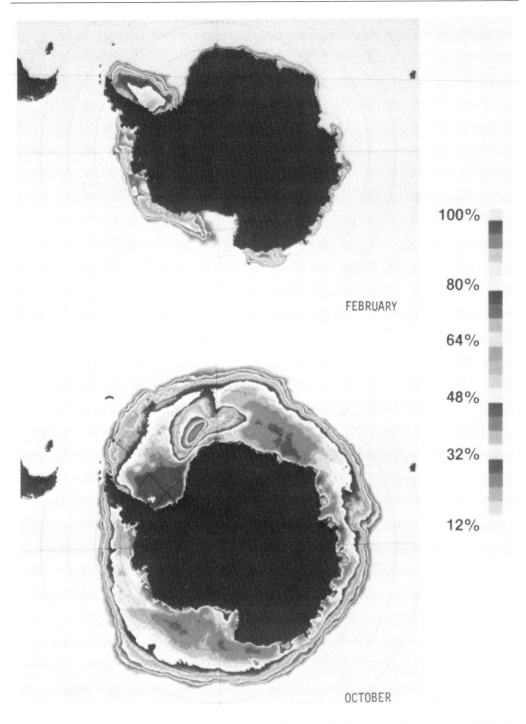

FEBRUARY

100%

80%

64%

48%

32%

12%

OCTOBER

Figure 4.15   Southern Ocean sea ice extent derived from Nimbus-5 ESMR data (Zwally and others 1983a). The months of minimum and maximum ice extent are shown; in each case, data are averaged over the indicated month then over the years 1973–6. The area of low mean concentration is called the Weddell polynya; it was observed in 1974–6, but not in 1973. From Untersteiner (1984).

The simplicity of non-iterative, real-time models permits reasonably accurate ice forecasts to be made with only modest computational capability and resources. As late as 1980, ESMR was the primary data source used to compile US Navy operational ice analyses. However, the use of a single frequency or polarisation creates serious ambiguities in data interpretation, particularly in regions of inhomogeneous ice cover such as the extensive MIZs of the Arctic.

Nimbus-6 ESMR

An upgraded, 37GHz (wavelength 0.81cm) and dual polarisation ESMR (1,270km swath width, ground resolution 20km) was launched on Nimbus-6 in 1976. This frequency change had the effect of roughly doubling the brightness temperature contrast between FY and MY ice, although some increase in ambiguity in ice concentration retrieval occurred. Unfortunately the data are not generally useful due to technical problems.

Second generation passive microwave sensors

Scanning Multi-channel Microwave Radiometers

In order classify sea ice more confidently, it is necessary to make use of the strong wavelength-(frequency-)dependence of the emission and scattering properties of the media within the IFOV, ie multi-frequency techniques are necessary. In effect, one needs to collect as many independent observations as possible, and ideally as many as there are unknown parameters. In an attempt to overcome the ambiguities caused by the lack of beam filling by a single ice type and to improve the accuracy of ice concentration calculations, identical Scanning Multi-channel Microwave Radiometers (SMMRs) were launched on Nimbus-7 in October 1978 and Seasat in June 1978. Seasat unfortunately only operated for 95 days, and its SMMR data are therefore of limited use compared to the Nimbus data set.

The SMMR sensor was fed by a single antenna and a calibration subsystem, and scanned conically (unlike the ESMR which scanned across-track). Two radiometers simultaneously measured the polarisation components of the received radiation, while another four alternated polarisation components during successive scans using a scanning antenna. With ten channels (corresponding to five dual-polarised signals), a swath width of 783km and pixel dimensions ranging from 27×32km at 37GHz to 148×151km at 6.63GHz, the Nimbus-7 SMMR offered complete coverage poleward of $72°$ latitude every day. Sea ice algorithms commonly remap these data into 25km pixel grids.

It was soon discovered that one of the major applications of multi-frequency passive microwave radiometry from space is not only the detection of sea ice but also its classification. The data are suitable for the application and development of inversion techniques to obtain significantly improved derivations of sea ice concentration, FY and MY fractions and information on critical variables such as snow cover state. The rationale behind these techniques is that the surface or near-surface characteristics of the ice cover interact with, and respond to, electromagnetic radiation differently at different frequencies and polarisations (Figure 4.16). For a given surface, different frequencies have different penetration depths and volume scattering characteristics. Emissivities of different target media, while similar at some microwave frequencies, are significantly different at others.

Combining measurements from the different channels is not a straightforward task, as both the size of the antenna FOV and atmospheric effects are different at different frequencies. The finite and coarse spatial resolution of the SMMR necessitates the use of an unfolding procedure to determine the fractional coverage of various surfaces within each IFOV. In spite of the inherent limitations, algorithms to extract both sea ice concentration

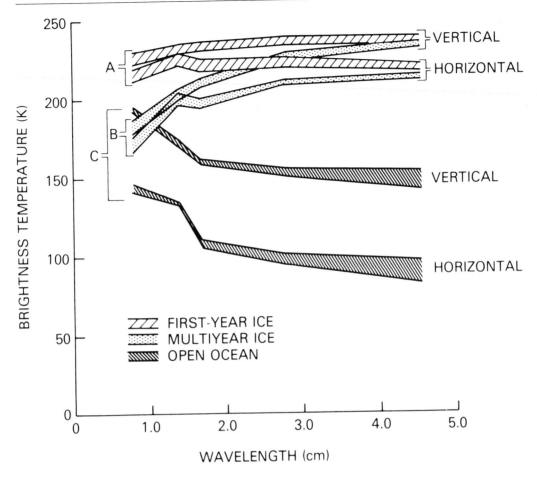

Figure 4.16 The microwave TB of sea ice and open ocean as measured by each of the ten Nimbus-7 SMMR channels in three regions of the Arctic, 3–7 February 1979. A corresponds to a region of consolidated FY ice in Baffin Bay, B to a region of largely MY ice in the Canadian Arctic, and C to ice-free ocean in the Norwegian Sea. Hatched bands indicate ±1 standard deviation about the mean. From Cavalieri and others (1984).

and type from these data have been developed and tested by researchers at various international institutions.

Comiso and Sullivan (1986) have estimated the overall error in ice concentration retrieval using the Comiso cluster analysis algorithm, for example, to be ±10% under freezing conditions (Figure 4.17). Similar accuracies are obtainable with other algorithms, eg the NASA 'Team Algorithm' (Cavalieri and others 1984). The error is greater during the spring/summer period, when uncertainties in the retrieval of ice concentration values occur due to the unpredictable nature of the snow/ice signature. In winter, the emissivity remains relatively stable and more predictable (particularly at frequencies of less than 37GHz), and the error is significantly smaller. Problems inevitably remain within the MIZ, where the widespread occurrence of pancake ice, combined with wave overwashing, may increase the error to ±20% (Comiso and others 1990) (Figure 4.18).

The accurate retrieval of MY ice extent remains a challenging problem. By examining the Arctic MY ice budget from early autumn to late winter, Comiso (1990) discovered than an

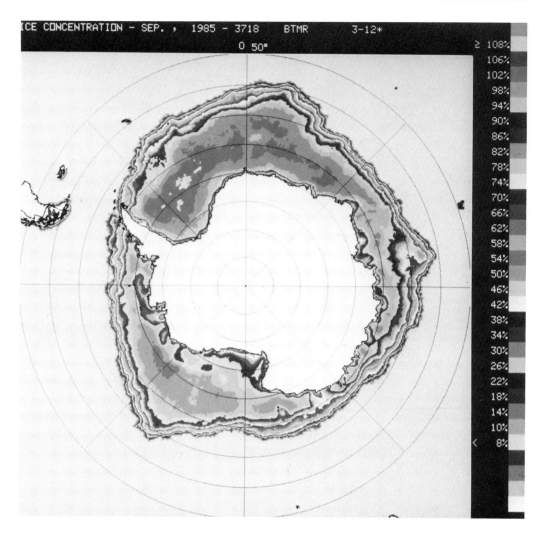

ICE CONCENTRATION - SEP. , 1985 - 3718   BTMR   3-12*

Figure 4.17 A monthly average (September 1985) map of Antarctic sea ice concentrations from Nimbus-7 SMMR data. The concentrations were retrieved from 37V and 37H channel data by applying the algorithm of Comiso (1983). The latter is a bootstrap technique, based upon the consistent distribution of clusters in scatter plots. It is employed to counteract the fact that (i) the absolute calibration of the satellite data is unknown, and (ii) radiative transfer models have yet to advance to the stage where the parameters affecting the radiometer measurements are adequately taken into account. Note that although concentrations of >100% are physically impossible, the amount is within the error limits of retrievals (c.10%) and is caused by variability in the emissivity of the surface.

inordinate amount of MY ice had disappeared in addition to that which could be reasonably exported through Fram Strait (see Figure 1.4). The latter forms the main outlet for ice advecting southwards from the central Arctic. He concludes that the flooding of floes under heavy snowcover loading, and the subsequent refreezing of this more saline saturated layer, would significantly alter (raise) the observed microwave emissivity. The net effect would be to 'mask' the identity of the MY ice, thereby explaining the deficit.

Figure 4.18   Wave overwashing of floes in the marginal ice zone of the Fram Strait, July 1984.

In spite of their inherent limitations, data from the Nimbus passive microwave instruments have contributed greatly to our knowledge of the large-scale distribution and behaviour of sea ice and its seasonal/interannual variations. Although Nimbus-7 was designed with a one-year life expectancy, the SMMR did not cease operation until September 1987. Analysis of ESMR and SMMR data of sea ice in the Southern Ocean for the period 1973–82 has revealed a number of surprising phenomena (Zwally and others 1983b). These include large regional variations, a significant decrease in mean ice extent during the mid-1970s (also reported by Kukla and Gavin 1981), and an increase in the extent of the ice cover in subsequent years. Investigators in the Oceans and Ice Branch of NASA GSFC are producing a combined atlas of SMMR data from both polar regions (Gloersen and others, in press). The analysis of passive microwave data has become reliable enough to warrant its routine use (with AVHRR data) on an operational basis by the US Navy and other maritime organisations in producing maps of sea ice concentration and extent. Imagery faxed in near real-time to ships participating in recent field experiments (eg MIZEX) has also been very useful in helping to plan those experiments from within the field.

As the emissivity of open ocean outside the sea ice margin is relatively well defined and largely constant, SST in cloud-free regions can be computed from passive microwave measurements in the lower frequency range (ie <10GHz) with an absolute accuracy of 1K and a relative accuracy of 0.2K (Bernstein 1982). It is also theoretically possible to determine the surface temperature of snow and ice masses using these data. In practice, however, this is complicated by the presence of intervening clouds and the as yet largely unquantified variability in the emissivity of snow and ice at microwave frequencies.

Sea surface emissivity increases with wind speed and foam cover. Surface roughness affects multiple surface scattering, whereas the major effect of a foam layer on the sea surface is to alleviate the discontinuity between air and water. For this reason, information on wind speed (but not direction) can be inferred from passive microwave data over open ocean (Wilheit 1979). Wilheit and others (1980) describe atmospheric correction techniques for passive microwave data collected over open ocean. Reviews of the methodology and applications of passive microwave radiometry to open ocean studies are given by McClain (1980) and Swift (1980).

Third generation passive microwave sensors

The Special Sensor Microwave/Imager
On 18 June 1987, the US Defense Meteorological Satellite Program (DMSP) initiated the launch of a series of spacecraft that will carry the Special Sensor Microwave/Imager (SSM/I), a scanning microwave radiometer that collects data from a 1,300km wide swath. Similar to the SMMR, the SSM/I differs in that it operates at a different set of frequencies (V and H polarisations at 19.35, 37 and 85.5GHz; V polarisation only at 22.235GHz), and is optimised for the measurement of sea ice and atmospheric water vapour, while lacking SST measurement capabilities (ie it lacks the lower frequency channels of SMMR). The addition of 85GHz is of paramount importance to the study of polar regions. The SSM/I is the first operational passive microwave sensor.

Surface measurements from all spaceborne passive microwave radiometer systems are affected by atmospheric water vapour and rain. Second- and third-generation instruments therefore include spectral frequency bands that are sensitive to these and correct for their effects. The 22.235GHz channel of SSM/I serves this purpose; it is close to a water vapour resonance line. The spatial resolution is frequency-dependent and ranges from about 50km at 19.4GHz to about 15km at 85.5GHZ. Although DMSP is primarily a military satellite series, many of the polar data have been made available to the wider scientific community through the Cryospheric Data Management System (CDMS) in Boulder, Colorado.

The SSM/I provides improved all-weather, all-season, day and night routine mapping of sea ice concentration and type at 25km resolution (once processed), and ice edge delineation to a resolution of 12.5km (a significant improvement over the SMMR) on a daily basis (as opposed to three days for the SMMR). Three-day average SMMR images (actually two days of data separated by one day when the sensor was inoperative) have been invaluable in time-lapse studies of the large-scale advance and retreat of the sea ice masses of both polar regions. Significant jumps in the position of the ice pack can, however, occur over the space of three days. In certain circumstances, such as during and immediately after the passage of a severe storm, the more frequent coverage offered by the SSM/I is desirable, particularly in MIZs.

Passive microwave remote sensing from space, in the form of a continuing programme of SSM/I launches well into the next century and the EOS polar platforms, may be fundamental in detecting and monitoring any global changes that may be occurring via their effects on sea ice extent and concentration (Figure 4.19). It may be too early to draw concrete conclusions from the 17-year time series of satellite passive microwave data already collected (Gloersen and Campbell 1988; Parkinson 1989).

MOS-1 and -1b Microwave Scanning Radiometer (MSR)
Since 1987, a dual frequency passive microwave radiometer, the MSR, has flown on Japanese MOS satellites. This sensor collects data over ice at 23.8 and 31.4GHz (317km swath width, ground resolution 23km). Sea ice concentrations are retrieved using an ESMR-like linear interpolation algorithm. Similar drawbacks exist.

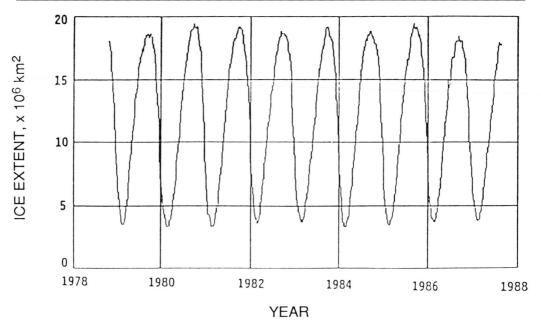

Figure 4.19   Southern Ocean sea ice extents for 1978–87, derived from Nimbus-7 SMMR data. From Gloersen and others (in press).

ERS-1 Along-Track Scanning Radiometer and Microwave Sounder (ATSR-M)
Important supplementary snow and ice data will be provided by the 36.5GHz channel of the ERS-1 ATSR-M (launch 1991), at a spatial resolution of 20km over a 500km swath, although it has already been seen that single-frequency passive microwave data must be interpreted with care. This instrument is doubly important in that it provides simultaneous TIR measurements. ERS-1 will also carry a suite of important active MW sensors (see later), which will produce sea ice products to both validate and complement SSM/I data. The problem of the registration of datasets collected simultaneously by different sensors is difficult to solve, but is not insurmountable.

Fourth generation passive microwave sensors

Advanced Microwave Sounding Units
The two Advanced Microwave Sounding Unit (AMSU) instruments (ie AMSU-A and AMSU-B), due to be launched on the NOAA 'Next' Series of operational meteorological satellites in the early 1990s and the EOS-A polar platform in 1998, will use even higher frequencies, ranging from 23.8–183GHz and operating over a 2,300km swath, and should provide valuable supplementary data to the SSM/I (Heacock 1985; Hollinger and others 1984). Microwave sounders provide data on the vertical profiles of temperature and molecular constituents in the atmosphere by making measurements near the molecular resonance frequencies. Sounders differ from imagers in that they have a lower spatial resolution and do not necessarily offer contiguous coverage. Sounder data (eg from the Stratospheric Sounding Unit or SSU of the current NOAA satellites) are exploited operationally to supplement the horizontal measurements supplied by traditional passive sensors operating in the VIS and IR.

Although designed primarily for meteorological application, certain channels of AMSU-B in particular, with a 15km spatial resolution and a 20-year life expectancy, theoretically have the potential to become an important source of operational sea ice data (Croom 1985). AMSU channels potentially valuable for sea ice studies include those operating at 23.8GHz, 36.5GHz, 90GHz, and 166GHz. The former three will stand direct comparison with the similar frequencies of the SSM/I. FY sea ice, due to its high brine content, has an emissivity greater than 0.9 at all frequencies. MY ice, on the other hand, is good volume scatterer, and has emissivities (in the Arctic) that decrease with increasing frequency to at least 90GHz, where it can be as low as 0.5. The gains in using higher frequency data must, however, be traded off against an increasing contribution from atmospheric and surface effects.

Passive microwave applications: (b) land ice and snow research

For sea ice applications, the land masses on the polar stereographic, mapped passive microwave image products are usually masked out. When not concealed in this fashion, the great ice sheets and shelves of Greenland and Antarctica show remarkably stable patterns of emissivity which have allowed extrapolation from the sparse network of ground observations. Their unexpectedly low brightness temperatures, as measured initially by the ESMR, have been the subject of several theoretical investigations (eg Comiso and others 1982). Unfortunately, no routinely applicable algorithms have been developed for extracting higher level geophysical ice sheet parameters from passive microwave data.

The penetration depth for dry, non-saline snow is of the order of metres at microwavelengths. As a result, the passive microwave properties of ice sheets tend to be influenced by the grain size/structure and mean temperature profile within approximately the upper 10–30m of firn; the physical temperature is controlled by topography/elevation and the meteorological transport of heat. Volume scattering is a dominant factor affecting the microwave emission of a uniform snow medium (Chang and others 1980). The emissivity of dry polar firn is sensitive to grain size variation with depth, which is in turn determined by the accumulation rate (ie larger grains for lower accumulation rates). At the single frequency (19.35GHz) and polarisation (H) of the Nimbus-5 ESMR, the emissivity varies from 0.8 to 0.9 in regions of Antarctica known to have high accumulation rates (eg parts of West Antarctica), and from 0.65 to 0.75 in areas of East Antarctica where accumulation rates are low (Zwally 1977). Consequently, it is theoretically possible to measure accumulation rates indirectly (to an accuracy of about ±20%) in certain regions using brightness temperature data. In practice, other factors come into play; wind redistribution processes, for example, appear to play a dominant role in determining the thermal history, and thus the grain size distribution, of land ice surface layers; other effects may account for <10% of the emissivity variance observed on SMMR imagery (Remy and others 1988).

Other data, collected by aircraft- and satellite-borne sensors at frequencies of 1.4–36GHz, have clearly delineated the position of the ice-firn boundary on the Greenland ice sheet. Moreover, regions of summer melting are delineated by the increased emissivity from wet versus dry snow (Zwally and Gloersen 1977), thereby offering the potential to monitor changes in the ice sheet ablation zones (eg S and W Greenland) (Figure 4.20). The brightness temperature of snow increases significantly with wetness (Stiles and Ulaby 1980), to values of up to 260K at the ESMR wavelength, corresponding to an emissivity of 0.95 at melting point. Conversely, Zwally (1977) observed very low brightness temperatures in the percolation zone of southern Greenland in winter, and this is due, it is thought, to the scattering by large grains formed during the previous summer melt period and subsequent refreezing. Zwally and Gloersen (1977) were able to detect ice stream drainage channels extending up to 500km inland from major outlet glaciers on ESMR data.

An independent determination of the physical temperature, possibly derived from more accurate and better calibrated satellite TIR data, would greatly assist in both the interpretation and mapping of ice sheet accumulation rates. The TIR band of the ATSR-M of ERS-1 may prove to be useful in this respect, possibly used in conjunction with data from its own microwave radiometers or from the DMSP SSM/I. Important limitations in the use of TIR and passive microwave-derived physical temperatures of ice masses are the uncertainty in determining the depth for which the data are representative, and the lack of accurate data on surface emissivities.

The analysis of SMMR and SSM/I data over the great ice sheets is at an early stage, and offers an exciting prospect in harness with theoretical studies and meteorological data (Jezek and others 1990). Brightness temperature data can be used to study classes of meteorological variation over such surfaces. With ice sheets, the choice of V polarisation minimises the effect of surface reflection and maximises the ability to observe radiation from the volume layer of the snow. Further systematic analysis is required. The study of finer-scale phenomena, such as mountain glaciers, using these techniques is effectively ruled out by the poor ground resolution of passive microwave sensors.

Scanning Microwave Spectrometers

Other research has focused on variations in the brightness temperatures of polar firn using data from the dual-frequency (22.235GHz and 31.65GHz) Nimbus-6 Scanning Microwave Spectrometer (SCAMS). SCAMS, along with the Nimbus-5 (Nimbus-E) Microwave Spectrometer (NEMS), was designed to determine atmospheric temperature profiles and liquid water/vapour over the ocean surface. The NEMS was the first microwave instrument to be launched into space to measure temperature profiles, and was the precursor of the current NOAA Microwave Sounding Unit (MSU). Although maps of sheet ice accumulation rates have been assembled using these data (Figure 4.21), both NEMS and SCAMS are severely limited by their poor spatial resolution (200km for NEMS, 150–300km for SCAMS) and the severity of the atmospheric contribution.

Passive microwave measurements of snowpacks

SMMR data have been used with some success to determine regional and global parameters of mid-high latitude seasonal snowpacks (Foster and others 1984; Robinson and others 1984). During the accumulation season, satellite coverage is desirable once every five to six days, corresponding to the time taken for individual weather systems to develop and move along preferred storm tracks. As snow melt begins, the coverage interval should ideally be reduced to three to four days. Prior to their melt season, snowpacks typically exhibit an increase in the average dimension of their crystals (from 0.1–0.2mm to 4–5mm). This metamorphosis has been observed to change the brightness temperature of snowpacks observed at 19 and 37GHz, an effect which is currently being studied in an effort to measure snow-water equivalent (SWE) from space as an aid to increasing the accuracy of snow melt runoff forecasts. Areas with rugged terrain and much vegetation present a greater challenge to algorithm retrieval techniques, although this problem can be partially overcome by using higher spatial resolution (ie 85.5GHz) data.

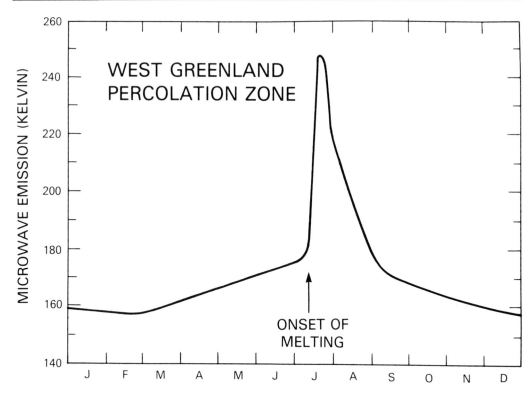

Figure 4.20   Regions of ice sheet surface melt in summer can be mapped by passive microwave remote sensing, as the brightness temperature of the snow/firn varies slowly with seasonal temperature until wetness causes an abrupt increase in the emissivity. From Thomas and others (1985).

Even for the SSM/I, however, the precision of the snowline retrieval on a regional scale is limited by the resolution of the instrument, although the latter does not adversely affect the snowline determination on hemispheric or global maps (at a scale of 0.5° × 0.5°). Until now, radiometers operating at 37GHz (0.8cm wavelengths) have been the most widely used sensors for snowpack monitoring. Scattering of radiation at this frequency is strong since the snow crystal sizes often surpass the wavelength. Although several algorithms are currently available (eg Hallikainen and Jolma 1986; Chang 1986; Foster and others 1984), a generally accepted, reliable SWE algorithm suitable for universal application has proved elusive. It is anticipated that the SSM/I 85.8GHz data will provide more realistic values for snow density and grain size, which are important parameters in radiative transfer models.

Active microwave applications

Whereas the great advantages of synoptic global coverage have been evident since the launch of the first meteorological satellites, it was the brief performance of Seasat (6 June– 10 October 1978) which demonstrated that active microwave instrumentation had reached a point where scientifically useful accuracies and resolutions could be obtained from space.

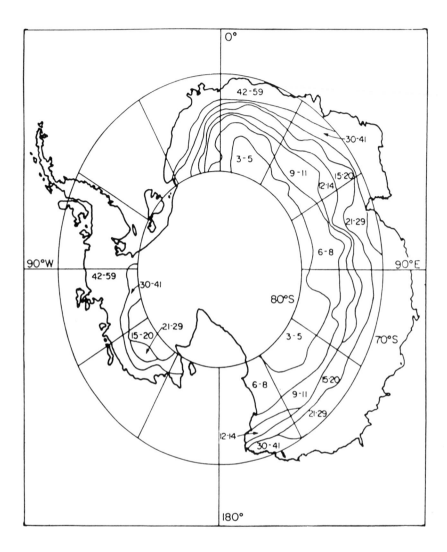

Figure 4.21 Retrievals of snow accumulation rate (g cm^{-2}yr^{-1}) in Antarctica based on 31.6GHz brightness temperature data from Nimbus-6 SCAMS in 1975–6. From Staelin (1980).

Seasat was the first research and development satellite to carry a suite of both active and passive microwave sensors dedicated to the study of the world's oceans. For this reason, it was not required to fly in a sun-synchronous orbit. Although it failed well before its anticipated two-year lifetime and provided no coverage poleward of latitude 72°, Seasat yielded a wealth of unique data which are still keeping scientists gainfully employed.

The Seasat project evaluation teams did not develop algorithms specifically for the extraction of ice parameters from the Seasat SMMR data. Most of our present knowledge on satellite-borne active microwave sensors in the polar regions has however been gained from the Seasat Radar Altimeter (ALT), Scatterometer (SCATT) and Synthetic Aperture Radar (SAR) data. Due to its limited mission lifetime, however, many of the planned series of supporting airborne and 'surface truth' experiments were never carried out. Polar coverage is largely limited to the Beaufort Sea (more than 100 passes). No Seasat SAR data were collected in Antarctica.

The Soviet satellite COSMOS–1500, launched on 28 September 1983, carried an X-band, real-aperture radar (a sideways-looking airborne radar or SLAR) with a spatial resolution of 1km, plus a scanning microwave radiometer and a multi-band VIS scanner (satellite inclination 82.5°, altitude 630km, period 97.3 minutes). The SLAR images, covering a swath width of 450km, are processed at three main receiving stations and supplied to over 500 subsidiary reception points at a reduced resolution (2km). These data are routinely used to assist in ship-routing in the Soviet Arctic; their practical value was demonstrated in December 1983 when images were used to map escape routes for dozens of Soviet ships trapped in heavy sea ice near Wrangel Island. Multi-year ice can also be distinguished from FY ice under most conditions, although the coarse resolution causes problems in interpretation (Bushuyev and others 1985). Promising studies of the Antarctic ice sheet have also been made with this imagery (Burtsev and others 1985). SLAR use elsewhere is mainly limited to aircraft reconnaissance. The Soviet Union has been using space-borne SAR operationally since the summer of 1987, when COSMOS–1870 was launched to monitor ice and oceanographic conditions.

This satellite, the first in the Almaz series, carried an S-band (10cm wavelength) SAR with a ground resolution of 25–30m over an image of 20 × 250km. From its operational inclination of 72°, the SAR can reach a maximum latitude of 78°. Almaz-1, launched in March 1991, carries a 15m resolution SAR (image size 45 × 40–300km). Almaz digital data are commercially available within the USA from the Space Commerce Corporation, 504 Pluto Drive, Colorado Springs, CO 80906, telephone 719–578–5490. For further details regarding current Soviet space-borne radars, please see the earlier section entitled *The Russian Resurs series*. Okean-2 is the present active Russian satellite. NASA has agreed to a cooperative scientific programme with the USSR, and it is hoped that Soviet radar data will be received by the Alaska SAR Facility. Future Okean-O satellites will include a new sideways-looking radar which will provide coverage on both sides of nadir, sweeping out a wider swath while retaining virtually the same resolution.

(a) Synthetic Aperture Radar (SAR)

SAR is arguably the single most powerful satellite-borne sensor available for ice sheet and polar oceans research, due to its high resolution and almost all-weather, all-season, day and night imaging capability. Tradeoffs, however, exist for all satellite sensors, and SAR is no exception. The collection of SAR data (ie the effective spatial and temporal coverage that can realistically be achieved from space) is limited by the following practical constraints:

i)   The narrow swath width (80–150km). The data represent a 'snapshot' sample of a limited area of the surface at any one time, and as such are unsuitable for studies requiring synoptic observations.
ii)   The very high data rate (up to 110 million bits per second [Mbps]) and power requirement leads to possible incompatibility with other sensors on the same bus.
iii) The complexity and expense of data processing.

Sea ice concentration
Nevertheless, high resolution SAR data contain a wealth of detail which offer considerable benefits to a number of critical research areas in polar regions. The analysis of Seasat L-band (frequency 1.275GHz, wavelength 23.5cm) data revealed that the major morphological features of sea ice, including individual leads, polynyas, pressure ridges, fast ice and MY floes, can be readily identified from SAR data (Fu and Holt 1982). It also showed that SAR may provide improved estimates of ice concentration and floe size distribution in areas of low concentration (eg coastal, polynya and ice edge regions) (Figure 4.22). As such, the data are very useful in the validation of sensors with poorer spatial resolutions but wider swaths; they can be used in conjunction with the latter, *in situ* observations and coupled air-sea-ice interaction models to tackle complex process-oriented problems. Commonly used radar band names and their respective frequencies and wavelengths are given in Table 4.2.

Figure 4.22   A typical band of FY ice near the largely diffuse ice edge of the Bering Sea, as seen from the NOAA ship *Discoverer*, March 1983. Such features are difficult to resolve on low resolution (ie passive microwave) satellite data, but are more easily discernible using SAR.

Sea ice kinematics and dynamics

The use of sequential SAR imagery is, along with satellite-tracked buoys, the best method of monitoring mesoscale sea ice motion (kinematics) and deformation, especially in areas far from coastlines and points of reference. This technique relies upon the highly accurate information that can be obtained using geometrically-corrected and Earth-located SAR imagery (Leberl and others 1983). Seasat data have been geolocated to an accuracy of 50–100m (Curlander 1982).

Accurate, SAR-derived ice motion vectors are, with related meteorological and buoy data, an essential input to, and validation for, kinematic and dynamic models. The latter describe how the ice responds to a given velocity field/set of atmospheric and oceanic forcing conditions. Such models are a prerequisite to better understanding ice mass balance, heat, moisture, salt and momentum fluxes on the regional scale, and how these relate to forcing fields; these quantities have been among the most poorly measured in the entire air-sea-ice system. Moreover, SAR-derived kinematics can be used to map, by inference, mesoscale ocean structure within the interior MIZ (Manley and others 1987).

The estimation of sea ice motion from satellite imagery, first applied to SAR imagery by Hall and Rothrock (1981), has in the past been a manual procedure, and thus very time-consuming. Consequently, research is proceeding towards the inauguration of automatic techniques to extract motion vectors from the stream of SAR data to be received from ERS-1, JERS-1 and Radarsat by stations such as NASA's Alaska SAR Facility (ASF), which are currently preparing for the data deluge anticipated in the early 1990s (Figure 4.23). The use of motion algorithms is computationally very intensive; to operate within practical time limits and permit interactive experimentation, high-speed parallel processors have been built at NASA's Jet Propulsion Laboratory (JPL) for installation at the ASF. Similar developments have been taking place in Canada and Europe.

Two procedures have been developed for the tracking of sea ice:

i)   cross-correlation (Fily and Rothrock 1986), a technique which has been further developed to require no operator intervention by Collins and Emery (1988); and
ii) feature tracking (Vesecky and others 1988).

Both methods treat the tracked floes as Lagrangian drifters responding to the combined effects of wind stress and ocean currents. In highly dynamic MIZs, Hall and Rothrock (1981) have concluded that ice rotation and drift rates tend to be too great to permit high density ice feature tracking. The chosen repeat period of the satellite is a critical factor here; periods of greater than about three days may be unsuitable for the tracking of sea ice with SAR (even though they may be most suitable for the monitoring of sea ice by radar altimetry). This problem is offset if the orbits 'drift' in the same direction as the ice.

Sea ice type classification

The imminent data downpour has also prompted research into the development of fully automated algorithms to produce sea ice type classification maps and concentration estimates from digital SAR imagery (Wackerman and others 1988). The rationale behind these techniques, which assign tonal and textural variations observed on the image to different ice classes, is once again based upon the concepts of penetration depth and volume scattering, described earlier (Figure 4.24). SAR may prove to be particularly important in the detection of thin ice in leads and polynyas. The problems experienced by passive microwave sensors, due to melt and snow cover effects, etc, also apply to SAR (Drinkwater 1989; Livingstone

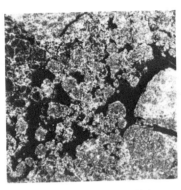

**SEASAT REV 1481
OCT 8, 1978**

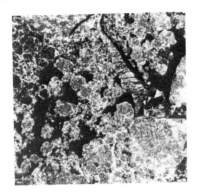

**SEASAT REV 1438
OCT 5, 1978**

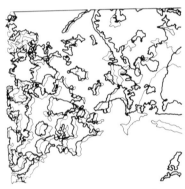

EXTRACTED EDGES
------ **1438**
——— **1481**

DERIVED ICE MOTION

Figure 4.23 Alaska SAR Facility Geophysical Processor System automated ice motion tracking from Seasat SAR data in the Beaufort Sea. The drift of sea ice, as it relates to the mass balance of the pack in space and time, is a central topic of air-sea-ice interaction studies. Such products will be produced routinely from future SAR data. Courtesy of Ronald Kwok, NASA JPL.

and others 1987). Research issues remain regarding the separation of thin ice, smooth water and wind-roughened water signatures.

Other sea ice variables
Fily and Rothrock (1988) have developed an algorithm, based on the techniques described above, for measuring both the opening and closing of leads to an estimated accuracy of ±10–20% by comparing two sequential digital SAR images. Lead distribution, orientation and behaviour can give valuable insight into the dynamics of the pack, and are essential inputs to heat, moisture and salt flux computations. Burns (1988) has proposed the use of SAR data to distinguish sea ice regions with different drag coefficients. The latter form a critical input to sea ice modelling, and at present encompass a wide range of values derived from very few surface measurements.

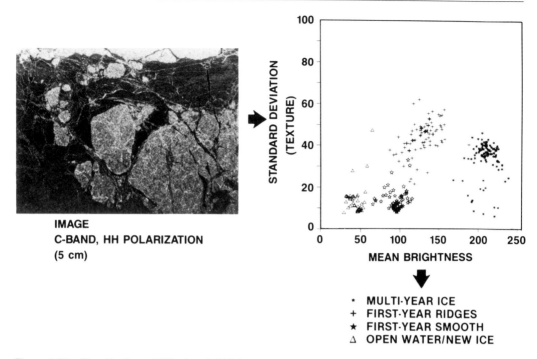

IMAGE
C-BAND, HH POLARIZATION
(5 cm)

* MULTI-YEAR ICE
+ FIRST-YEAR RIDGES
★ FIRST-YEAR SMOOTH
△ OPEN WATER/NEW ICE

Figure 4.24 Classification of JPL aircraft SAR image of sea ice, using tone and texture. Courtesy of Ronald Kwok, NASA JPL.

Waves and oceanography

Over open ocean, an SAR obtains instantaneous maps of the short (1–30cm) gravity wave field present, as well as variations in these short waves induced by longer waves, currents and winds (NASA 1987b). Understanding how these variations are produced and interact with the electromagnetic radiation is a fundamental prerequisite to retrieving quantitative information from SAR data (Figure 4.25). Present experience is based more on theoretical research than experiments in the field (Alpers and others 1986; Hasselmann and others 1985).

The launch of a number of important SARs in the 1990s will provide a unique opportunity to verify and refine models with reliable measurements of directional wave spectra (Lyden and others 1988). Waves and swell are important in sculpting not only the MIZ (Wadhams and others 1988) but also the inner pack (Liu and Mollo-Christensen 1988). Accurate data on wave penetration into, and attenuation by, the ice are essential to the study of acoustic noise levels in the MIZ (Johannessen and others 1988; Rottier 1989).

SAR-derived wave direction and wavelength data will, when synthesised with radar altimeter (ALT) wave height and scatterometer (SCATT) wind velocity data, significantly improve wind-wave forecast and ocean circulation models. These fields are critical to the success of the World Ocean Circulation Experiment (WOCE), to be conducted in the mid-1990s with an important Southern Ocean element (WMO 1987b). SAR also has the unique ability to detect and monitor internal waves in ice-free regions (Alpers 1985), which are most commonly found near coastal regions with a sharp, shallow seasonal thermocline. The convergent and divergent current components of shallow internal waves modulate the short surface waves and accumulate surface films to produce both dark and light bands on the

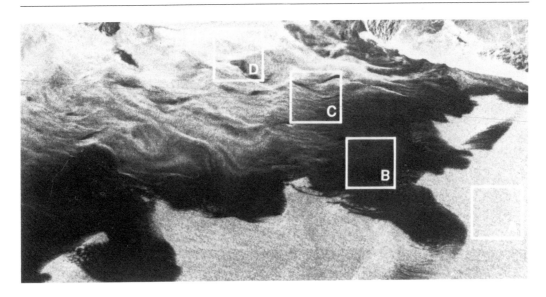

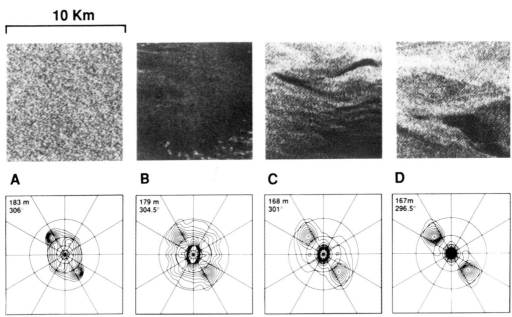

Figure 4.25   Seasat SAR imagery of surface waves penetrating into new ice in the Chukchi Sea, 8 October 1978. The boxes indicate the location where wave spectra were obtained; A is open ocean, B is frazil ice (appears dark due to high frequency wave damping), C and D contain pancake ice (appears bright due to increasing roughness). The corresponding wave spectra are shown below each subscene. The wavelength decreases with increasing distance into the ice field, and wave direction indicates counter-clockwise rotation. The edge of the pack is seen in the upper right of the image. The change in wavelength and direction has been compared to a theoretical model of waves penetrating a frazil ice cover in Wadhams and Holt (1990). Courtesy of Ben Holt, NASA JPL.

ocean surface which are detectable by SAR. SAR can observe fine scale structures in the oceanographic circulation field that would otherwise remain undetected, eg eddies, currents and rings. Methods used to detect these phenomena are described in NASA (1987b).

Iceberg detection

The large amount of airborne data collected over Canadian Arctic waters by organisations such as the Canada Centre for Remote Sensing, the International Ice Patrol and Intera Consultants Inc. has shown conclusively that SAR can both detect icebergs (Pearson and others 1980) and measure their velocity. Such data enable a better understanding of calving and deterioration rates. X-band (and C-band) SAR may be preferable to lower frequencies (ie L-band) for iceberg mapping and detection, due to the decreasing penetration into dry snow/firn with increasing frequency (Rott and others 1985).

Ice sheets and glaciers

It was stated earlier that land ice exhibits a very low dielectric loss (compared to sea ice), and has a correspondingly high penetration depth at microwavelengths. However, although the penetration depth of pure ice is roughly 10m at 6GHz (Ulaby and others 1982), the radar return signal from glacier ice tends to be dominated more by surface scattering, the effects of which depend on the scale of surface roughness relative to wavelength. In contrast, the penetration depth of snow is of the order of 10m at 10GHz, decreasing to about 1m at 40GHz (see Figure 3.3). Under dry, freezing conditions, backscattering properties of firn are largely influenced by the snow stratigraphy, although the surface contribution becomes more important with increasing radar frequency and decreasing incidence angle (Rott and others 1985). The dry accumulation zones of glaciers and ice sheets typically exhibit a strong return signal, as absorption is low and a thick layer contributes to backscattering (Rott and Mätzler 1987). The presence of wet snow significantly increases the dielectric losses, and scattering accordingly emanates largely from the surface and near-surface layers.

Table 4.2 Commonly used radar band names. The C-band, 3.9–6.2GHz, overlaps the S- and X-bands. K_u-band is 12.5–18GHz.

Frequency band	Approximate frequency (GHz)	Approximate wavelength (cms)
P-band	0.225 – 0.39	140 – 76.9
L-band	0.39 – 1.55	76.9 – 19.3
S-band	1.55 – 5.20	19.3 – 5.77
X-band	5.20 – 10.90	5.77 – 2.75
K-band	10.90 – 36.00	2.75 – 0.834
Q-band	36.00 – 46.00	0.834 – 0.652
V-band	46.00 – 56.00	0.652 – 0.536

Over ice sheets, the few SAR data collected so far from space suggest that one of the advantages of SAR relative to other sensors is its ability to accurately map and monitor, on an all-weather, all-season basis, coastal margins, ice shelves, crevasse fields, tidewater glacier advance and retreat, ice motion and rate of discharge (iceberg calving rates), particularly where the latter is rapid, eg the Jakobshavn Glacier in West Greenland (Figure 4.26). It is eminently well suited to carrying out routine detailed surveys of regions which have been

Figure 4.26 Jakobshavn Glacier, West Greenland. With a terminus velocity of 7km/yr, this is the fastest glacier on Earth.

noted to be critical. Other applications include the study of permafrost and glacial/proglacial geomorphological features (NASA Alaska SAR Facility Prelaunch Science Working Team 1989). SAR data collected over terrestrial ice sheets complement those from VIS and NrIR sensors (Vornberger and Bindschadler, in press) and altimeters. However, terrain distortion and shadow effects largely preclude the use of SAR in regions of steep elevation, eg valley glaciers.

The very limited Seasat SAR dataset collected over ice sheets reveals a number of interesting features not detectable from the ground, eg anisotropic returns and wave-like patterns in regions which are known to have featureless, apparently isotropic snow surfaces. The exact cause of these unusual patterns is, as yet, unknown (Swift and others 1985). Theoretical calculations of scattering from snow and ice are of immense importance to the unambiguous interpretation of radar data. Unfortunately, this remains a difficult, although not insurmountable, task as the typical dimensions of surface roughness are often similar in magnitude to the radar wavelengths of interest. Although ice surfaces are generally rougher than snow surfaces and thus have a higher radar return, glacier ice often appears smooth and is characterised by low backscattering at longer wavelengths (eg L-band) (Rott and Mätzler 1987).

Bindschadler and others (1986), by studying Seasat (L-band) and airborne X-band SAR images of the Greenland Ice Sheet, suggest that textural and tonal variations may be due to the presence of lakes, streams, residual snow patches, topographic (orientation) effects and ablation zones (Figure 4.27). X-band and higher frequencies are particularly sensitive to refrozen crust layers, formed by diurnal freeze-thaw cycles (Rott and Mätzler 1987). Before SAR data can be used to delineate land ice features, however, corrections must be applied to

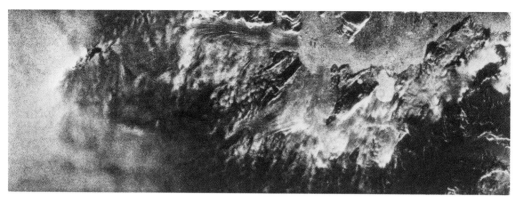

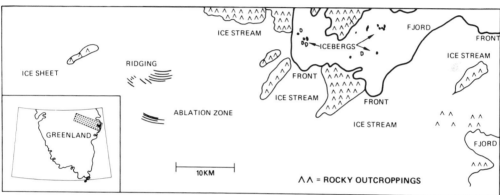

Figure 4.27 A Seasat SAR image of an area on the SW coast of Greenland. Many interesting features are evident. SAR produces data over ice sheets to complement radar altimeter data. This is an optically processed image; a significantly improved resolution can be obtained by digital processing. From Thomas and others (1985).

compensate for sensor attitude, terrain, and slant range distortions, which intensify with surface elevation; automatic procedures are currently being developed and will be operational for the launch of ERS-1 (R. Bindschadler, personal communication).

Wet snow (ablation) regions show up as areas of decreased backscatter. Rott (1985) found that X-band is suitable for mapping wet snow. C-band should also prove useful in this respect; the recent study of Vornberger and Bindschadler (in press) suggests that it should be possible to apply the ERS-1 SAR to the accurate mapping of seasonal changes in the boundary between melting and dry snow, especially when the data are combined with VIS and IR data. Quantitative analyses of airborne data suggest that the contrast between a wet snow cover and a snow-free surface decreases from X-band to longer wavelengths (Rott and Mätzler 1987). Claims that accumulation rates can be derived from SAR data may be premature, as many unknown parameters within the snow/firn volume affect its scattering behaviour. Unfortunately, few direct calibrated SAR (or scatterometer) measurements have been collected over ice and snow in polar regions against which scattering theories and models can be tested. Some quarters have expressed a desire to have a P-band (80cm wavelength) instrument on the EOS SAR, although its usefulness in glaciological studies is subject to speculation as few systems operating at this frequency have been tested to date, and such a system would require a huge antenna and much power.

Snowpacks and hydrology

Snow water equivalent (SWE) has a unique relationship to the radar backscatter coefficient, dependent on polarisation, frequency, snow wetness (Figure 4.28) and incidence angle (Figure 4.29). The measurement of SWE should ideally be made at night, when diurnal effects are less drastic and snowpacks are dryer. Inversion algorithms for the retrieval of SWE from SAR data are described in NASA (1987b). Airborne and surface experiments have shown that the SWE and extent of seasonal mid-high latitude snowpacks are best observed at frequencies >15GHz, which show an increase in volume scattering due to snow grain and internal layering effects; unfortunately, no space-borne SARs are planned at such high frequencies. An L-band instrument will often penetrate a dry snow cover to the underlying soil. Thick snowpacks are essentially invisible at L-band, marginally visible at C-band and quite visible at X-band or K_u-band. More information may be retrievable by combining SAR with optical, TIR and passive microwave data. The theory of radar remote sensing of snow is described by Rott and others (1985).

Future reception of SAR data

The onboard recording capacity of ERS-1 is insufficient to cope with the high-bit rate SAR data. Thanks to the signing of international agreements, however, dedicated receiving stations at Kiruna (Sweden), Gatineau (Canada), Fairbanks (Alaska) and West Freugh (Scotland) should ensure that most of the Arctic is covered by ERS-1, Radarsat (from 1995) and the Japanese JERS-1 satellite (from 1992) (Figure 4.30). The ground segments of ERS-1 and Radarsat are designed to be compatible. In addition to receiving JERS-1 L-band SAR data, plans are afoot for the reception of VNIR data at the ASF; this would provide a unique opportunity to analyse and compare simultaneous optical and SAR data.

The ASF at Fairbanks, for example, consists of a receiving station, SAR processor system, archive and operations system, and a geophysics product system. SAR lends itself readily to the application of automatic, unsupervised processing. The Geophysical Processing System can process over 200 100 × 100km (Seasat-like) frames (or 52 minutes of data) per day from the raw SAR data, at a ground resolution of about 30 × 30m with 4-looks (NASA Alaska SAR Facility Prelaunch Science Working Team 1989). The products from ERS-1 will be available within one week of data reception, are of high geometric accuracy and geolocated to within 500m, and will be archived on-site. Similar provisions have been made by the European Space Agency.

Future capabilities of the ASF facility and its follow-on will include the integration of data from other sources (eg meteorological and other satellite data) in the generation of higher level geophysical products (B. Holt, personal communication). These augmentative data will serve both to improve the performances of the basic data production by increasing the accuracy and to extend the data product level by allowing the computation of key fluxes in the sea ice environment, eg. heat, momentum, brine, freshwater, radiation balance, and melt pond coverage and thermodynamics. This architecture will be in place for the launch of EOS in 1998.

The ASF will obtain nearly simultaneous L- and C-band SAR data (from 1992 onwards). Data processing and analysis can be undertaken using either an optical or a digital processor. The optical method produces images which are suitable for qualitative and quick-look analysis, and provides a fast, relatively inexpensive way of surveying large-area data sets. The digital processor, on the other hand, converts SAR data into imagery that is suitable for quantitative analysis, and is accordingly more time-consuming and expensive. Selective transmission of SAR data will take place via a SARCOM link from the ASF to the US

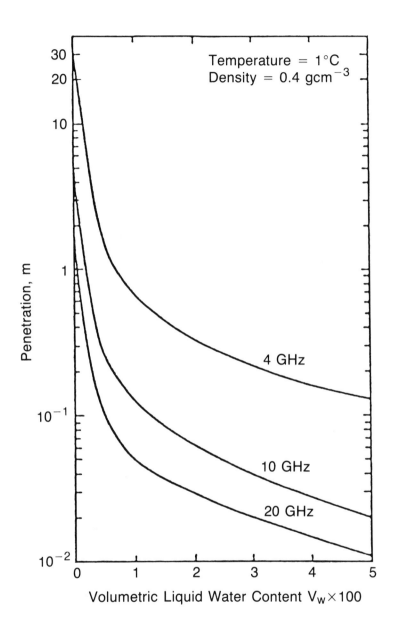

Figure 4.28 Penetration depth in snow as a function of liquid water content. From Stiles and Ulaby (1982).

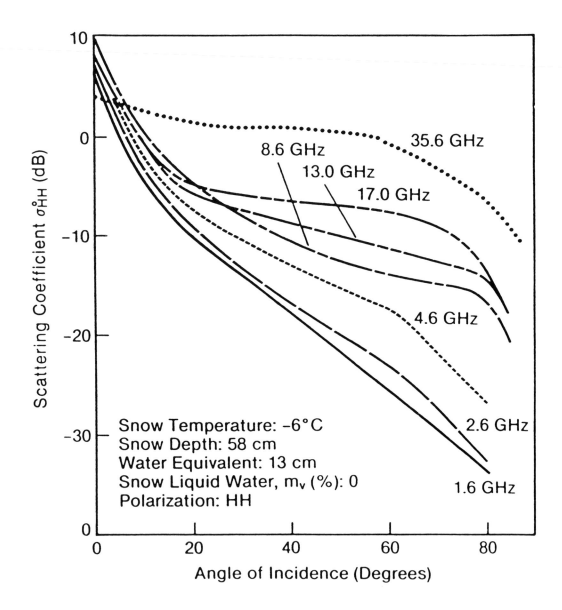

Figure 4.29 Angular response of the backscatter coefficient for ground covered with dry snow, mid-afternoon. From Stiles and others (1981).

Navy/NOAA Joint Ice Center in Suitland, Maryland for use in operational ice forecasting and the production of ice charts.

Coverage of Antarctica at the launch of ERS-1 will be restricted to that area within the masks of two receiving stations: i) a German receiving station at the Chilean base Bernardo

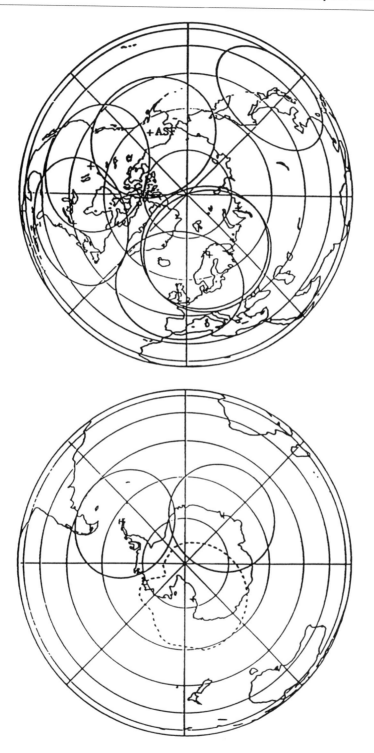

Figure 4.30 Northern and southern hemisphere SAR receiving stations and their data acquisition masks. The dashed line represents the mask for a possible future station at McMurdo Sound base.

O'Higgins, at the tip of the Antarctic Peninsula; and ii) the facility at Syowa, which Japan has upgraded to enable it to receive ERS-1 and JERS-1 data (see Figure 4.30). Australia has also been looking into the possibility of constructing an SAR receiving station at either Casey or Davis bases; Macquarie Island and Tasmania are also under consideration. Such a facility, if followed through, will not be ready in time for the launch of ERS-1. Further information is available from Dr A. Finney, Convenor, Tasmanian Earth Resources Satellite Station Project Working Party, TASUNI Research, Box 252C GPO, Hobart, Tasmania 7001, Australia. These facilities would together enable high resolution surveillance of a large proportion of the Antarctic coastline. However, a large gap remains in the very important Ross Sea sector, which could be filled by a suitable receiving facility at the US McMurdo Sound station (negotiations are underway; if construction began soon, this station may be ready in time for the launch of Radarsat).

Radarsat, with a planned mission lifetime of five years, has an onboard recorder with a capacity for a limited amount of SAR data. Although designed primarily to give operational information on ice conditions in the Canadian Arctic, it will offer the opportunity for periodic mapping of much of the Antarctic ice sheet margin. Moreover, it is uniquely able to swing the SAR beam from the right to the left side by a yaw manoeuvre of the spacecraft; this will enable periodic complete coverage of central Antarctica. This beam switching capability gives it the advantage of flexibility over fixed swath systems such as ERS-1 and Seasat; coverage of the Antarctic in its entirety is planned on two occasions. ERS-1, while flying in a 99° orbit, has a right-looking SAR which will restrict southern hemisphere coverage to north of about 78°S.

Radarsat has the ability to operate at a reduced resolution (100m), wide swath mode that will contain significantly more data and offer better coverage than the low resolution ERS-1 product, an important factor in the study of ice dynamics. An additional benefit of the Radarsat wide swath mode is that it has many similarities to the NASA EOS sensor to be launched in 1999 (at the earliest), and it will therefore act as a perfect testbed for both data systems and scientific modelling work (Carsey and Weeks 1988).

One crucial consideration regarding future SAR systems, and other systems relevant to polar research, is to work out sensible duty cycles in advance to ensure a balanced coverage of all regions of interest, and to avoid duplication. An integrated international research programme using SAR in polar regions is taking place from 1991–6 in the form of the European Space Agency (ESA) Programme for International Polar Oceans Research (PIPOR) (ESA 1985); a Polar Ice Sheet Programme has also been established with similar aims. Both underline the crucial need for satellite sensor calibration and validation, requiring carefully coordinated simultaneous airborne and surface measurements.

NASA's Shuttle Imaging Radar programme

An important advance in recent remote sensing technology has been NASA's Shuttle Imaging Radar Programme, brought to a tragic standstill by the *Challenger* disaster of January 1986 and subsequently by technical difficulties. The successful launch of *Discovery* from Cape Canaveral on 29 September 1988, marked the re-emergence of a programme which will, when fully operation, play a unique role not only as a launch (and possibly maintenance) vehicle for future satellites (for example Radarsat and Landsat) but also as a primary testbed for sensors, particularly multi-frequency and multi-polarisation SAR.

The SIR-B mission STS-17 of October 1984, carrying an L-band HH polarised instrument, failed to gain much polar coverage (apart from the Antarctic sea ice margin in the vicinity of the South Sandwich Islands [Carsey and others 1986]) due to communications problems, but showed the potential of flying, retrieving and modifying SAR. SIR-B data were digitally recorded and processed. SIR-C/X-SAR, comprising three flights in different seasons of 1993–6, will provide the first space-borne high resolution imaging sensor with a multi-spectral (L- and C- and X-bands) and a multi-polarisation capability. It will also incorporate new distributed solid-state technology, and electronic beam steering. Unfortunately, the orbital inclination will be only 57°, thereby severely limiting polar coverage. Polar-orbiting Shuttles must be launched from Vandenburg Air Base; those launched from Cape Canaveral cannot obtain a polar orbit. Nevertheless, this and possible future missions, will lay the groundwork for the EOS SAR (see below).

Potential improvements
Surface and air-borne measurements suggest that, for a given incidence angle, the higher C-, X- and K_u-band frequencies allow improved ice type discrimination (compared to L-band), due to the greater effects of volume scattering at shorter wavelengths, as well as more information on small-scale surface roughness characteristics (Onstott and Gogineni 1985). Onstott and Shuchman (1988) report that FY and MY Arctic ice separate by as much as 10dB at X-band like-polarisation, although much variability exists within each ice type.

K_u-band frequencies appear to be even better suited to the study of ice type than those in the C- and X-bands. Under freezing conditions, K_u-band signatures of FY and younger ice are dominated by surface scattering properties; volume scattering appears dominant for MY ice (Livingstone and others 1987). These higher frequencies, however, are more sensitive to atmospheric and possibly ionospheric effects. Moreover, low density frost flowers on the surface of newly refrozen leads have a centimetre-scale roughness which, combined with their high salinity, can lead to a relatively high radar backscatter from new ice at C-, X- and K_u-bands. The upward migration of brine onto the new ice surface, and its subsequent wicking into the snowcover, is a critical issue in microwave remote sensing (Drinkwater and Crocker 1988). Such processes can confound the unambiguous discrimination of new ice and open water. Regarding older ice, the effects of internal ice structure and air/gas bubble distribution and diameter on the observed variability of volume scattering and emissivity are complex issues that require further research.

Higher frequency radars have yet to be flown in space as, compared to C- and L-band systems, they are more expensive to construct and operate, require more power, a more accurate antenna flatness and are heavier. An X-band instrument, for example, is greater than L-band in its required transmitting power by a factor of 7, in weight by a factor of 1.8, and in power consumption by a factor of 7 (Horikawa 1985).

The EOS SAR to be launched, possibly as a dedicated free-flier in 1999, is a three-frequency (L-, C- and X-band), multi-polarisation instrument (NASA 1990b). The orbit will be polar, sun-synchronous, with an afternoon equator crossing time, at an altitude of 620km. The mission lifetime will be five years, with two replacement satellites subsequently being launched. Swath width and incidence angle will be variable (over the range of 15–40°), with three different modes of operation:

  i)   20–30m ground resolution, 30–50km swath width (Local High Resolution mode).
 ii)   50–100m resolution, 100–200km swath width (Regional Mapping mode).
iii)   250–500m resolution, swath width up to 500km (Global Mapping mode).

The received image amplitudes will be the four basic linear polarisations, with the received image phases being the phase differences between the linear polarisations. It will be possible to synthetically transmit a right circularly polarised signal and receive a left circularly polarised signal. This cross polarisation synthesis technique is known as polarimetry, and can be used to enhance the response of certain discrete targets (Figure 4.31) (M. Drinkwater, personal communication). This 15-year mission will produce a huge amount of very important follow-on data. The EOS SAR will be capable of acquiring multi-incidence angle data, using electronic beam steering and other imaging geometries by mechanically pitching, yawing and rolling the antenna.

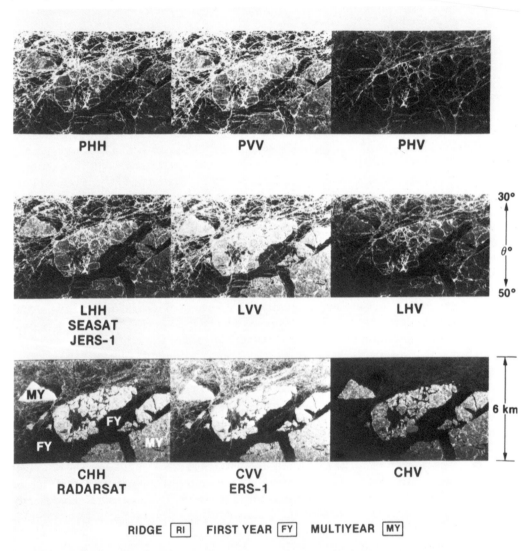

Figure 4.31 Aircraft images of sea ice in the Beaufort Sea, acquired using the JPL multi-frequency polarimetric SAR. Individual panels indicate frequency and polarisation responses of various ice types. Note that the ERS-1 SAR operates at C-band (VV), Radarsat at C-band (HH), and Seasat and JERS-1 at L-band (HH). Courtesy of Mark Drinkwater and Ben Holt, NASA JPL.

Even taking into account the inherent limitations and present ambiguities in interpretation, the level of detail obtained from SAR in polar regions is far greater than that obtained by any other satellite sensor. Intensive research involving both surface and aircraft experiments is continuing with the aim of formulating routinely applicable techniques for the interpretation of radar backscatter data collected over polar ice and oceans. Improved models describing the physics of microwave scattering under a variety of conditions will play a major role in this respect.

(b) Radar altimeters (ALTs)

The ALT is a nadir-viewing instrument which transmits short duration (narrow) radar pulses with known power in a pencil beam towards the Earth's surface, then measures reflected energy in a number of time gates. The signal amplitudes in successive gates create a waveform. The incidence angle is only a fraction of a degree from nadir. The measurement of the return energy determines three parameters, which can then be used to derive a variety of important geophysical quantities:

i)   the time delay between pulse transmission and return (receipt) of the backscattered energy;

ii)   the absolute value of the backscatter coefficient; and

iii)   the shape of the returned pulse, and its leading edge in particular.

The time delay until receipt of the reflected signal (or two-way travel time of the pulse), when coupled with a knowledge of the velocity of propagation through the ionosphere and wet troposphere, can be converted to a highly accurate measurement of the altitude of the satellite, and therefore a measurement of surface topography (assuming that the orbit ephemeris is accurately determined).

Over the ocean, the ALT waveform profile is sufficiently well understood to permit real-time estimates of ocean parameters to be carried out on board the satellite. Coupled with knowledge of the geoid and the ocean density field, satellite altimetry provides the only feasible means of determining both the wave climate and absolute geostrophic currents in the open ocean on a global scale (Fu and others 1988). Altimeters are also designed to yield data on ocean height relative to the geoid and surface wind speed (but not direction) (Wunsch and Gaposhkin 1980). With the growing concern over possible global change, scientists have also realised the unique value of the ALT as the only practical means of gaining an accurate database of ice sheet elevations (Allison 1983). Due to its narrow swath width, the ALT requires a longer repeat period than the SAR to ensure a dense ground coverage; ground tracks converge, however, towards the poles.

It was noted earlier that the vast polar ice sheets remain largely uncharted, with a near complete absence of geodetic control points rendering conventional surface mapping virtually impossible. The accurate, reliable and consistent topographic mapping and monitoring of these remote regions over long time periods is essential to obtain an inventory of the fundamental features and processes affecting ice sheet mass balance and dynamics. Time series of altimeter measurements will allow an assessment of the input into the system (accumulation rates), the advection of ice from the interior to the periphery (ice dynamics, the study of which requires the definition of elevation gradients), and the output (ablation rate in terms of ice sheet thinning and iceberg calving). The non-trivial problem of comparing tracking over non-ocean surfaces by different satellite altimeter missions is covered by Brenner and others (1990).

Skylab

For ice surfaces, the waveform shape does not always conform to a simple model, and the application of specialised algorithms is necessary to extract geophysical information from ALT data. Analysis of returns over both sea ice and ice sheets has, however, demonstrated the unique versatility and precision of the ALT. Although the first ALT (designated S193) flew onboard Skylab in 1974 (with an altitude accuracy of ±1m), it only orbited to a maximum latitude of 50°. The Skylab ALT was designed to investigate the accuracy, precision and overall practicality of satellite ALTs to i) determine mean sea levels, ii) monitor dynamically the mean surface slopes, and iii) measure small-scale departures of the ocean surface from overall mean sea level. The range measurement precision was quite crude (McGoogan 1975), and was further degraded by perturbations in the spacecraft's orbit. Nevertheless, the data served to tantalise the scientific community as to the potential applicability of ALT data.

GEOS-3

The first ALT measurements of polar ice from space were not acquired until the launch of Geodetic Earth Orbiting Satellite-3 (GEOS-3) in April 1975. The intensive operating mode of this ALT (at 13.0GHz) provided ±20cm precision with an overall height accuracy of ±60cm (over open ocean). Principal error sources again included orbital uncertainty, altitude error, and unmeasured geophysical effects (eg ionospheric electron content, wet and dry tropospheric errors). The GEOS-3 ALT provided a wealth of data over its almost three-and-a-half-year lifetime. Brooks and others (1978b) first noted that the returned power and echo waveform shapes obtained over sea ice differ distinctly from those observed over the open ocean, and that these differences could be used to delineate ice-ocean boundaries. Based upon observations of the return signal strength of GEOS-3 ALT data, Dwyer and Godin (1980) developed an 'ice index', which yielded positive values over sea ice and negative values over open ocean. Direct comparison with contemporary ice charts, constructed from VIS and passive microwave data, showed good agreement in ice edge location.

Although GEOS-3 obtained no coverage poleward of latitude 65°, its data were also used to map the topography of the south Greenland Ice Sheet to an estimated accuracy of about ±50cm (Brooks and others 1978a). Consistently accurate mapping and monitoring of the surface elevation of ice sheets, hitherto impossible by conventional means, is critical on two accounts:

i) No other method can repetitively observe the slight variations that occur in polar ice caps and ice shelves. The early ALT missions showed that the lower the accuracy of an ALT, the longer the time span of measurements required to be certain of changes in elevation. An accuracy of ±10cm in surface elevation over flat terrain is believed to be adequate to detect major changes in volume over a ten-year period (Bindschadler and others 1987).

ii) Details of the undulating ice sheet/glacier surface contain information on the character of the ice flow, ie its dynamics. Since shear between bedrock and ice provides the principal resistance to motion, the surface slope is a proxy indicator of basal glaciological conditions and gross aspects of subglacial relief (McIntyre and Drewry 1984). Drainage basins can be delineated with a knowledge of flow direction (Bindschadler 1983). An elevation accuracy of ±3m is sufficient to measure mean slopes of 0.002 over a distance of 30km to an accuracy of 10% (Bindschadler and others 1987).

Smaller-scale surface undulations (three to five times the ice thickness) are proxy indicators of conditions at the base of the ice sheet. Smooth, horizontal surfaces often demarcate the presence of sub-glacial lakes, whereas a slightly rougher surface with a low mean slope indicates a well-lubricated and fast-flowing glacier or ice stream. Even rougher surfaces correspond to slow or thin ice which is often frozen to the underlying bedrock. The undulations commonly have amplitudes of a few tens of metres and wavelengths of a few tens of kilometres. To measure these features accurately, a footprint of <100m in diameter on a flat surface is required.

Although not optimised for use over rougher surfaces (ie ice sheets), the GEOS-3 mission confirmed the ALT as a primary sensor for the study of ice sheet dynamics and volume, providing data to unique levels of accuracy.

Seasat

The Seasat ALT, which had only a three-month lifespan, extended coverage of the Greenland and Antarctic ice sheets to 72°. It tracked the half-power point on the leading edge of a composite return pulse, formed by summing 50 consecutively received pulses. This technique worked well over the ocean, where measured ranges change very slowly. Frequent loss of track of the pulse occurred over ice, however, and even over the quite gentle regional slopes of the interior portions of the great ice sheets, which are usually within the ±0.43° design range of the ALT. This behaviour may be attributed to a combination of causes which are documented by McIntyre and Drewry (1984).

Although both the Greenland and Antarctic ice sheets are parabolic in shape and virtually flat in the centre (50% of the Antarctic continental surface has gradients of less an 0.03%), slopes increase to as much as 3% towards the coastal margins, where ice flows down via huge outlet glaciers and ice streams. Such regions are of sufficient relief that the conventional wide bream altimeters tend to be inappropriate for topographic measurement of the ice surface.

It is for these reasons that elevation accuracies over the steep outer margins of the Greenland ice cap exceed 100m (Brenner and others 1983). In spite of these problems, Seasat's ALT acquired 100,000 useful observations of the Antarctic and Greenland ice sheets; topographic data sets have been created with an average precision of between about 2 to 3m over smooth terrain (Bindschadler and others 1990). Seasat data have been used to construct ice sheet topographic maps with contour intervals of ±5m (Brooks 1983). Bindschadler and others (1990) have isolated individual drainage basins from gridded (20km) data in Greenland.

In theory, therefore, the ALT is ideally suited to monitoring the subtle changes that occur in ice sheet elevation over time (of the order of centimetres per annum) by taking the difference in range between successive measurements at orbit crossover points. In practice, however, crossover differences are often unrealistically large; recent research has attacked this problem in a number of ways, one of which is to edit the crossover differences that are larger than a defined threshold value. Lingle and others (1990) compute the elevation at a given crossover point on the ice sheet surface by linearly interpolating between the closest elevations on either side; each crossover difference is weighted in proportion to the inverse square of the noise level in the data in the vicinity of the crossover point. Semivariograms are used to estimate the noise level as a function of the position on the ice sheet. This approach is suitable for use near ice sheet margins, where altimeter noise levels tend to be high and variable due to rough topography and steep slopes. It also allows the comparison of data sets from different satellites, a critical factor if a time series is to be built up that is of sufficient length to enable meaningful studies of possible global change effects.

Another promising application of altimetry is the accurate detection and delineation of the ocean-sea ice and sea ice-ice shelf or ice sheet boundaries, due to the convergence of orbits towards the poles and the narrow pulse-limited footprint diameter of the ALT (Thomas and others 1983). Monitoring of ice front positions to an accuracy of ±100m on a routine basis offers a direct indication of ice shelf growth, and provides advance early warning of iceberg calving events. Similarly, the location of shear zones and the grounding line, both theoretically discernible from ALT data, are sensitive indicators of mass balance conditions in the grounded portion of glacial drainage basins.

Seasat further highlighted the potential ability of ALTs to obtain important measurements over sea ice. Signals from the latter are varied and much stronger than those from the open ocean; the wind-roughened sea surface is a diffuse reflector, whereas sea ice is generally specular in its reflectance characteristics. Moreover, the shape of the scattered return signal, and especially its leading edge, can yield important information on ice surface and near-surface characteristics (ie roughness and reflectivity), but only if the instrument operation and normal incidence backscatter from natural surfaces are understood.

GEOSAT

Radar altimeter evolution has been characterised by a continuing improvement in instrument design, measurement precision and understanding. The launch of the military-dedicated ALT on GEOSAT in March 1985 provided dense coverage of the polar ice masses up to latitude 72° with a precision of 5cm for $1s^{-1}$ averages over open ocean. Although its primary objective was to generate maps of mean sea surface topography with high spatial resolution for operational naval purposes, GEOSAT produced exciting data in ice-covered regions, which have been made available to the polar research community (Figure 4.32).

GEOSAT has established that, although ALTs are limited in terms of spatial coverage, they are eminently suitable for the uninterrupted production, rapid processing and subsequent dissemination of operational sea ice products, such as ice/ocean boundary location (sea ice extent). The US Naval Ocean and Atmosphere Research Laboratory (NOARL) developed the GEOS-3 ice index to provide an operation, all-weather, day/night GEOSAT product indicating ice edge location to a few kilometres' accuracy (Fetterer and others 1988; Laxon 1989). The processed data were transmitted to the Joint Ice Center in Maryland, where they were used in conjunction with NOAA AVHRR and DMSP SSM/I data to create ice charts. The ice index used is a ratio of the energy returned in the early gates to that returned in the later gates, and is a simple parameterisation of waveform shape.

Recent research using Seasat ALT data from the Beaufort Sea and GEOSAT ALT data from the Bothnian Sea suggests that normal-incidence backscatter coefficients may be used to differentiate sea ice types (Ulander 1988). Moreover, smooth surfaces (melt ponds and new ice) are the most likely sources of observed high power peaked returns, whereas older ice gives diffuse, low power returns (Laxon 1989). Further research is necessary in relating pulse echo waveforms to dielectric, penetration depth and surface roughness characteristics (over both sea ice and ice sheets) (Jezek and Alley 1988). It should be noted that GEOSAT did not carry a passive microwave instrument suitable for measuring atmospheric water vapour, spatial variations of which lengthen the electromagnetic propagation path and can cause variations in computed ALT height (Monaldo 1988). Water vapour corrections have been applied using DMSP SSM/I data (for the latter part of the mission).

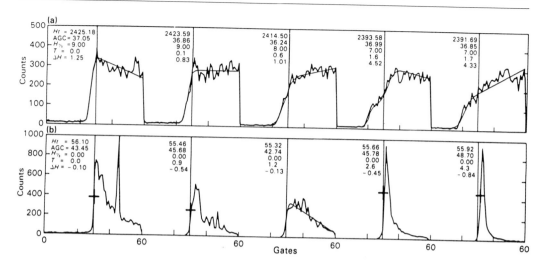

Figure 4.32 (a) The characteristics of a diffuse reflecting surface on sample GEOSAT waveforms from snow-covered ice sheets. The smooth curve is the function fitted to the waveform data to obtain the range correction for deviation of the midpoint of the ramp from the central gate. (b) Sample waveforms over sea ice, showing the characteristics of specular reflecting surfaces mixed with diffuse reflections of varying strengths. Ht is the uncorrected surface height relative to the ellipsoid in meters; AGC is the automatic gain control (larger values indicate stronger signals); $H_{1/3}$ is the altimeter significant wave height value; T is the relative time in seconds; and ΔH is the height correction in metres. Courtesy of Jay Zwally, NASA, GSFC.

Over land ice sheets, the precision of elevation retrievals is about ± 1.5m overall and may be as good as ± 30cm in areas of smooth topography (Zwally and others 1987). NASA scientists have established that the standard deviation of the elevation difference at 17,161 crossovers on the Greenland Ice Sheet for 110 days' worth of data is 1.61m, with a mean difference of -0.5m between descending and ascending passes (Figure 4.33). An important comparison with the Seasat dataset is theoretically possible.

Future ALT missions
In order for an ALT to operate reliably over ice surfaces without loss of track lock while retaining the ability to collect high precision ocean height measurements, it should contain the following enhanced design specifications (Bindschadler and others 1987):

i)   Adaptive control of sampling gate widths (that constitute the range window). These will be required to accommodate high relief ice surfaces.
ii) Multiple track modes are required to ensure that the complex backscattered waveform shapes and the rapid rate of change from waveform to waveform do not cause a loss of lock in the tracking loops.
iii) The pulse repetition frequency should be variable to maximise the noise reduction that is achieved with increasing the number of decorrelated sampled waveforms.

These improvements will largely be incorporated into future designs.
    ERS-1, due for launch in 1991, is of particular interest to ice sheet glaciologists, as this satellite will cover the latitudinal band of 78–82° for the first time, thereby monitoring 80% of the polar region not reached by Seasat, ie the entire Greenland Ice Sheet and a major part

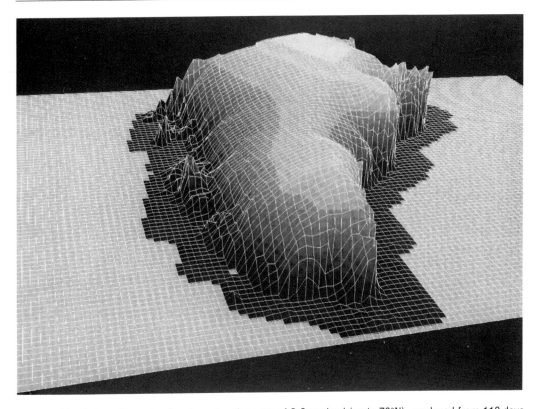

Figure 4.33   A three-dimensional surface elevation map of S Greenland (up to 70°N), produced from 110 days of retracked GEOSAT data; each grey level interval represents 500m elevation. Surface slope indicates the direction of flow. The N–S running divide, and the outlines of several drainage basins, are discernible. Courtesy of Jay Zwally, NASA GSFC.

of the climatologically sensitive and apparently unstable areas of Antarctica. This ALT is the first to be provided with an 'ice mode' of operation, whereby a centre-of-gravity tracker will be adopted in an attempt to overcome the tracker lock problems encountered by previous satellite ALTs over ice; this modification should assist in the recovery of irregular waveforms. Altimetry may enable detection of the equilibrium line, the boundary between the accumulation and melt zones, through the associated change in return signal. ERS-1 data should enable more precise studies of the mass balance and dynamics of almost all of Antarctica's major ice shelves, which are an important source of tabular icebergs and Antarctic bottom water. Since floating ice shelves are in hydrostatic equilibrium, measurement of their surface elevation to accuracies of ±1m should enable their thickness to be determined to ±10m (McIntyre and Drewry 1984).

Any inaccuracies that still occur over sloping portions of the ice sheet (Brenner and others 1983) are likely to be compensated for by the effect of more substantial thickness changes at the margins, which are due to higher accumulation rates and more active flow. It has been estimated that a network of transects with a minimum spacing of 5km would be required to map the surface adequately (Bindschadler and others 1987). This exercise would take six months to complete, and should be repeated once every two years to look for statistically significant changes in volume.

Radar altimetry from ERS-1 may also provide critical data on the wind and wave climates of MIZs. The interaction of ocean swell with sea ice can be readily detected from ALT data (Rapley 1984). Measurement of ocean significant waveheight within sea ice-covered regions is, however, only possible where the shape of the ALT return waveform corresponds closely to that observed over the open ocean (Allan 1987); some correction for onboard tracking may be necessary at these times. The interpretation is more difficult at times of quasi-specular return; short (<10m) wavelength components in the wave spectrum may be damped out by the ice. The effect of longer wavelengths – which can penetrate deep into the pack – on the ALT signal is poorly understood, and provides an exciting research topic.

Another important application will be the mapping of the geoid in sea ice regions; mapping of mean surface topography has previously been impossible in permanently ice-covered areas, thereby excluding a large proportion of the world's oceans. Due to their inherent sampling problems, ALTs are unlikely to compete with passive microwave radiometers in terms of providing global sea ice extent data, although they will make a unique contribution to the accurate delineation of sea ice edges (Laxon 1989). Precision of location is theoretically the same as that of the ERS-1 waveform sampler (ie 330m), depending on the precision of knowledge of the satellite's along-track position. Sensitivity to the presence of sea ice depends on prevailing wind and weather conditions. Moreover, data from the ALT may indirectly yield invaluable information on sea ice thickness (averaged over a coarse resolution cell), derived from measurements of the freeboard in relation to open water.

Unlike the SAR, the ALT is a low bit-rate sensor which can operate virtually continuously. Consequently, the ALT may have an important role to play gathering data on the nature of MIZs when the SAR is unable to operate. In the case of ERS-1, this research may be supplemented by Active Microwave Instrument (AMI) scatterometer mode data within the pack, and AMI wave mode data outside it (and within the MIZ). The ALT may prove useful in the detection of polynyas and large leads within the pack/close to the coast. Sea ice operational products produced at the Earth Observation Data Centre in Farnborough include 'quick dissemination' ice-ocean boundaries, which can be transmitted to ships operating in ice-infested regions.

Future proposed ALT missions are the Japanese MOS-2 satellite, scheduled for a 1992 launch (a Seasat-class ALT), possibly a future SPOT satellite, and the dedicated TOPEX/Poseidon satellite. The latter, with a 1992 launch date, will offer a ±13cm sea level height measurement accuracy, but no coverage poleward of latitude 66.02° (an inclination that is optimised for ocean tide determination). This ALT is a primary sensor of WOCE, a major component of the World Climate Research Program (WCRP). The ALT originally planned for launch on NROSS was to be included in the payload of the dedicated US Naval operational ALT mission Spinsat Altimeter (SALT) in 1991. At the time of going to press, however, it appears that SALT has been scrapped. Plans are now afoot to launch a series of GEOSAT follow-on satellites every three years, starting in approximately 1994. These will carry slightly improved GEOSAT-class ALTs; passive microwave radiometers will also be included for atmospheric correction. Into the next century, ALTs may be included in the payloads of DMSP satellites. The US Navy intends that these data should be available to the scientific community.

Probable future ALT developments are reviewed in a volume describing the altimetric system to be flown on EOS-B in 2000 (Bindschadler and others 1987). The EOS instrument will be dual frequency, operating at 13.6 (K_u-band) and 5.3GHz (C-band); the former is the primary altitude sensor, whereas the latter will correct for ionospheric pulse delay effects (NASA 1990b). The European polar platform will also carry a TOPEX-class ALT. Future missions carry sophisticated range and orbit ephemeris-tracking experiments (including laser) which will greatly improve the accuracy of altitude retrievals (the EOS ALT is aiming for a ±2cm precision over open ocean). Altimeter missions will also use satellites of the new US Department of Defense Global Positioning System (GPS) to track geocentric altitude variations independently to a high accuracy.

All spaceborne altimeters flown to date have been wide-beam pulse-limited systems, which are characterised by small antennae and relatively long wavelengths. The dimensions of their ground footprint, and thus accuracy, is determined mainly by the pulse duration. This is appropriate for open ocean and smooth terrain where side echoes from off-track topography are not easily confused with the desired nadir return. As discussed earlier, however, the pulse-limited footprint is physically meaningless in higher relief terrain (Figure 4.34). In such regions, it is desirable to produce a narrower radar beam (ie use a beam-limited approach) by employing larger antennae and/or higher frequencies (shorter wavelengths). This physical constraint of antenna size has discouraged the deployment of narrow-beam altimeters into space. Similarly, the highest frequency and high reliability radars corresponding to an atmospheric transmission window are limited by the state of technology to around K_a-band (37GHz). One novel solution to this problem involves the use of coherent processing or synthetic aperture techniques.

(c) Radar Scatterometers

The Radar Scatterometer (SCATT) yields a detailed quantitative observation of radar scattering behaviour, expressed as the normalised backscatter cross-section per unit area, and surface roughness off-nadir. Data are collected within a fan-beam antenna pattern in swaths on both sides of the satellite sub-track (Figure 4.35), providing spot measurements with a relatively coarse spatial resolution (the Seasat SCATT or SASS had a spatial resolution of 50km).

SCATTs are primarily designed to measure surface wind speed and direction over areas of open ocean by observing ocean capillary waves, which form due to wind stress and scatter the radiation in a Bragg-like manner (Barrick and Swift 1980). Accuracies for the SASS have been assessed at $1.5ms^{-1}$ in speed and 20° in direction for winds in the range of $5–20ms^{-1}$. The instrument samples the surface at a wide variety of incidence angles, and thus acquires much information about the scattering function. To date, K_u-band SCATTs have flown on two platforms, namely Skylab (1973) and Seasat (1978). The NASA NSCAT, originally earmarked for the defunct NROSS satellite, will be included in the payload of the Japanese satellite ADEOS (1995), and a C-band (5.3GHz) instrument is being flown on ERS-1 (from 1991).

In order to determine wind direction unambiguously, three observations are required at different values of the look angle. The NSCAT and the Wind SCATT Mode of the ERS-1 AMI therefore mark distinct design improvements over the SASS, having three look angles as opposed to two (Figure 4.36i). This change reduces the number of possible directional ambiguities from four to two.

SCATTs, by measuring surface wind speed and direction over open ocean, provide important inputs into the computation through models of the drag, stress and forcing coefficients of wind on ice, heat flux transfer and wave fields (Long and Mendel 1988), which in turn are essential inputs to both drive and test coupled dynamic-thermodynamic sea ice models (Hibler 1979; Preller and others 1990). It has been estimated that a satellite-borne SCATT can deliver measurements of vector winds at the ocean surface equivalent to 20,000

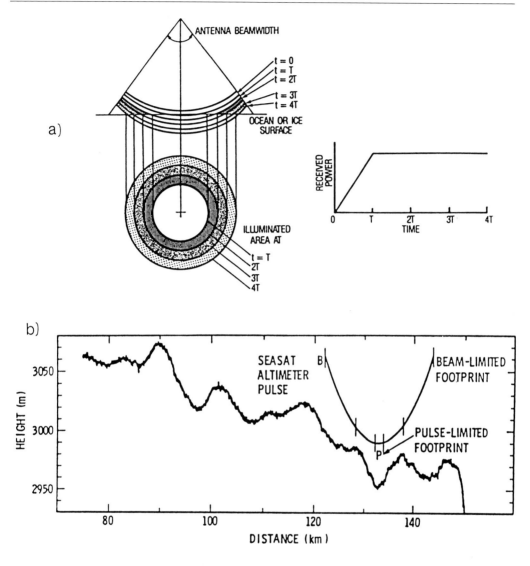

Figure 4.34 (a) The illumination of the Earth's surface by an altimeter pulse of width T, where the surface relief is small compared to T. This figure shows the surface area illuminated at time increments equal to the duration of the transmitted pulse width, and illustrates how the reflected signal received by the ALT changes with time for the case of diffuse scattering, eg from the open ocean surface (specular reflection is characteristic of ice surfaces). From Hawkins and Lybanon (1989). (b) Comparison of the Seasat radar altimeter pulse shape to typical topography of the East Antarctic Ice Sheet. Echoes from hill slopes (marked by the vertical bars) adjacent to smooth topography return to the instrument before the nadir echo, thereby becoming virtually indistinguishable from the 'correct' return signal. The beam-limited footprint diameter of the Seasat pulse (approximately 20km) is a more realistic indicator of the true spatial resolution of this instrument than the pulse-limited footprint (several metres) over such terrain. Adapted from Robin and others (1983).

ship observations per day with a high accuracy at least equal to that specified by the World Meteorological Organization (Gower 1986).

The NSCAT, for example, should provide accurate wind data over at least 90% of the global, ice-free ocean with a sampling frequency of at least every two days and for a three-

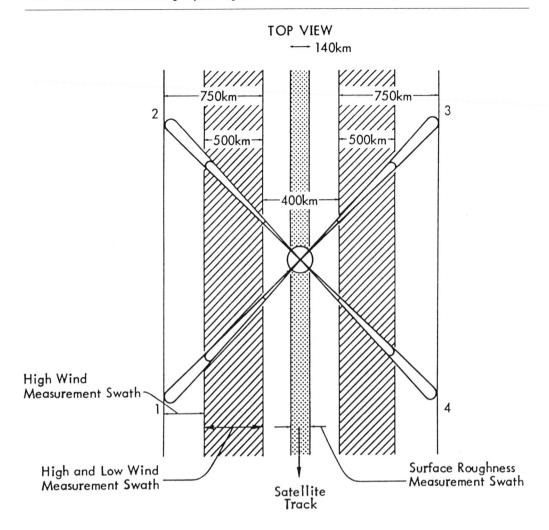

Figure 4.35 Antennae surface illumination patterns for the Seasat SASS.

year period. SCATTs cannot measure winds when ice is present. However, although Seasat SASS data have not been widely applied directly to ice research, they indicate an exciting potential in terms of ice surface characteristics and properties information content (Figure 4.37) (Gray and others 1982; Onstott and Schuchman 1986). The resolution and swath width of SCATTs (25km and 1,200km respectively for NSCAT, for example) are similar to those of passive microwave radiometers. Most present experience of SCATTs operating over sea ice has been based upon surface, aircraft and helicopter experiments (eg Onstott and Shuchman 1988). At high incident angles (in the range of 40°–70°), open ocean has a lower backscatter than sea ice, and the two can be readily discriminated. Problems in interpretation are caused by melt effects. Few similar experiments have been conducted over land ice.

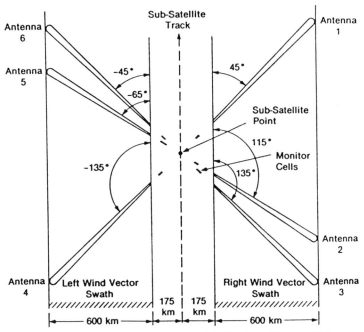

Figure 4.36 (i)   Antennae surface illumination patterns for NSCAT. From Elachi (1988).

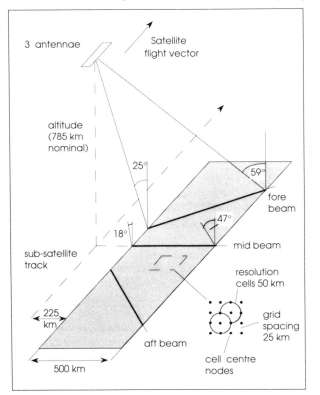

Figure 4.36(ii)   Wind Scatterometer geometry. Courtesy of Chris Rapley, MSSL and UK Earth Observation Data Centre.

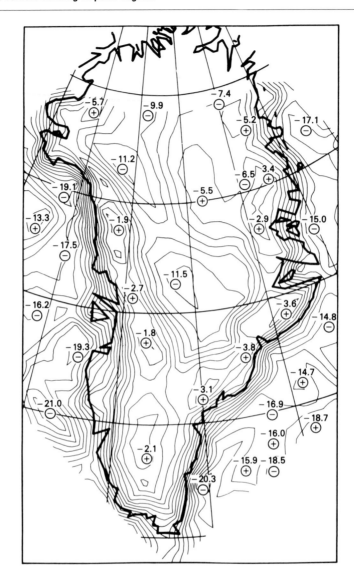

Figure 4.37 Contours of radar backscatter (in dB) over Greenland derived by averaging 30 days of Seasat SASS data obtained during September and October 1978. Regions of strong backscatter along the coast (marked with +) appear to be centred over areas of summer melting. The latter, followed by refreezing, leads to the subsequent formation of subsurface ice lenses, which act as significant scatterers. From Thomas and others 1985).

5. Future developments

After accuracy and consistency, continuity is the most important requirement of studies using satellite datasets of the polar regions. Even though mission longevity has improved significantly since the early 1960s, there is still a need to launch replacement spacecraft at two to five-year intervals, as satellites commonly become inoperable within their life expectancy. Moreover, worldwide networks of ground stations require constant maintenance. In an effort to overcome onboard tape recorder malfunction problems and offer continuous coverage, many future missions will rely upon geostationary data relay satellites (eg the Tracking and Data Relay Satellites [TDRS]) for onward data transmission to the ground when out of sight of dedicated receiving stations. Until the relaunch of the Shuttle programme on 29 September 1988, only one TDRS satellite was operational; TDRS-C was launched on this first re-flight. TDRS-D was launched from *Discovery* on 13 March 1989.

The Earth Observing System (EOS) polar platforms

An agreement was signed on 29 September 1988, between the USA, ESA, Canada and Japan to build a permanent manned Space Station 'Freedom', although the precise nature of this project has yet to be laid down. To be launched in 1998, it will have a life expectancy of 30 years. This programme includes up to six co-orbiting polar platforms (two to be supplied by ESA [EPOP-M1 and -N1 (Benz and others 1989)], one by Japan [JPOP] and three series by NASA [EOS-A and -B and a dedicated SAR satellite, EOS SAR]) (Figure 5.1). Exact orbital configurations are yet to be determined, although the following baseline parameters have been proposed (for the NASA elements): 705km altitude, 98.2° inclination, sun-synchronous, two-day exact repeat, ascending node equator crossing 13:30 LST. The planned launch dates are: EPOP-M1 and EOS-A 1998, JPOP 1998, EOS SAR 1999, and EOS-B and EPOP-N1 2000. All will be large (about four times the size of Landsat), multi-disciplinary, platforms with essentially complementary payloads. The NASA platforms will be replaced every five years to achieve the 15-year mission lifetime goal.

The first EOS will be part of a three satellite EOS-A series, each with a design lifetime of five years. The central focus of EOS-A will be the physical-climate system as it relates to potential global warming. It will be accompanied by a progressive shift from discipline-orientated to interdisciplinary research programmes. The payload of the first EOS-A platform includes:

i) Clouds and Earth's Radiant Energy System (CERES): measurement of incoming and emitted radiation at the top of the atmosphere;

Figure 5.1 An artist's impression of one of NASA's EOS platforms, which will carry continuous, global observations of the Earth into the twenty-first century. From NASA Earth System Sciences Committee (1986).

   ii)   MODIS-N/-T: cloud phenomena (-N), and biological processes (-T);

  iii)   Earth Observing Scanner Polarimeter (EOSP): aerosols and clouds as they heat and cool the Earth;

  iv)   Multi-angle Imaging Spectro-Radiometer (MISR): global observations of the directional characteristics of reflected light;

   v)   Atmospheric Infrared Sounder (AIRS)/AMSU-A/AMSU-B: atmospheric temperature profiles plus data on atmospheric water vapour, cloud and sea- and land-surface temperatures. Supplemental information on snow and ice;

  vi)   High-Resolution Dynamics Limb Sounder (HIRDLS): levels of trace gases that contribute to the greenhouse effect, thereby extending the measurements of UARS;

 vii)   Stick Scatterometer (STIKSCAT): surface wind speed and direction over open ocean;

viii)   Advanced Spaceborne Thermal Emission and Reflection (ASTER): high resolution (15 to 90m) data of surface and clouds;

  ix)   Measurement of Pollution in the Troposphere (MOPITT): global measurements of atmospheric carbon dioxide and methane;

   x)   Wide-Band Data Collection System (WBDCS): collection of global *in situ* data; and

  xi)   Multifrequency Imaging Microwave Radiometer (MIMR): atmospheric water content, rain rate over open ocean, sea surface temperature, and ice and snow cover parameters.

All of the above apart from HIRDLS have been tentatively selected for flight on the remaining satellites of the EOS-A series. In addition, the High-Resolution Imaging Spectrometer (HIRIS) has been selected for flight on the second and third EOS-A satellites. The complementary EOS-B series, also consisting of three satellites over 15 years, will address the other elements of the US Global Change Research Program, including ocean circulation and glaciology (it will carry a laser altimeter/ranger and a radar altimeter). Details of these and the other candidate sensors are given in NASA (1990b; 1990c). The final payload allocation for EOS-B (and JPOP) will be made in September 1991. For further information, contact Dr Stan Wilson, EOS Program Scientist, NASA HQ, 600 Independence Avenue SW, Washington DC 20546, telephone (202)–453–1725.

EPOP-M1 will have a morning equatorial crossing time, and both this satellite and the continuing operational NOAA series will carry AVHRR/4, AMSU, High Resolution Infrared Sounder/4 (HIRS/4, for global atmospheric water vapour and temperature profile measurements) and ARGOS systems; the ESA platform will also carry a CERES. The former will also include a SAR, SCATT and ALT. Final payload details are yet to be determined.

The platforms will carry larger instruments of advanced design, with scope for large antennae PMW radiometers offering improved spatial resolutions of possibly 2–10km. Large antennae may allow wider use of lower frequency passive microwave channels (ie 1–10 GHz), which are less affected by surface and atmospheric variability but which are currently limited by poor spatial resolution (Kostiuk and Clark 1984). However, an increase in antenna aperture tends to affect sensitivity for high scanning rates, as the noise level of the receiver for a given bandwidth is dependent on integration time. Alternatively, the synthetic aperture approach used with radar may be extended to encompass passive microwave instruments, thereby allowing improved resolution without the need for huge antennae. How applicable such an instrument would be to studying polar regions remains to be seen.

Although its frequencies are similar to those employed by the SMMR and SSM/I, the spatial resolution of the Multi-frequency Imaging Microwave Radiometer (MIMR), a probable candidate for inclusion on EOS-A, are better by a factor of three (ie at 37GHz, the MIMR will have a footprint size of 11.6km as opposed to 38×30km for the SSM/I). This is partly a function of the larger antenna employed (1.6m for MIMR, 60×65cm for the SSM/I). The radiometric sensitivity of the corresponding channels of MIMR and SSM/I are similar due to technological improvements with respect to the noise temperature of the receivers.

As expected, the sensors under consideration for future deployment are not radically different from existing and proven systems, with one or two notable exceptions, ie a laser altimeter (Bufton and others 1982) and the alternative multibeam radar ALT (Bindschadler and others 1987). It is unlikely that microwave radiometry will completely replace VIS and IR radiometry. An improved system may be one that combines improved spatial resolution with wide swath coverage. It must be ensured that future optical instruments (eg the MODIS-T/-N) have dynamic ranges which are high enough for snow and ice masses not to saturate the sensors; these hyperspectral sensors can sample the electromagnetic spectrum over far narrower wavebands than has hitherto been practical. Excellent reviews of possible new directions in the next generation of sensor technology are provided by Bindschadler *et al.* (1987) and NASA (1986; 1987a–e).

Sounding of the three-dimensional temperature field of the atmosphere is one area where improvements will undoubtedly be made by exploiting different regions of the electromagnetic spectrum (Smith 1989). This field is currently mapped at a coarse spatial resolution (20km) by multi-channel IR and microwave radiometers; the resolution is dictated by the need for a dwell time sufficient for an adequate detector signal-to-noise ratio. Unfortunately, the performance is substantially degraded by a rapidly varying surface background and the presence of clouds within the IFOV. The next generation of sensors (eg the two AMSU instruments of EOS-A) will measure directly through the clouds in the microwave region through a wide selection of channels on the wings of the 90GHz oxygen and 180GHz water vapour lines, thereby yielding soundings of both temperature and water vapour concentrations to a higher resolution and accuracy. Advanced platforms in geostationary orbit will include microwave sounders, and will form another important element of EOS. The report of the NASA Earth System Sciences Committee (1988) provides an excellent review of these and related matters.

Lasers

The deployment of a laser altimeter or ranger in space will be an important innovation with significant implications (Curran 1989; Degnan 1985); both are under consideration for inclusion on EOS-B series satellites (Figure 5.2). The first spaceborne laser altimeter was flown in lunar orbit on an Apollo mission in the early 1970s. Light detection and ranging instruments (lidars) operate in the ultraviolet, VIS and IR regions of the electromagnetic spectrum. The new active laser remote sensing system has been designated the Geodynamics Laser-Ranging System (GLRS) (NASA 1987c). GLRS incorporates three separable remote sensing techniques; it will operate in both ranging (short pulse, precision 1cm) and altimetry (long pulse, precision 10cm) modes, as well as an atmospheric profiling mode. The laser and optics will transmit three colours to enable the accurate determination of the atmospheric delay. The altimeter mode will be nadir-pointing, whereas the range component will operate up to 50° from nadir in two colours; the diameter of the spot illuminating the surface will vary from 80 to 290m. Its primary role will be to sense the vertical and horizontal distribution of atmospheric constituents to a hitherto unprecedented resolution; products will include aerosol boundary and planetary boundary layer height, and cloud height, base and optical depth.

Glaciological research addressed by the GLRS will have two principal components, namely to determine the time-dependent topography and mass of ice sheets, and the measurement of ice strain and flow velocity. In ranging mode, the sensor will range to a number of ground-based retroreflectors. The latter will provide an accurate measure of ice sheet motion and deformation at a number of specified locations, and will enable verification of ALT data (Thomas and others 1985). The lidar altimeter will yield precise range measurement to the Earth's surface by acquiring data on laser pulse time-of-flight and reflected pulse energy. The GLRS will also measure crustal deformation and isostatic effects in polar regions by targeting corner cube retro-reflectors mounted on exposed rock outcrops around the periphery of the ice sheets. Lidar techniques may also provide improved measurements of surface wind speed and direction over the open oceans, but only in cloud-free regions. Additional information may be provided on sea ice roughness and freeboard, both of which affect drag coefficients.

In areas of complex ice sheet topography, particularly near coastlines/ice shelves, conventional ALTs are frequently unable to resolve the subtle smaller-scale variations present. Because they depend on multiple pulse averaging due to target-induced speckle modulation and the consequent low SNR for an individual pulse measurement, the optical wavelength of the laser permits even a relatively small transmitter to provide a sub-milliradian beam

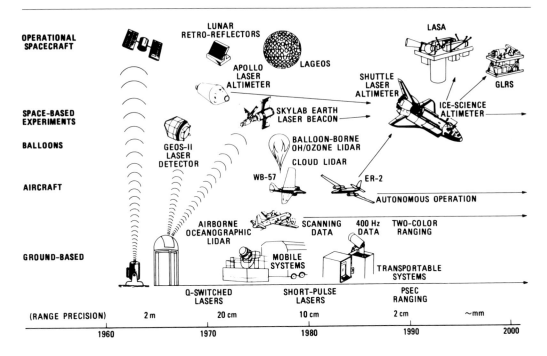

Figure 5.2 NASA laser-ranging evolution, culminating in the launch of GLRS on EOS. Laser atmospheric sounding will measure near-surface winds over ice. From NASA (1987c).

divergence and a small beam-limited footprint (about 70–100m as opposed to about 20km for the ALT) (NASA 1987c). Due to this factor, combined with a short, high-power pulse and freedom from speckle modulation, lasers/lidars could potentially provide more detailed surface profiles of sloping and undulating areas of ice sheets without the location ambiguities and biases inherent to conventional radar altimetry, and with a range precision of ±10cm (Zwally and others 1981; Thomas and others 1985).

Another potential application is the measurement of discontinuities in range, such as ice shelf margins, with the full horizontal resolution provided by the altimeter repetition rate. Land and sea ice roughness could be measured via laser-pulse waveform analysis. The resulting highly detailed surface elevation maps could be used both to delineate and monitor the time-dependent behaviour of ice flowlines, streams, grounding lines, zones of rapid basal sliding, sastrugi, and the seaward margins of the land ice. Laser altimetry, possibly in combination with narrow-beam radar or conventional radar altimetry (EOS contains both instruments), would also reveal the ice divides that define drainage basins and smaller glacier units (Swithinbank 1985).

Laser systems are restricted to clear (non-cloudy) sky conditions, and impose tighter attitude control requirements than less precise systems. Other technological limitations exist, including the present short lifetimes of the flashtubes, although the design of a laser capable of 5×10^7 shots appears to be feasible in the not too distant future (Bufton and others 1981). This should provide sufficient sampling to complete at least one survey of the major ice sheets.

It is hoped that future satellites, carrying a suite of microwave instruments, will abandon the sun-synchronous orbit in favour of the true polar orbit in order to offer more complete coverage of polar regions (hopefully in their entirety). Unfortunately, no missions on the horizon are planning to do this. Consequently, large data gaps will continue to be present within 7–8° latitude of the poles. The ice sheet dynamics and mass balance of central Antarctica are highly significant yet poorly understood, and this is likely to remain the case for some time to come. The problem is alleviated to some extent by wide swath sensors (eg passive microwave radiometers) that scan towards the poles.

Technological improvements

One major development in radiometer technology is the gradual replacement of mechanical scanning techniques by electronic (pushbroom) solid-state scanners; with the former, the field of view is changed by mechanical rotation of the sensor system or angular movement of its scan mirror. In present opto-mechanical imaging systems (eg the Landsat MSS and TM, and the MOS-1 VTIR), the spectral bands are broad because of the small signal obtained. The latter results from the short residence time of the detector in each IFOV. Pushbroom scanners, on the other hand, rely on the satellite motion to provide one direction of scan and electronic sampling of the detectors in the cross-track direction to provide the orthogonal scan component to form an image. They allow an increased residence time in each IFOV, and can thus operate in narrower spectral bands. No scanning mirror is required, as the entire swath width is covered by the detector elements, or charge-coupled devices (CCDs). Each detector is dedicated to one IFOV, and the image is created by scanning the detector array once during the period that the detector array moves forward by one IFOV along-track.

CCDs, which consist of arrays of photosensitive elements, are a dramatic innovation in space-borne remote sensing as they shift a group of signal electrons from input to output without distorting the signal itself. The fixed geometry results in high geometric accuracies in the line direction, which simplifies data processing. Finally, a time delay and integration (TDI) technique can be applied, whereby the signals from individual detector elements in the scan direction are delayed and added in phase to achieve signal enhancement by increasing the number of elements in the serially scanned line (Chen 1985). With TDI, the signal-to-noise ratio is significantly improved in the scan direction, and an averaging effect smooths out element-to-element nonuniformities.

As the technology develops, large and more sensitive mosaic array sensors could replace scanning systems. This field has become a principal area for research and future development. However, present CCDs, including those used in the MESSR and HRV sensors currently orbiting on MOS-1 and SPOT-1 respectively, are limited to a spectral bandwidth ranging from 0.4–1.1μm, ie VIS and NrIR. In order to accomplish photon detection at wavelengths greater than about 1.1μm, a detector with a narrower bandgap than that of pure silicon is required (Goetz and others 1985). Moreover, pushbroom technology to date does not yet allow the simultaneous acquisition of data from many contiguous spectral bands. It is for this reason that the request by potential users of MESSR data that an additional blue band (0.45–0.52μm) be added to the sensor could not be met. Geometric fidelity and spatial co-registration of spectral bands are additional constraints placed on sensor design.

The use of longer-wave IR spectral bands is also limited by the cooling requirements of the system, which preclude the use of a passive cooler. The sensitivity of a radiometer, which is a measure of its ability to detect changes in brightness temperature, generally decreases as the system temperature increases. In most TIR sensors, thermal noise limits the sensor performance, and cooling the detectors, amplifiers or optics greatly enhances the sensitivity. Virtually all instruments operating in the IR are dependent on cryogenic cooling, but this technique is difficult to employ in space and remains a challenge. Plans are afoot, however, to include IR and TIR CCDs on the next generation of SPOT and Landsat satellites.

The near-future availability of large orbiting space platforms and their associated transport systems has greatly influenced the current research and development of remote sensing philosophy and technology (Kostiuk and Clark 1984). There is a general trend towards the replacement of independent single purpose instruments by multi-functional systems capable of some degree of autonomous operation and decision-making. This trend requires multiple detector and receiver systems to obtain large spatial coverage in the minimum amount of time, involving heavier instrumentation with higher power consumption. The size and efficiency of solar panels in space has been increasing steadily. Artificial intelligence and telerobotics have obvious practical applications in space. Moreover, satellites will play significant role in the location of, and data relay from, instrument packages on the surface (Figure 5.3).

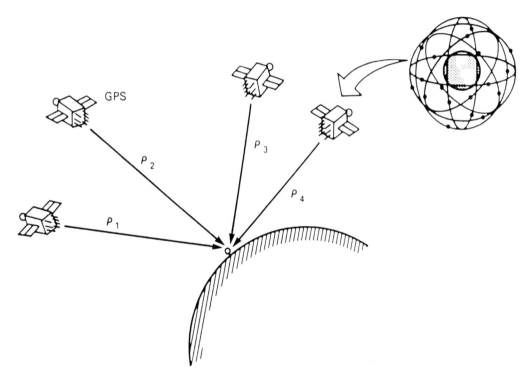

Figure 5.3 Radio signals transmitted via polar orbiting navigational satellites such as TRANSIT/NAVSAT and the new Navstar satellites in the Global Positioning System (GPS), can, when recorded at sites on the surface of an ice sheet, give accurate flow velocity and strain/deformation data (Drew and Whillans 1984). Simultaneous phase measurements from at least two satellites with at least two ground receivers enables precise tracking of the differential range change of the satellites. The location of a ground receiver is determined with respect to the satellite ephemeris by using the Doppler shift in frequency of signals transmitted from the passing satellite. Accuracies of ±1.5m in all three coordinates are obtainable using translocation techniques to geodetically determine ground control. GPS will employ 21 high altitude (20,200km) satellites when fully operational (in 1993). It will give precise tracking of satellite orbits, a factor of great importance to altimetry missions.

Substantial improvements in the overall retrieval accuracies from passive microwave radiometers are likely to result from instrument upgrading, the use of more frequencies, and the reduction of ambiguities inherent to the algorithms designed to extract geophysical parameters from the data. Multi-frequency, multi-incidence angle and multi-polarisation (polarimetric) radar imaging will feature more prominently in future space-borne programmes over snow and ice as our ability to handle the anticipated flood of data improves. Recent experiments carried out in the Arctic by NASA JPL, using a multi-frequency and polarisation SAR, have yielded a wealth of comparative and interesting results, eg an improved distinction between open water and new ice (M. Drinkwater, personal communication) (Figure 5.4). It would be enlightening to generate images showing the phase differences between HH and VV since for reasons that are unclear, volume scattering produces almost no relative phase shift, while surface scattering can produce a 180° phase shift (for sea ice).

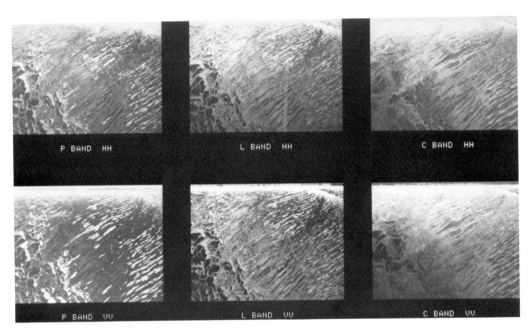

Figure 5.4 JPL aircraft multi-frequency polarimetric SAR images of the early stages of ice growth in the St Lawrence Island Polynya, Bering Sea at P-, L- and C-band frequencies, HH and VV polarisations. Streamers of frazil ice organised by Langmuir circulation are discernible. Courtesy of Ben Holt, NASA JPL.

The primary use of multi-polarised SAR data is in separating volume from surface scattering effects, thereby gaining further insight into the mechanisms involved and allowing interference of the surface properties. At large viewing angles (>30°), surface (Bragg) scattering is stronger for VV than for HH polarisation. The potential value of cross-polarised data is poorly quantified at present due to difficulties in the details of modelling the backscattering process. The few experimental data available suggest, however, that cross-polarisation images should provide additional information about the distribution of volume scatterers (particularly at C- and higher wavebands), and therefore about sea ice type. Cross-polarisation for snowpacks is sensitive to volume scattering by liquid water and enlarged snow grains (Rott and others 1985). It is not anticipated that microwave sensors with a wide areal coverage will be able to resolve details smaller than approximately 15–20m.

Before powerful, multi-frequency/polarisation SARs can be included on the EOS platforms, however, a number of technological problems must be addressed and solved. For example, more accurate measurements require end-to-end calibration. Current advances in sold-state amplifier technology will enable the use of SAR systems operating at higher frequencies than are presently possible. One probable scheme for high-power radiation is the use of a distributed approach, whereby the amplifying elements are distributed across the aperture and co-located with the radiating elements (Carver and others 1985). This technique will be tested for the first time in space when SIR-C/X-SAR is launched in 1993, albeit in a non-polar orbit. Both this and subsequent flights will use electronic beam scanning, and will be calibrated.

The near-simultaneous acquisition of multi-polarisation data will allow easy registration of the imagery and the derivation of the relative phase on a pixel-by-pixel basis. The spectral region range being considered for future space-borne SARs is from 500MHz to 60GHz, allowing for gaps due to water vapour and oxygen absorption. The lower end of the region is limited to SAR by frequency-allocation regulations and ionospheric effects. The upper end is limited more by technological factors such as the need to generate high power and the weight of the instrument, as well as by atmospheric opacity. Such developments will be accompanied by dramatic increases in data rates. Higher frequency SARs, operating in the range of 10–12GHz, are, however, desirable for investigations of snow cover and volume scattering properties of ice.

Consequently, the next decade will be a crucial transition period from the preceding 20 years of research and development with air- and space-borne microwave sensors to the routine application of satellite data to polar research. The simultaneous use of passive and active microwave systems with VIS and IR radiometers will yield more information than would come from these sensors used individually. Campbell and others (1975) highlighted the merits of an integrated approach to polar remote sensing well over a decade ago, although an optimum sensor combination for ice research applications (both sea and land) has yet to be determined. The integration of different data sets must overcome the problem of melding data with different spatial resolutions on to common grids (map projections). Only then can detailed inter-comparisons be made and the full scientific value of the data be realised.

6. Data processing, archiving and dissemination

The crucial importance of the ground segment of satellite missions cannot be over-emphasised; there is no point in generating mountains of data if no provision is made for their processing, archiving and rapid dissemination. It has become increasingly clear that traditional channels for handling scientific data are totally inadequate when faced with the huge volumes of data generated by current satellites. For example, a single Landsat TM scene covering a surface area of 170×185km contains 273 Mbytes of data, enough to fill $7 \times$ 1600bpi computer-compatible tapes (CCTs). Each SPOT HRV multi-spectral scene covering an area of 60×60km contains 27Mb of data. The problem will be compounded in the EOS era, when the scale of requirements will be many orders of magnitude larger than they are at present. The reasons for this include the need for quantisation of 12 rather than 8 bits, spectral coverage of many more bands, reprocessing of complete datasets, merging of imagery with topographic and other geophysical data, and directional image acquisition rather than nadir-looking only. The MODIS instrument, for example, will produce data volumes of approximately 10^{12} bits per day.

The data rate of SAR, at over 100 million bits per second (Mbps), is monumental. Indeed, the proposed multi-polarisation and multi-frequency EOS SAR will require a data rate of 200 to 300Mbps (Figure 6.1). If it were to operate continuously for 24 hours, it would generate up to 2.5×10^{13} bits, or the equivalent of about 250,000 CCTs of data. Such frightening volumes will necessitate the development of onboard selective processing, improved data transmission links and high density storage media. With several very important satellites due to be launched in the next decade, it has been (and is) crucial to prepare, test and establish systems to handle, in a streamlined fashion, the expected data deluge before the event; only then can the full benefits be attained and the information content understood. Data compression and subsequent de-compression techniques will play an important role in handling such huge data rates and volumes. Hardware and software designed for this purpose are becoming standard components in digital communications and storage devices.

Such large data rates inevitably come at a cost as follows:

i)   Very large data transmission channel bandwidths are required.
ii) The processing of raw data into usable data in near real-time requires ground processors with speeds of the order of 1 GFlop s^{-1} (a Flop is a floating point operation per second).

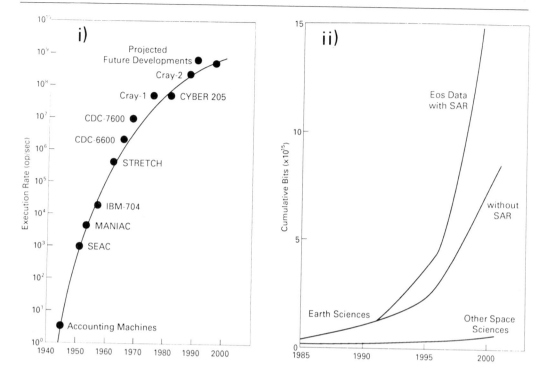

Figure 6.1 (i) Trends in computational capacity demonstrate the advancing power of super-computers; the cost per unit computation has fallen steadily, and is expected to continue doing so. (ii) In spite of technological advances, the dramatic increase in the volume of data to be processed, validated and made readily available over the next decade (from EOS) will greatly challenge the ground segment. The actual EOS data rate will depend primarily on the duty cycle of the SAR. After NASA Earth System Sciences Committee (1988).

iii)   After processing, adequate storage and retrieval mechanisms must be developed for easy access to the data.

iv)   The enormous volume of data threatens to totally overwhelm the ability of human interpreters to extract the necessary information. Consequently, a considerable effort is being made to develop advanced techniques for automated image segmentation, co-registration and change detection. These tasks require the training and education of additional skilled staff before the launch of each satellite, a factor which is often overlooked by the powers that be when writing their mountains of proposals. Moreover, the communication gateways between system designer and data user are open to improvement; the former may be motivated to create a sensor with the highest spatial resolution in the world, but is this what the user wants and needs?

The usefulness of satellite data has often been limited by the speed with which they can be obtained from the data processing, dissemination and archival centres, even if these are known to exist. Frequently, there are long delays between the receipt of raw data on the ground and the dissemination of preprocessed data to the user. Furthermore, these centres may not be geared towards handling specific requests for specially processed data. Consequently, even where data have been archived and can be located, the storage and duplication costs passed onto the user may be prohibitive. Polar data have often been stored, if retained at all, in a highly fragmented fashion, and dispersed throughout various national archives.

It is unlikely, for practical reasons, that any single archive can satisfy the many requirements of the polar research community. Although there is a healthy trend toward archive centralisation, the inevitable continued existence of independent archives necessitates the adoption of uniform standards and archival practices, a situation that does not exist at present. The proliferation of data formats, image analysis techniques and processing algorithms in the last decade has only served to underline the critical need for standardisation on an international basis. These and related questions should be addressed sooner rather than later. Compatibility of both hardware and software is crucial, as are improved guidelines/documentation regarding their availability. Space agencies such as ESA and NASA are fully aware of the problem, and international negotiations are taking place in a concerted effort to rectify the situation.

Uniformity is best achieved by having each archive serve as a node of an easily accessible, electronic network, along the lines of the Cryospheric Data Management System (CDMS) at the US National Snow and Ice Data Center (NSIDC) in Boulder, Colorado, which has developed to disseminate polar passive microwave products; this archive also contains a wide range of other polar data. Similarly, NOAA has developed an electronic catalogue system not only to streamline the distribution of critical meteorological data (AVHRR and TOVS) but also to improve access to archives for individual users through interactive communication networks. The NASA Master Directory (NMD) is an on-line multi-disciplinary computer directory of ocean, land and atmospheric (and space) datasets; it contains descriptions of the data, and provides mechanisms for searching for data by criteria such as geophysical parameter, time and spatial coverage. Further details are available from Dr Joy Beier, Code 633, NASA Goddard Space Flight Center, Greenbelt, MD 20771, USA, telephone (301)–786–5289. Similarly, an important goal of the ERS-1 mission is to have a ground segment capable of providing operational fast delivery products (ie within three hours of reception). NASA's ASF will provide users with an electronic catalogue of raw and processed data and data products as well as an electronic browse file of lower resolution imagery for access on line. Great progress is therefore being made, although much work remains to be done before the application of satellite data to polar research becomes routine.

The integration of complementary data from future missions is essential, as is the need to avoid duplication of coverage by similar sensors (duplication of instruments on different satellites is desirable, however, as an insurance against the failure of one). As applications proliferate and new sensors evolve, it is clear that a high degree of international coordination and cooperation is required, not so much to design more efficient sensors as to make sure that orbits and data acquisition and delivery systems are chosen to satisfy both the broad requirements of data users groups and the various government and commercial programme objectives.

With so many potentially conflicting sensors on board some future satellites, mission control and efficient scheduling of observations will become crucial, with some allowance to cope with unforeseen events such as large iceberg calving. Careful advance planning within the polar community will involve a rigorous definition of scientific objectives and a thorough understanding of spacecraft and sensor capabilities and orbits. An important example of this is the PIPOR programme, and its ice sheet equivalent the Polar Ice Sheet Program, which are designed to coordinate field experiments to calibrate and validate ERS-1 data (Gudmandsen and others 1989). Vast amounts of *in situ* data have already been collected by a number of important international experiments (eg MIZEX and CEAREX) (Luther 1988). These data have greatly improved our ability to interpret satellite data, and have identified problem areas (a special issue of the *Journal of Geophysical Research (Oceans)*, 92, C7, 1987, is devoted to MIZEX). The detection of possible global change using satellite data requires an absolute calibration, control of data quality, and careful attention to sensor drift and the stability of calibration. Moreover, if product retrieval algorithms are to evolve, then the changes incorporated must be consistent if a data set collected in, say, 1975 is to be compared with one from 1990.

Geographical Information Systems and data management

There is an increasing call for the generation of readily-usable, gridded geophysical products from level 1b data, as well as the archiving of the original format raw data. One of the main difficulties in implementing remote sensing data as a standard tool for polar research is the current need for investigators to have a substantial specialised knowledge of satellite data processing; their efforts would be more usefully employed if they were able to concentrate on the interpretation of geophysical data products rather than their generation. The potential user should not be deterred by a) feeling that he/she needs to be competent at computer programming to post-doctoral level, and b) by being confronted by a bewildering array of both image processing hardware and software. Simplicity should be a major consideration.

The real power of remote sensing emerges when data from more than one source are merged. Geographical Information Systems (GIS) is a powerful technique that combines elementary manipulations of raster images into overlays, areal measures and ever more powerful inferential tools (Figure 6.2). Recent advances in GIS technology may provide the most efficient means of improving data cataloguing and retrieval by providing a method for combining ice and complementary data from different sensors into a common and readily accessible data base (Marble and Peuquet 1983). For example, the Ice Centre Environment Canada (ICEC) in Ottawa, Canada, is developing an Ice Data Integration and Analysis System (IDIAS), in conjunction with MacDonald, Dettwiler and Associates, to aid users by synthesising information from various data sources, all corrected to one geographic coordinate system, and available on menu-driven work stations (Ramsay and others 1988). IDIAS will automate the ingestion and processing of the large volume of digital radar data from ice reconnaissance aircraft, NOAA AVHRR data, Landsat data, and future input using ERS-1, Radarsat and DMSP data. Data from other sensors may be added at a later date. ICEC are also developing an Iceberg Analysis and Prediction System. Further details are available from Ice Centre, Environment Canada, 373 Sussex Drive, Ottawa, Ontario K1A OH3, Canada, telephone (613)–996–4297.

Similarly, the primary purpose of the CDMS (NSIDC) is to extract polar data from the SSM/I data stream, create gridded data sets, provide interactive data cataloguing with remote access via standard telecommunications circuits, distribute data to users, and implement revised algorithms to extract improved ice and snow products. The software utilises a menu-driven user interface, and has an on-line browse facility. Recent technological innovations have opened up an exciting new era, one in which satellite and relevant ancillary data will be held and transmitted in both off- and on-line forms (Figure 6.3).

By performing many of the preliminary data processing activities and effectively reducing the data to a range of standard, ready-to-use products, NASA hopes to encourage a wider use of the data among scientists. The feedback of ideas and recommendations from the latter will

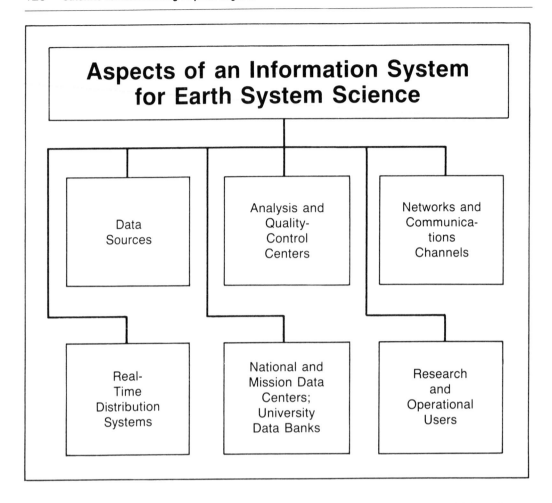

Figure 6.2 Principal aspects of an information system for Earth System Science. After NASA Earth System Sciences Committee (1988).

be mutually beneficial. The CDMS is also ingesting, in cooperation with the NASA Goddard Space Flight Center (GSFC), Nimbus-7 SMMR data (sea ice data re-binned into the 25km resolution SSM/I grid) and gridded, georeferenced Seasat and GEOSAT ALT ice sheet data to allow detailed intercomparison between different sensors, past, present and future. Moreover, not only original brightness temperatures and geophysical data in gridded format will be archived; original satellite swath format data will also be retained, thus enabling researchers to develop their own algorithms for the extraction of specific parameters of their choice. Similarly technologically sophisticated, but conceptually simple, innovations have been incorporated into the NASA's National Space Science Data Center (NSSDC) and Climate Data System (NCDS). Users can better than ever before determine which data are specifically relevant to their needs before acquiring those data. Such facilities greatly improve the efficiency of data archiving and dissemination, thereby reducing costs and improving product quality and availability.

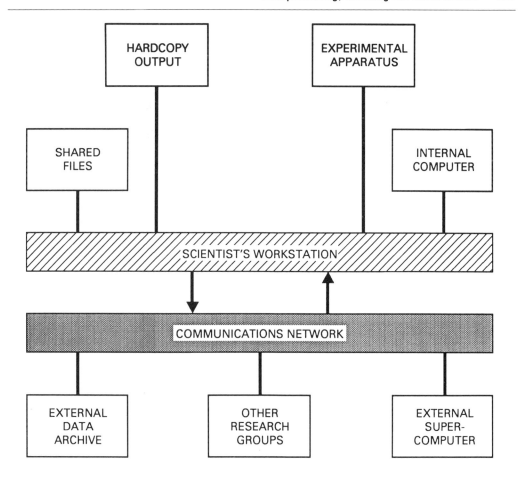

Figure 6.3 Modern scientific workstations, linked to data archives, experiments and a variety of computational aids, have revolutionised the ability of the scientist to both access and use remote sensing data. They combine high speed processing with extensive core/virtual-core memories that can drive large, high resolution colour displays, eg 1024 × 1024 pixels. Networking allows communication with remote mass storage access devices/other computers/on-line data archives. Communications capabilities range from voice-line speeds of $1.2kbs^{-1}$ up to current typical mass data transfer speeds of $56kbs^{-1}$. After NASA Earth System Sciences Committee (1988).

EOSDIS

The EOS programme is also placing great emphasis on the ground segment, called the Data and Information System (EOSDIS). This comprises a number of Distributed Active Archive Centers (DAACs), each of which are subdivided into 3 sectors: i) product generation; ii) data storage and distribution; and iii) information and management. To date, seven centres have been designated as archives responsible for a specific range of environmental data from EOS; for example: i) NASA GSFC will be responsible for data related to climate and global change, SST and ocean colour (ie this will be an extension of the existing NSSDC); ii) NASA JPL will handle oceans data (by extending the NASA Ocean Data System or NODS); and iii) the University of Colorado will handle all sea ice and snow cover data collected by low bit-rate EOS and non-EOS instrument, including *in situ* data steams (as an extension of the

NSIDC). Specific products are yet to be determined. EOSDIS will incorporate electronic browse systems, allowing remote access to, and exploration of, data files via communications networks. This will help bring the new technology, and its advantages, to a broader user community. Quick-look imagery on-line will allow the investigator to determine cloud cover characteristics and to determine whether the dataset suits his/her specific needs. For further information, contact the EOS Project Science Office, NASA Goddard Space Flight Center, Greenbelt, MD 20771, USA.

Optical disc storage and data dissemination

Since the data base being produced is useful beyond the lifetime of its parent sensor, the use of optical discs will become increasingly prevalent as the primary medium for both the storage and dissemination of data (Mass and others 1987). Many potentially important polar satellite data have in the past been discarded because of the cost of archiving them. Even high-density (6250bpi) magnetic tapes can hold only 130–80 Mbytes of data. Conversely, all of the data from the future TOPEX/Poseidon mission, amounting to about 3 $\times$ 10^9 bits of data over a three-year period, should fit onto just three optical discs.

Two types of optical disc system are well suited as storage/dissemination media, namely Write-Once-Read-Many (WORM) and Compact Disc-Read Only Memory (CD-ROM). WORM technology enables users/data dissemination centres to write data from their tapes or other storage media to optical discs. Twelve-inch WORMS can hold up to 1GByte of data on each side of a two-sided disc. The lack of widely accepted standards is a problem. The CD-ROM, on the other hand, uses virtually the same technology as the music compact disc, but employs superior error-correcting electronics. Indeed, CD-ROM drives with interfaces for various personal computers, or computers with Small Computer System Interface (SCSI) ports, are readily available. These drives can provide data from the discs at rates of up to 150,000 bytes per second, and a single 4.72-inch compact disc can hold over 600Mbytes of digital information (equivalent to about 20 $\times$ 1600bpi CCTs). Alternative high density storage media include Digital Audio Tapes (DAT), with a 1,300 Mbyte capacity (9.3 times that of a 9-track 6250bpi magnetic tape), and 8mm tapes, with a 2,100 Mbyte capacity.

The CDMS distributes Nimbus-7 SMMR sea ice and DMSP SSM/I brightness temperature datasets on CD-ROM. Hardware specifications can be found in the Nimbus-7 section in Part 2. The recent reprocessing of Nimbus-7 CZCS data onto optical digital discs by scientists at the NASA GSFC and the University of Miami is another prime example. Such systems, with their user friendly software, greatly facilitate archive browsing and data accessibility.

The NASA JPL node of NODS has compiled an invaluable set of references to help those users receiving data and data products on CD-ROM to learn more about the technology and the hardware specifications necessary to access and use the data. For further information, contact Elizabeth Smith, Mail Stop 300–323, JPL/NODS, 4800 Oak Grove Drive, Pasadena, CA 91109, USA, telephone (818) 354–6980; telemail NODS.JPL on OMNET; telex 675429 (attention NODS); FAX 818–393–6720 (attention NODS); or via SPAN as EAS::STANS.

Commercialisation of satellite data

International cooperation is essential in that it improves access to data, aids verification and helps to avoid duplication. These considerations are of paramount importance at a time when financial restrictions are increasingly prohibitive. With this in mind, the recent commercialisation of the Landsat system by the US Remote Sensing Commercialisation Act of 1984 (transferring its operations to the EOSAT Corporation) in an effort to recoup the expense of the system, and its direct competition with the SPOT system since 1986, should theoretically improve the quality of both the products and the service offered. It is a sign of the times that projects must be shown to be economically viable if they are allowed to continue (NRC 1985). The commercial success of Landsat and SPOT would at least provide justification for the launch of further satellites with improved sensors onboard.

If prices continue to rise at such an alarming rate, however, the result will be simple: the user will not be able to afford the product. It is no use producing the highest quality imagery in the world if no one can afford to benefit from it. EOSAT and SPOT Image have monopolised commercial high resolution remote sensing. Russia has recently entered the fray, announcing in December 1987 that it is selling images of 6m spatial resolution of any area apart from its own and its allies' territories, with a sales outlet for data in the USA. The spirit of glasnost and perestroika may further benefit the scientific community, as the USSR has a long experience in the remote sensing of polar regions from space. The exchange of ideas and data between East and West would be a major development.

Other competitors, including Japan, Canada, ESA, Brazil, China and India will follow by launching polar-orbiting Earth observation satellite systems in the 1990s. These developments could have a serious and detrimental impact on data availability and accessibility. The other side of the coin should, however, be stated: even though the recent trend has been towards drastic increases in the cost of satellite data, the cost of good quality imagery is likely to be minimal compared to the cost of conducting experiments on the ground.

Also under threat is the 'open skies' policy of remote sensing mentioned earlier; different national space agencies may not continue to allow such ready access to data from their satellites as they fly overhead. Indeed, current policies appear to challenge the 'Principles relating to remote sensing of the Earth from space' laid down by the General Assembly of the United Nations (United Nations 1987). In this respect, the memoranda of understanding currently being signed by different national agencies are of crucial importance in ensuring long-term availability of data from remote polar regions. The European Space Agency (ESA), NASA, the Canadian Space Agency and the Japanese National Space Development Agency have recently signed agreements to allow access to data from each other's satellites, ie ERS-1, JERS-1 and Radarsat. Technical developments are being carried out to ensure full compatibility of the various crucial satellite receiving stations, eg Kiruna, Ottawa, Syowa base in Antarctica and the ASF.

On a less pessimistic note, it is encouraging to note that a primary provision of EOS is to make operational data available to any user from participating countries at a cost not to exceed that of reproduction and transmission. Principal research investigators will be unable to reserve a period of exclusive use. This and other matters regarding the availability of satellite data and interagency/international cooperation in the EOS era are well covered by Baker (1990).

The threat to future data availability

The unique value of, and the huge benefits offered by, satellite remote sensing of polar regions is beyond doubt. The future of a great deal of important data yet to be received, however, is far from assured, as is the continued maintenance of existing data archives. The threat comes mainly in the form of budget cuts. If these are to be made, it is imperative that i) the polar-user community should first be consulted to select digital data for important areas and for periods of special interest, eg those collected during surface experiments, and ii) the most useful levels of data to be retained should be specified.

Level-1b data products are the lowest data level of interest to investigators, and it is these data that should be placed into permanent archive. Routine systematic and specific processing can then be implemented downstream from the fundamental level products as required.

Satellite data are an international, dynamic resource, and hold the key to a better comprehension of the impact and causes of possible global change. Part 2 underlines the fact that a vast reservoir of unique satellite data, largely untapped by the polar research community, has already been collected and archived. Understanding these data will greatly aid our interpretation of data from future sensors. They form a beginning of a priceless time series, and as such their existence should not be forgotten in the imminent data deluge.

References and bibliography to Part 1

Ackley, S. 1981a. A review of sea ice-weather relationships in the southern hemisphere. *Proceedings of Symposium on Sea level, Ice and Climatic Change*, 7–8 December 1979, Canberra. *IAHS Publication 131*: 127–59. International Association of Hydrological Sciences, Wallingford, UK.

Ackley, S. 1981b. Sea ice-atmosphere interactions in the Weddell Sea using drifting buoys. *Proceedings of Symposium on Sea level, Ice and Climatic Change*, 7–8 December 1979, Canberra. *IAHS Publication 131*: 177–91. International Association of Hydrological Sciences, Wallingford, UK.

Ackley, S., Smith, S. and Clarke, D. 1982. Observations of pack ice properties in the Weddell Sea. *Antarctic Journal of the United States* 17: 91–3.

Ad Hoc Panel on Remote Sensing of Snow and Ice. 1989. Prospects and concerns for satellite remote sensing of snow and ice. *Special Report, Committee on Glaciology, Polar Research Board, Commission on Physical Sciences, Mathematics and Resources, National Research Council*. Polar Research Board, NAS, Washington, DC 20418.

Allan, T. 1987. Study of ERS-1 mission options. *Special Report of the Deacon Laboratory of the Institute of Oceanographic Sciences*, Wormley, Surrey, UK.

Allison, I. (ed.). 1983. *Antarctic climate research*. SCAR, Cambridge, UK.

Alpers, W. 1985. Theory of radar imaging of internal waves. *Nature* 314: 245.

Alpers, W., Bruening, C. and Richter, K. 1986. Comparison of simulated and measured Synthetic Aperture Radar image spectra with buoy-derived ocean wave spectra during the Shuttle Imaging Radar B Mission. *IEEE Transactions of Geoscience and Remote Sensing* GE-24: 559.

Andersen, T. and Odegaard, H. 1980. Application of satellite data for snow mapping. *Report Number 3*, Norwegian National Committee for Hydrology, Oslo.

Andreas, E. and Ackley, S. 1982. On the differences in ablation seasons of Arctic and Antarctic sea ice. *Journal of Atmospheric Sciences* 39: 440–7.

Arking, A. and Childs, J. 1985. Retrieval of cloud cover parameters from multispectral satellite images. *Journal of Climatology and Applied Meteorology* 24: 322–33.

Asrar, G. (ed.). 1989. *Theory and applications of optical remote sensing*. John Wiley and Sons, New York.

Barath, F., Barrett, A., Copeland, J., Jones, D. and Lilley, A. 1964. Mariner 2 microwave radiometer experiment and results. *Astronomical Journal* 69, 1.

Baker, D.J. 1990. *Planet earth: the view from space*. Harvard University Press, Cambridge, Massachusetts and London.

Barkstrom, B., Harrison, E., Smith, G. and Cess, R. 1989. Results from the Earth Radiation Budget Experiment (ERBE). *Remote sensing of atmosphere and oceans. Proceedings of COSPAR 27th Plenary Meeting*, Espoo, Finland, 18–29 July 1988: 75–82.

Barnett, T. 1983. Long-term changes in dynamic height. *Journal of Geophysical Research* 88: 9,547–52.

Barrick, D. and Swift, C. 1980. The SEASAT microwave instruments in historical perspective. *IEEE Journal of Oceanic Engineering* OE-5: 74–80.

Barry, R. 1985. The cryosphere and climate change. In MacCracken, M. and Luther, F. (eds), *Detecting the climatic effects of increasing carbon dioxide*. US Department of Energy Special Report *DOE/ER-0235*. Washington DC.

Barry, R., Crane, R., Weaver, R. and Anderson, M. 1984. Sea-ice and snow-cover data availability, needs and problems. *Annals of Glaciology* 5: 9–15.

Basharinov, A., Gurvich, A., Yegorov, S., Kurskaya, A., Matvyev, D. and Shutko, A. 1971. The results of microwave sounding of Earth's surface according to experimental data from the satellite COSMOS 243. *Space Research* 11. Akademie-Verlag, Berlin.

Beckman, J. 1983. Communication and data transmission systems. In Colwell, R. (ed.), *Manual of Remote Sensing*, Volume 1: 681–98. American Society of Photogrammetry.

Benson, C. 1989. Remote sensing of global snowpack energy and mass balance: *in situ* measurements on the snow of interior and Arctic Alaska. *Special Report of the Geophysical Institute*. University of Alaska, Fairbanks.

Benz, R., Gebauer, G., Tanner, T. and Zeiss, E. 1989. Multimission capability of the European Polar Platform. *Proceedings of IGARSS '89*, Vancouver, Canada, 10–14 July 1989: 570–4.

Berg, C., Wiesnet, D. and Legeckis, R. 1982. The NOAA-6 satellite mosaic of Antarctica: a progress report. *Annals of Glaciology* 3: 23–6.

Bernstein, R. 1982. Sea surface temperature mapping with the SEASAT microwave radiometer. *Journal of Geophysical Research* 87. C10: 7,865.

Bernstein, R. (ed.). 1983. Image geometry and rectification. In Colwell, R. (ed.), *Manual of Remote Sensing*, Volume 1: 873–922. American Society of Photogrammetry.

Billingsley, F. (ed.). 1983. Data processing and reprocessing. In Colwell, R. (ed.), *Manual of Remote Sensing*, Volume 1: 719–92. American Society of Photogrammetry.

Bindschadler, R. 1983. Jakobshavn glacier drainage basin: a balance assessment. *Journal of Geophysical Research* 89: 2,066.

Bindschadler, R. (ed.) 1990. SeaRISE; a multidisciplinary research initiative to predict rapid changes in global sea level caused by collapse of marine ice sheets. *NASA Conference Publication 3075.*

Bindschadler, R., Jezek, K. and Crawford, J. 1986. Glaciological investigations using the synthetic aperture radar imaging system. *Annals of Glaciology* 9: 11–19.

Bindschadler, R., Born, G., Chase, R., Fu, L., Mouginis-Mark, P., Parsons, C. and Tapley, B. 1987. Altimetric system. *Earth Observing System Panel Report*, Volume IIh, NASA HQ, Washington, DC.

Bindschadler, R., Zwally, J., Major, J. and Brenner, A. 1990. Surface topography of the Greenland ice sheet from satellite radar altimetry. *NASA SP-503*. NASA, Washington DC.

Bindschadler, R. and Scambos, T. 1991. Satellite-image-derived velocity field of an Antarctic Ice Stream. *Science* 252, 5003, 242–246.

Bogorodsky, V., Bentley, C. and Gudmandsen, P. 1985. *Radioglaciology*, D. Reidel Publishing Co., Dordrecht.

Bolle, H.-J. 1985. Assessment of thin cirrus and low cloud over snow by means of maximum likelihood method. *Advances in Space Research* 5, 6: 169–75.

Booth, A. and Taylor, V. 1969. Meso-scale archive and computer products of digitized video data from ESSA satellites. *Bulletin of the American Meteorological Society* 50: 431.

Brenner, A., Bindschadler, R., Thomas, R. and Zwally, H. 1983. Slope-induced errors in radar altimetry over continental ice sheets. *Journal of Geophysical Research* 88, C3: 1,617.

Brenner, A.C., Frey, H.V. and Zwally, H.J. 1990. Comparisons between GEOSAT and Seasat tracking over non-ocean surfaces. *Geophysical Research Letters* 17, 10: 1,537–40.

Brooks, R. 1983. Land science applications achievable from satellite altimetry. *Geoscience Corporation Report, NASA Contract NAS6–3232*. NASA, Washington DC.

Brooks, R., Campbell, W., Ramseier, R., Stanley, H., and Zwally, H. 1978a. Ice sheet topography by satellite altimetry. *Nature* 274, 5,671: 539–43.

Brooks, R., Roy, N. and Stanley, H. 1978b. Sea ice boundary determination from satellite radar altimetry. *EOS Transactions of the American Geophysical Union* 260.

Bufton, J., Robinson, J., Femiano, M. and Flatow, F. 1981. Satellite laser altimeter for measurement of ice sheet topography. *Proceedings of IGARSS '81, IEEE Digest*: 1,003.

Burke, S. and Anderson, J. 1989, Oceanographic measurements with drifting buoys. *ARGOS Newsletter* 37, June 1989: 13–15.

Burns, B. 1988. SAR image statistics related to atmospheric drag over sea ice. *Proceedings of IGARSS '88*, Edinburgh, 13–16 September 1988. *ESA SP-284 (IEEE 88CH2497–6)*, Volume 1: 409–12.

Burns, B., Cavalieri, D., Keller, M., Campbell, W., Grenfell, T., Maykut, G. and Gloersen, P. 1987. Multisensor comparison of ice concentration estimates in the marginal ice zone. *Journal of Geophysical Research* 92, C7: 6,843–56.

Burtsev, A., Krovottyntsev, V., Nazirov, M., Nikitin, P. and Spiridonov, Y. 1985. Radar maps of the Arctic and Antarctic from data of KOSMOS-1500 satellite and preliminary results of analyses. *Issledovaniye Zemli iz Kosmosa* 3: 54.

Bushuyev, A., Grishchenko, V. and Masanov, A. 1985. Interpreting sea ice on satellite radar images. *Issledovaniye Zemli iz Kosmosa* 3: 9.

Campbell, W., Gloersen, P., Nordberg, W. and Wilheit, T. 1974. Dynamics and morphology of Beaufort Sea ice determined from satellites, aircraft, and drifting stations. In Bock, P., Baker, F. and Ruttenberg, S. (eds), *Proceedings of the Symposium on Approaches to Earth Survey Problems through Use of Space Techniques*. Berlin, Akademie Verlag: 311–27.

Campbell, W., Ramseier, R., Weeks, W. and Gloersen, P. 1975. An integrated approach to the remote sensing of floating ice. *Proceedings of 3rd Canadian Symposium on Remote Sensing*: 39–72. Canada Centre for Remote Sensing, Ottawa.

Caprara, G. 1986. *The complete encyclopedia of space satellites*. Portland House, New York.

Carsey, F. 1980. Microwave observation of the Weddell polynya. *Monthly Weather Review* 108: 2,032–44.

Carsey, F. 1982. Arctic sea ice distribution at the end of summer 1973–1976 from satellite microwave data. *Journal of Geophysical Research* 87, C8: 5,809–35.

Carsey, F. 1985. Summer Arctic sea ice character from satellite microwave date. *Journal of Geophysical Research* 90, C3: 5,015–34.

Carsey, F. 1989. Review and status of remote sensing of sea ice. *IEEE Journal of Oceanic Engineering* 14, 2: 127–38.

Carsey, F., Holt, B., Martin, S., McNutt, L., Rothrock, D., Squire, V. and Weeks, W. 1986. Weddell-Scotia Sea marginal ice zone observations from space. *Journal of Geophysical Research* 91, C3: 3,920.

Carsey, F. and Zwally, H. 1986. Remote sensing as a research tool. In Untersteiner, N. (ed.). *The geophysics of sea ice*. Plenum Publishing Corporation, New York: 1,021–98.

Carsey, F. and Weeks, W. 1988. Plans for the development of EOS SAR systems using the Alaska SAR Facility. *Proceedings of IGARSS '88*, Edinburgh, 13–16 September 1988. *ESA SP-284 (IEEE 88CH2497–6)*, Volume 3: 1,491–4.

Carver, K., Elachi, C. and Ulaby, F. 1985. Microwave remote sensing from space. *Proceedings of the IEEE* 73, 6: 970–96.

Casassa, G., Jezek, K., Turner, J. and Whillans, I. In press. Relict flow stripes on the Ross Ice Shelf. *Annals of Glaciology* 15.

Cavalieri, D., Gloersen, P. and Campbell, W. 1984. Determination of sea ice parameters with the NIMBUS-7 SMMR. *Journal of Geophysical Research* 89, D4: 5,355–69.

Cavalieri, D., Gloersen, P. and Wilheit, T. 1986. Aircraft and satellite passive microwave observations of the Bering Sea ice cover during MIZEX West. *IEEE Transactions of Geoscience and Remote Sensing* GE-24, 3: 368–77.

Chang, A. 1986. NIMBUS-7 SMMR snow cover data. In Kukla, G., Barry, R.G., Hecht, A. and Wiesnet, D. (eds) Snow Watch '85, *World Data Center A for Glaciology Data Report GD-18*: 181–87. World Data Center A for Glaciology, University of Colorado, Boulder, Colorado.

Chang, A., Gloersen, P., Schmugge, T., Wilheit, T. and Zwally, H. 1976. Microwave emission from snow and glacier ice. *Journal of Glaciology* 16, 74: 23–39.

Chang, A., Choudhury, B. and Gloersen, P. 1980. Microwave brightness of polar firn as measured by

NIMBUS 5 and 6 ESMR. *Journal of Glaciology* 25: 85–91.

Charalambides, S., Hunt, G., Rycroft, M., Murgatroyd, R. and Limbert, D. 1986. Studies of the radiation budget anomalies over Antarctica during 1974–1983, and their possible relationship to climatic variations. *Advances in Space Research* 5, 6: 127–32.

Chen, H. 1985. *Space remote sensing systems: an introduction.* Academic Press, Orlando, Florida, USA.

Chetty, P. 1988. *Satellite technology and its applications.* TAB Books Inc., Blue Ridge Summit, PA 17294, USA.

Clarke, D. and Ackley, S. 1984. Sea ice structure and biological activity in the Antarctic marginal ice zone. *Journal of Geophysical Research* 89: 2,087–95.

Colbeck, S. 1979. Grain clusters in wet snow. *Journal of Colloid Interface Science* 72: 371.

Colbeck, S.C. 1982. An overview of seasonal snow metamorphism. *Reviews of Geophysics* 20: 45–61.

Colbeck, S.C. 1991. The layered character of snow covers. *Reviews of Geophysics* 29, 1: 81–96.

Collins, M. and Emery, W. 1988. A computational method for estimating sea ice motion in sequential SEASAT synthetic aperture radar imagery by matched filtering. *Journal of Geophysical Research* 93, C8: 9,241–51.

Colwell, R. (ed.). 1983. *Manual of Remote Sensing (second edition),* Volumes 1 and 2. American Society of Photogrammetry, Falls Church, VA 22046, USA.

Comiso, J. 1985. Remote sensing of sea ice using multispectral microwave satellite data. In Deepak, A., Fleming H. and Chahine, M. (eds), *Advances in remote sensing methods*: 349–70. A. Deepak Publishing, Hampton, Virginia.

Comiso, J. 1990. Arctic multi-year ice classification and summer ice cover using passive microwave satellite data. *Journal of Geophysical Research* 95, C8: 13,411–22.

Comiso, J. and Zwally, H. 1982. Antarctic sea ice concentrations inferred from NIMBUS-5 ESMR and LANDSAT imagery. *Journal of Geophysical Research* 87: 5,836–44.

Comiso, J., Zwally, H. and Saba, J. 1982. Radiative transfer modelling of microwave emission and dependence on firn properties. *Annals of Glaciology* 3: 54–8.

Comiso, J. and Sullivan, C. 1986. Satellite microwave and *in situ* observations of the Weddell Sea ice cover and its marginal ice zone. *Journal of Geophysical Research* 91, C8: 9,663–81.

Comiso, J., Maynard, N., Smith, W. and Sullivan, C. 1990. Satellite ocean color studies of Antarctic ice edges in summer and autumn. *Journal of Geophysical Research*, 95, C6: 9,481–96.

Cornillon, P. 1982. A guide to environmental satellite data. *Ocean Engineering NOAA Sea Grant Technical Report 79*, School of Oceanography, University of Rhode Island, Kingston, Rhode Island.

Crane, R. and Anderson, M. 1984. Satellite discrimination of snow surfaces. *International Journal of Remote Sensing* 5: 213.

Croom, D. 1985. Future passive microwave satellite data for sea ice research. *Proceedings of the Conference on the Use of Satellite Data in Climate Models,* Altbach, Austria, 10–12 June 1985. *ESA SP-244*: 125–7.

Curlander, J. 1982. Location of spaceborne SAR imagery. *IEEE Transcations of Geoscience and Remote Sensing* GE-20: 359–64.

Curran, P. 1985. *Principles of remote sensing.* Longman, London.

Curran, R. 1989. Satellite-borne LIDAR observations of the Earth: requirements and anticipated capabilities. *Proceedings of the IEEE 77,* 3: 478–80.

Degnan, J. 1985. Satellite laser ranging: current status and future prospects. *IEEE Transactions on Geoscience and Remote Sensing* GE-23: 398.

Dewey, K. and Helm, R. 1982. A digital archive of Northern Hemisphere snow cover, November 1966 through December 1980. *Bulletin of the American Meteorological Society* 63: 1,132–41.

Dey, B., Moore, H. and Gregory, A. 1977. The use of satellite imagery for monitoring ice break-up along the Mackenzie River, N.W.T. *Arctic* 30, 4: 234–42.

Diesen, B. and Reincke, D. 1978. Soviet Meteor satellite imagery. *Bulletin of the American Meteorological Society* 59, 7: 804.

Douglas, R. 1988. *Satellite communications technology.* Prentice Hall, New York.

Dowdeswell, J. and McIntyre, N. 1986. The saturation of LANDSAT MSS detectors over large ice masses. *International Journal of Remote Sensing* 7, 1: 151–65.

Dozier, J. 1980. Spectral signature of alpine snow cover from LANDSAT Thematic mapper. *Remote Sensing of Environment* 28: 9–22.

Dozier, J., Schneider, S. and MacGinnis, D. 1981. Effect of grain size and snowpack water equivalence on visible and near-infrared satellite observations of snow. *Water Resources Research* 17: 1,213–21.

Dozier, J. and Strahler, A. 1983. Ground investigations in support of remote sensing. In Colwell, R. (ed.), *Manual of Remote Sensing*, Volume 1: 959–86. American Society of Photogrammetry.

Dozier, J. and Warren, S. 1982. Effect of viewing angle on the infrared brightness temperature of snow. *Water Resources Research* 18: 1,424.

Drew, A. and Whillans, I. 1984. Measurement of surface deformation of the Greenland Ice Sheet by satellite tracking. *Annals of Glaciology* 5: 51–5.

Drewry, D. (ed.). 1983. *Antarctica: glaciological and geophysical folio*. Scott Polar Research Institute, Cambridge, UK.

Drinkwater, M. 1989. LIMEX '87 ice surface characteristics: implications for C-band SAR backscatter signatures. *IEEE Transactions on Geoscience and Remote Sensing* 27, 5: 501–13.

Drinkwater, M. and Crocker, G. 1988. Modelling changes in the dielectric scattering properties of young snowcovered sea ice. *Journal of Glaciology* 118: 274–82.

Dwyer, R. and Godin, R. 1980. Determining sea ice boundaries and ice roughness using GEOS-3 altimeter data. *NASA Contractor Report 156862*.

Ebert, E. 1987. A pattern recognition technique for distinguishing surface and cloud types in polar regions. *Journal of Climate and Applied Meteorology* 26: 1,412–27.

Elachi, C. 1987. *Introduction to the physics and techniques of remote sensing*. John Wiley and Sons, New York.

Elachi, C. 1988. *Spaceborne radar remote sensing*. IEEE Press, New York.

ESA. 1985. A programme for international polar oceans research (PIPOR). *ESA Report SP-1074*. Paris, European Space Agency.

Farman, J., Gardiner, D. and Shanklin, J. 1985. Large losses of total ozone in Antarctica reveal seasonal $CCIO_x/NO_x$ interaction. *Nature* 315: 207–10.

Ferrigno, J. and Gould, W. 1987. Substantial changes in coastline of Antarctica revealed by satellite imagery. *Polar Record* 23, 146: 577–83.

Ferrigno, J., Williams, R. and Kent, T. 1983. Evaluation of LANDSAT 3 RBV images. In Oliver, R., James, P. and Jago, J. (eds), *Antarctic Earth science, Proceedings of the Fourth International Symposium on Antarctic Earth Science*, 16–20 August 1982, University of Adelaide, South Australia: 446–9. Australian Academy of Science, Canberra.

Ferrigno, J., Williams, R., Lucchitta, B. and Molnia, B. 1990. Recent changes in the coastal regions of Antarctica documented by Landsat images. *International Conference on the Role of the Polar Regions in Global Change*, 11–15 June 1990. University of Alaska, Fairbanks: 24.

Fetterer, F., Johnson, D., Hawkins, J. and Laxon, S. 1988. Investigations of sea ice using coincident GEOSAT altimetry and synthetic aperture radar during MIZEX-87. *Proceedings of IGARSS '88*, Edinburgh, 13–16 September 1988. *ESA SP-284 (IEEE 88CH2497–6, Volume 2: 1,125–6.*

Filey, M. and Rothrock, D. 1986. Extracting sea ice data from satellite SAR imagery. *IEEE Transactions of Geoscience and Remote Sensing* GE-25, 6: 849–54.

Filey, M. and Rothrock, D. 1988. Measuring lead area changes in sea ice imagery. *Proceedings of IGARSS '88*, Edinburgh, 13–16 September 1988. *ESA SP-284 (IEEE 88CH2497–6), Volume 2: 799–800.*

Foster, J.L., Hall, D.K. and Chang, A.T.C. 1984. An overview of passive microwave snow research and results. *Reviews of Geophysics and Space Physics* 22, 2: 195–208.

French, J. 1988. Power budget improvements for miniature PTTs. *ARGOS Newsletter* 34, July 1988: 13.

Fu, L. and Holt, B. 1982. SEASAT views oceans and sea ice with SAR. *JPL Publication 81–120*. Jet

Propulsion Laboratory, Pasadena, California.

Fu, L., Chelton, D. and Zlotnicki, V. 1988. Satellite altimetry: observing ocean variability from space. *Oceanography* 1, 2: 4–11.

Giovenetto, M. and Bentley, C. 1985. Surface balance in ice drainage systems of Antarctica. *Antarctic Journal of the United States* 20, 4: 6–13.

Gloersen, P. and Campbell, W. 1988. Variations in the Arctic, Antarctic, and global sea ice covers during 1978–1987 as observed with the NIMBUS-7 Scanning Multichannel Microwave Radiometer. *Journal of Geophysical Research* 93, C9: 10,666–74.

Gloersen, P., Campbell, W., Cavalieri, D., Comiso, J., Parkinson,C. and Zwally, H. In press. Arctic and Antarctic sea ice, 1978–87: satellite passive microwave observation and analysis. *NASA Special Publication*.

Gloersen, P., Nordberg, W., Schmugge, T. and Campbell, W. 1973. Microwave signatures of first-year and multi-year sea ice. *Journal of Geophysical Research* 78, 18: 3,564–72.

Gloersen, P., Wilheit, T., Chang, T., Nordberg, W. and Campbell, W. 1974. Microwave maps of the polar ice of the Earth. *Bulletin of the American Meteorological Society* 55: 1,442–8.

Gloersen, P., Zwally, H., Chang, A., Hall, D., Campbell, W. and Ramseier, R. 1978. Time-dependence of sea ice concentration and multi-year ice fraction in the Arctic Basin. *Boundary Layer Meteorology* 13: 339–59.

Goetz, A., Wellman, J. and Barnes, W. 1985. Optical remote sensing of the Earth. *Proceedings of the IEEE* 73, 6: 950–69.

Gordon, A. 1988. The Southern Ocean and global climate. *Oceanus* 31, 2: 39–46.

Gordon, H. and Castaño, D. 1987. The Coastal Zone Color Scanner atmospheric correction algorithm: multiple scattering effects. *Applied Optics* 26: 2,111–22.

Gow, A., Ackley, S., Weeks, W. and Govoni, J. 1982. Physical and structural characteristics of Antarctic sea ice. *Annals of Glaciology* 3: 113–17.

Gower, J. (ed.). 1986. Opportunities and problems in satellite measurements of the sea. *UNESCO Technical Papers in Marine Science 46*. UNESCO, Paris.

Gray, A., Hawkins, R., Livingstone, C., Arsenault, L. and Johnstone, W. 1982. Simultaneous scatterometer and radiometer measurements of sea ice microwave signatures. *IEEE Journal of Oceanic Engineering* OE-7, 1: 20–32.

Grenfell, T. 1979. The effects of ice thickness on the exchange of solar radiation over the polar oceans. *Journal of Glaciology* 22: 305–20.

Grenfell, T. 1983. A theoretical model of the optical properties of sea ice in the visible and near infrared. *Journal of Geophysical Research* 88, C11: 9,723–35.

Grenfell, T. and Comiso, J. 1986. Multifrequency passive microwave observations of first-year sea ice grown in a tank. *IEEE Transaction of Geoscience and Remote Sensing* GE-24: 826–31.

Grenfell, T. and Lohanick, A. 1985. Temporal variability of microwave signatures of sea ice during late spring and early summer near Mould Bay, NWT. *Journal of Geophysical Research* 90, C3: 5,063–74.

Grenfell, T. and Maykut, G. 1977. The optical properties of ice and snow in the Arctic Basin. *Journal of Glaciology* 18: 445–63.

Gudmandsen, P., Carsey, F. and McNutt, L. 1989. PIPOR – a programme for international polar oceans research. *Remote sensing of atmosphere and oceans. Proceedings of COSPAR 27th Plenary Meeting*, Espoo, Finland, 18–29 July 1988: 31–8.

Hall, D. 1988. Assessment of polar climate change using satellite technology. *Reviews of Geophysics* 26, 1: 26–39.

Hall, R. and Rothrock, D. 1981. Sea ice displacement from Seasat synthetic aperture radar. *Journal of Geophysical Research* 86, C11: 11,078–82.

Hallikainen, M. and Jolma, P. 1986. Retrieval of snow water equivalent from NIMBUS-7 SMMR data. In Kukla, G., Barry, R., Hecht, A. and Wiesnet, D. (eds) Snow Watch '85, *WDC A for Glaciology Data Report GD-18*: 173–9. WDC A for Glaciology, Boulder, Colorado.

Hasselmann, K., Raney, R., Plant, W., Alpers, W., Shuchman, R., Lyzenga, D., Rufenach, C. and

Tucker, M. 1985. Theory of Synthetic Aperture Radar ocean imaging: a MARSEN view. *Journal of Geophysical Research* 90, C3: 4,659.

Hawkins, J.D. and Lybanon, M. 1989. GEOSAT altimeter sea ice mapping. *IEEE Journal of Oceanic Engineering* 14, 2: 139–48.

Heacock, E. 1985. US remote sensing of the Earth from space – a look ahead. In Schnapf, A. (ed.), *Monitoring Earth's ocean, land and atmosphere from space – sensors, systems, and applications*. New York, American Institute of Aeronautics and Astronautics: 713–45.

Hibler, W. 1979. A dynamic thermodynamic sea ice model. *Journal of Physical Oceanography* 9: 815–46.

Hibler, W. and Ackley, S. 1983. Numerical simulation of the Weddell Sea pack ice. *Journal of Geophysical Research* 88: 2,873–87.

Hollinger, J., Troy, B., Ramseier, R., Asmus, K., Hartman, M., and Luther, C. 1984. Microwave emission from high Arctic sea ice during freeze-up. *Journal of Geophysical Research* 89, C5: 104–22.

Houghton, J.T., Jenkins, G.J. and Ephraums, J.J. (eds) 1990. *Climate change: the IPCC scientific assessment*. Cambridge University Press.

Hyvärinen, T. and Lammasniemi, J. 1987. Infrared measurement of free-water content and grain size of snow. *Optical Engineering* 26: 342.

ICSU/SCAR. 1989. *The role of Antarctica in global change. Scientific priorities for the International Geosphere-Biosphere Programme (IGBP)*. SCAR, Cambridge, UK.

Jensen, J. 1986. *Introductory digital image processing – remote sensing perspective*. Prentice-Hall, Englewood Cliffs, New Jersey.

Jezek, K. and Alley, R. 1988. Effect of stratigraphy on radar altimetry data collected over ice sheets. *Annals of Glaciology* 11: 60–3.

Jezek, K., Hogan, A. and Cavalieri, D. 1990. Antarctic ice sheet brightness temperature variations. In Ackley, S. and Weeks, W. (eds), Sea ice properties and processes. *CRREL Monograph 90–1*: 217–23.

Johannessen, O., Starke, K. and Payne, S. 1988. Ambient noise in the marginal ice zone. *Nansen Remote Sensing Center Technical Report Number 14*. Nansen Remote Sensing Center, Edvard Griegsvei 3a, N-5037 Solheimsvik, Norway.

Johnson, N. 1989. The Soviet year in space, 1989. Teledyne Brown Engineering, 1250 Academy Park Loop, Suite 240, Colorado Springs, CO 80910, USA.

Keller, L., Weidner, G., Stearns, C. and Sievers, M. 1988. Antarctic automatic weather station data for the calendar year 1988. Report of the Department of Meteorology, University of Wisconsin, Madison, WI 53706, USA.

Kostiuk, T. and Clark, W. 1984. Space-borne sensors (1983–2000 AD): a forecast of technology. *NASA Technical Memorandum 86083*. NASA GSFC, Greenbelt, Maryland.

Kukla, G. and Gavin, J. 1981. Summer ice and carbon dioxide. *Science* 214: 497–503.

Lange, M. 1988. Basic properties of Antarctic sea ice as revealed by textural analysis of ice cores. *Annals of Glaciology* 10: 95–101.

Lange, M., Ackley, S., Wadhams, P., Dieckmann, G. and Eicken, H. 1989. Development of sea ice in the Weddell Sea, Antarctica. *Annals of Glaciology* 12: 92–6.

Laxon, S. 1989. Sea ice operational products from space-borne radar altimeters. In Barrett, E. and Brown, K. (eds), *Remote sensing for operational applications, Proceedings of the 15th Conference of the Remote Sensing Society*, Bristol, 13–15 September 1989: 237–43.

Leberl, F., Raggam, J., Elachi, C. and Campbell, W. 1983. Sea ice motion displacements from SEASAT SAR images. *Journal of Geophysical Research* 88, C3: 1,915–28.

Lingle, C. 1984. A numerical model of interactions between a polar ice stream and the ocean: application to ice stream E, West Antarctica. *Journal of Geophysical Research* 89, C3: 3,523–49.

Lingle, C., Brenner, A. and Zwally, H. 1990. Satellite altimetry, semi-variograms and seasonal changes in the ablation zone of West Greenland. *Annals of Glaciology* 14.

Liu, A. and Mollo-Christensen, E. 1988. Wave propagation in a solid ice pack. *Journal of Physical*

Oceanography 18: 1,702–12.

Livingstone, C., Singh, K. and Gray, A. 1987. Seasonal and regional variations of active/passive microwave signatures of sea ice. *IEEE Transactions of Geoscience and Remote Sensing* GE-25, 2: 159–73.

Long, D. and Mendel, J. 1988. Model-based estimation of wind fields over the ocean from wind scaterometer measurements. *Proceedings of IGARSS '88*, Edinburgh, 13–16 September 1988. *ESA SP-284 (IEEE 88CH2497–6)*, Volume 1: 553–6.

Luther, C. 1988. Remote sensing in a marginal ice zone: a brief overview. *Proceedings of IGARSS '88 Symposium on Remote Sensing: Moving towards the 21st century*, Edinburgh, 13–16 September 1988. *ESA SP-284 (IEEE 88CH2497–6)*, Volume 2: 1,107–9.

Lyden, J., Schuchman, R., Zago, C., Rottier, P., Wadhams, P. and Johannessen, O. 1988. SAR imaging of ocean waves in the marginal ice zone. *Proceedings of IGARSS '88*, Edinburgh, 13–16 September 1988. *ESA SP-284 (IEEE 88CH2497–6)*, Volume 3: 1,435–7.

MacCracken, M. and Luther, F. (eds) 1985. Detecting the climatic effects of increasing carbon dioxide. *Report DOE/ER-0235*. US Department of Energy, Washington DC.

Macqueen, A. 1988. *Radio echo-sounding as a glaciological technique: a bibliography*. World Data Centre 'C' for Glaciology, SPRI, Cambridge, UK.

Mairs, R. 1970. Oceanographic interpretation of Apollo photos. *Photogrammetric Engineering and Remote Sensing* 36, 10: 1,045–58.

Manabe, S. and Stouffel, R. 1979. A CO_2-climate sensitivity study with a mathematical model of the global climate. *Nature* 282: 491–2.

Manley, T., Schuchman, R. and Burns, B. 1987. Use of synthetic aperture radar-derived kinematics in mapping mesoscale ocean structure within the interior marginal ice zone. *Journal of Geophysical Research* 92, C7: 6,837–42.

Marble, D. and Peuquet, D. (eds). 1983. Geographic Information Systems and Remote Sensing. In Colwell, R. (ed.), *Manual of Remote Sensing*, Volume 1: 923–58. American Society of Photogrammetry.

Martinson, D., Killworth, P. and Gordon, A. 1981. A convective model for the Weddell polynya. *Journal of Physical Oceanography* 11: 466–88.

Mass, C., Edmon, H., Friedman, H., Cheney, N. and Recker, E. 1987. The use of compact discs for the storage of large meteorological and oceanographic data sets. *Bulletin of the American Meteorological Society* 68, 12: 1,556–8.

Massom, R. 1988. The biological significance of open water within the sea ice covers of the polar regions. *Endeavour* 12, 1: 21–7.

Massom, R. 1989. *The study of Weddell Sea ice using passive microwave and buoy data*. Unpublished PhD thesis, University of Cambridge, UK.

Mather, P. 1987. *Computer-processing of remotely-sensed images. An introduction*. John Wiley and Sons, Chichester, UK.

Matson, M. and Wiesnet, D. 1981. New data base for climate studies. *Nature* 289: 451.

Matveev, D. 1976. Antarctic ice characteristics according to Kosmos 243 satellite measurements. *Meteorological Research, Transemantics*, 5–12. Washington DC.

Mätzler, C., Aebnischer, H. and Schanda, E. 1984. Microwave dielectric properties of surface snow. *IEEE Journal of Oceanic Engineering* OE-9, 5: 366–71.

Maykut, G. 1985. Large-scale heat exchange and ice production in the Central Arctic. *Journal of Geophysical Research* 87, C10: 7,971–84.

McClain, E. 1980. Passive microwave radiometry of the ocean from space – an overview. *Boundary Layer Meteorology* 18: 7–24.

McClain, E. and Baker, D. 1969. Experimental large-scale snow and ice mapping with composite minimum brightness charts. *ESSA Technical Memorandum NESCTM 12*. Washington DC.

McGinnis, D. and Schneider, S. 1978. Monitoring river ice breakups from space. *Photogrammetric Engineering and Remote Sensing* 44, 1: 57–68.

McGoogan, J. 1975. Satellite altimetry applications. *IEEE Transactions of Microwave Theory and*

Technology MTT-23: 970.

McGuffie, K., Barry, R., Schweiger, A., Robinson, D. and Newell, J. 1988. Intercomparison of satellite-derived cloud analyses for the Arctic Ocean in spring and summer. *International Journal of Remote Sensing* 9, 3: 447–67.

McIntyre, N. and Drewry, D. 1984. Modelling ice-sheet surfaces for ERS-1's radar altimeter. *ESA Journal* 8: 261–74.

Meier, M. 1973. Evaluation of ERTS imagery for mapping and detection of changes of snowcover on land and on glaciers. *Symposium on Significant Results Obtained from the ERTS-1 Satellite*, Volume 1, section A: 863–75.

Meier, M. 1980. Remote sensing of snow and ice. *Bulletin of Hydrological Sciences* 25, 3: 307–30.

Meier, M. (ed.). 1985. Glaciers, ice sheets and sea level: effect of a CO_2-induced climatic change. *Report DOE/ER/60235–1*. US Department of Energy, Washington DC.

Mitchell, J. 1989. The 'greenhouse' effect and climate change. *Reviews of Geophysics* 27, 1: 115–24.

Monaldo, F. 1988. The influence of water vapour on the detection of ocean mesoscale fronts and eddies by the GEOSAT altimeter. *Proceedings of IGARSS '88*, Edinburgh, 13–16 September 1988. *ESA SP-284 (IEEE 88CH2497–6)*, Volume 2: 643–6.

Moore, R. 1983a. Radar fundamentals and scatterometers. In Colwell, R. (ed.), *Manual of Remote Sensing*, Volume 1: 369–427. American Society of Photogrammetry.

Moore, R. 1983b. Imaging Radar Systems. In Colwell, R. (ed.), *Manual of Remote Sensing*, Volume 1: 429–74. American Society of Photogrammetry.

Moore, R., Fung, A., Dome, G. and Birrer, I. 1978. Estimate of oceanic surface wind speed and direction using orthogonal beam scatterometer measurements and comparison of recent sea theories. *NASA Contractor Report 158908*, Langley Research Center, Hampton, Virginia.

Muller, J.-P. (ed.). 1988. *Digital image processing in remote sensing*. Taylor and Francis, London and Philadelphia.

Nagaraja Rao, C.R., Stowe, L.L. and McClain, E.P. 1989. Remote sensing of aerosols over the oceans using AVHRR data. Theory, practice and applications. *International Journal of Remote Sensing* 10, 4 and 5: 743–9.

NASA. 1986. MODIS: Moderate-Resolution Imaging Spectrometer. *Earth Observing System Instrument Panel Report Volume IIb*. NASA, Washington DC.

NASA. 1987a. HIRIS: High-Resolution Imaging Spectrometer. Science Opportunities for the 1990s. *Earth Observing System Instrument Panel Report Volume IIc*. NASA, Washington DC.

NASA. 1987b. SAR: Synthetic Aperture Radar. *Earth Observing System Instrument Panel Report Volume IIf*. NASA, Washington DC.

NASA. 1987c. LASA: Lidar Atmospheric Sounder and Altimeter. *Earth Observing System Instrument Panel Report Volume IId*. NASA, Washington DC.

NASA. 1987d. HMMR: High-Resolution Multifrequency Microwave Radiometer. *Earth Observing System Instrument Panel Report Volume IIe*. NASA, Washington DC.

NASA. 1987e. LAWS: Laser Atmospheric Wind Sounder. *Earth Observing System Instrument Panel Report Volume IIg*. NASA, Washington DC.

NASA. 1989. *Ocean color from space*. NASA GSFC, Greenbelt, Maryland.

NASA. 1990a. *Upper Atmospheric Research Satellite: a program to study global ozone change*. NASA, Washington DC.

NASA. 1990b. *Earth Observing System Reference Manual*. NASA GSFC, Greenbelt, Maryland.

NASA. 1990c. *The Early Earth Observing System Reference Handbook: Earth Science and Applications Division Missions 1990–1997*. NASA Goddard Space Flight Center, Greenbelt, Maryland.

NASA Alaska SAR Facility Prelaunch Science Working Team. 1989. Science plan for the Alaska SAR Facility Program. Phase 1: data from the first European Remote Sensing Satellite, ERS-1. *JPL Publication 89–14*. NASA JPL, Pasadena, California.

NASA Earth System Sciences Committee. 1988. *Earth System Science: a closer view*. NASA, Washington DC.

NASA Science and Applications Working Group. 1979. Ice and Climate Experiment (ICEX). *NASA Goddard Space Flight Center Report*, Greenbelt, Maryland.

Nichols, D. 1983. Digital hardware. In Colwell, R. (ed.), *Manual of Remote Sensing*, Volume 1: 841–71. American Society of Photogrammetry.

Ninnis, R., Emery, W. and Collins, M. 1986. Automated extraction of pack ice motion from AVHRR imagery. *Journal of Geophysical Research* 95: 10,725–34.

Njoku, E. and McClain, E. 1985. Report of the COSPAR International Workshop on Satellite-derived Sea Surface Temperatures for Global Climate Applications. *World Climate Research Programme Report WCP-110, WMO/TD-No. 93*. World Meteorological Organization, Geneva.

NOAA/NASA/NSF. 1989. Airborne geoscience: the next decade. *Report of the Special Interagency Task Group on Airborne Geoscience*. Washington DC.

NRC. 1985. Remote sensing of the Earth from space: a program in crisis. *Report of the Space Applications Board, Commission of Engineering and Technical Systems, National Research Council*. National Academy Press, Washington DC.

Onstott, R. and Gogineni, S. 1985. Active microwave measurements of Arctic sea ice under summer conditions. *Journal of Geophysical Research* 90, C3: 5,035–44.

Onstott, R. and Schuchman, R. 1986. SAR/SCAT intercomparison of microwave signatures of Arctic sea ice. *Proceedings of the Open Symposium on Wave Propagation: Remote Sensing and Communications*. Durham, HH, 28 July–1 August 1986: 1.7.1–1.7.3.

Onstott, R. and Schuchman, R. 1988. Radar backscatter of sea ice during winter. *Proceedings of IGARSS '88, Edinburgh, 13–16 September 1988. ESA SP-284 (IEEE 88CH2497–6)*, Volume 2: 1,115–8.

Orheim, O. and Lucchitta, B. 1987. Snow and ice studies by Thematic Mapper and Multispectral Scanner Landsat images. *Annals of Glaciology* 9: 109–18.

Parkinson, C. 1983. On the development and cause of the Weddell polynya in a sea ice simulation. *Journal of Physical Oceanography* 13: 501–11.

Parkinson, C. 1989. On the value of long-term satellite passive data sets for sea ice/climate studies. *GeoJournal* 18: 9–20.

Parkinson, C., Comiso, J., Zwally, H., Cavalieri, D., Gloersen, P., and Campbell, W. 1987. Arctic sea ice, 1973–76: satellite passive-microwave observations. *NASA Special Publication SP-489*. NASA, Washington DC.

Pearson, D., Livinstone, C., Hawkins, R., Gray, A., Arsenault, L., Wilkinson, T. and Okamoto, K. 1980. Radar detection of sea ice ridges and icebergs in frozen oceans at incidence angles from 0° to 90°. *Proceedings of the Sixth Canadian Symposium on Remote Sensing*. Halifax, Nova Scotia: 231–7.

Perovich, D. 1983. On the summer decay of a sea ice cover. PhD thesis, 176pp. Geophysics Program, University of Washington, Seattle.

Perovich, D. and Grenfell, T. 1982. A theoretical model of radiative transfer in young sea ice. *Journal of Glaciology* 28: 341–56.

Preller, R., Cheng, A. and Posey, P. 1990. Preliminary testing of a sea ice model for the Greenland Sea. In Ackley, S. and Weeks, W. (eds), Sea ice properties and processes. *CRREL Monograph 90–1*: 259–77.

Price, J. 1984. Land surface temperature measurements from the split window channels of the NOAA-7 Advanced Very High Resolution Radiometer. *Journal of Geophysical Research* 89: 7,231–7.

Ramanathan, V. 1987. Role of Earth Radiation Budget studies in climate and general circulation research. *Journal of Geophysical Research* 92, D4: 4,075–95.

Ramsay, B., Henderson, D. and Carson, L. 1988. Real-time processing of digital image data in support of the Canadian sea ice analysis and prediction programme. *Proceedings of IGARSS '88*, Edinburgh, 13–16 September 1988. *ESA SP-284 (IEEE 88CH2497–6)*, Volume 3: 1,707–11.

Rango, A. and Martinec, J. 1979. Application of a snowmelt-runoff model using LANDSAT data. *Nordic Hydrology* 10: 225.

Rapley, C. 1984. First observations of the interaction of ocean swell with sea ice using satellite radar altimeter data. *Nature* 307: 150–2.

Reeburgh, W. and Whalen, S. 1990. The role of tundra and taiga systems in the global methane budget. *Abstracts of International Conference on the Role of the Polar Regions in Global Change*, 11–15 June 1990, University of Alaska, Fairbanks: 102.

Rees, W. 1990. The physical principles of remote sensing. Cambridge University Press, Cambridge, UK.

Rees, W. and Dowdeswell, J. 1988. Topographic effects on light scattering from snow. *Proceedings of IGARSS '88*, Edinburgh, 13–16 September 1988. *ESA SP-284 (IEEE 88CH2497–6)*, Volume 1: 161–4.

Remy, F., Anderson, M. and Minster, J. 1988. Comparison between active and passive measurements over Antarctica. *Proceedings of the 4th International Colloquium on Spectral Signatures in Remote Sensing, April 1988*: 69–72. ESA, Paris.

Richards, J.A. 1986. *Remote sensing digital analysis: an introduction.* Springer-Verlag, Berlin.

Robin, G., Drewry, D. and Squire, V. 1983. Satellite observations of polar ice fields. *Philosophical Transactions of The Royal Society London, Series A*, 309: 447–61.

Robinson, D., Kunzi, K., Kukla, G. and Rott, H. 1984. Comparative utility of microwave and shortwave satellite data for all-weather charting of snow cover. *Nature* 312, 5,993: 434–5.

Ropelewski, C. 1983. Spatial and temporal variations in Antarctic sea ice (1973–82). *Journal of Climate and Applied Meteorology* 22: 470–3.

Rosenberg, T. and Jezek, K. 1987. Polar communications: status and recommendations. *Report of the NASA Science Working Group.* NASA GSFC, Greenbelt, Maryland.

Ross, B. and Walsh, J. 1987. A comparison of simulated and observed fluctuations in summertime Arctic surface albedo. *Journal of Geophysical Research* 92, C12: 13,115–25.

Rott, H. 1985. Monitoring of snow and land ice parameters: possibilities and constraints of SAR and microwave imagining systems. *Proceedings of the Conference on the Use of Satellite Data in Climate Models*, Alpbach, Austria, 10–12 June 1985. *ESA SP-244*: 133–5. European Space Agency, Paris.

Rott, H., Domik, G., Mätzler, C., Miller, H. and Lenhart, K. 1985. Study on the use and characteristics of SAR for land snow and ice applications. Institut für Meteorologie und Geophysik, Universität Innsbruck, *Mitteilung Nr. 1 (1985)*. European Space Agency Report, Contract 5441/83/D/IM(SC).

Rott, H. and Mätzler, C. 1987. Possibilities and limits of synthetic aperture radar for snow and glacier surveying. *Annals of Glaciology* 9: 195–9.

Rottier, P. 1989. Observations of wave-ice interactions and ambient noise generation in the marginal ice zone. *Scott Polar Research Institute Sea Ice Group Special Report 89–5.* Scott Polar Research Institute, Cambridge, UK.

Sabins, F. 1987. *Remote sensing: principles and interpretation.* W.H. Freeman and Co., New York.

Sarmiento, J. and Tottweiler, J. 1984. A new model for the role of the ocean in determining atmospheric PCO_2. *Nature* 308: 621–3.

SCAR Steering Committee for the IGBP. 1989. *The role of Antarctica in global change.* ICSU Press/SCAR, Cambridge, UK.

Schanda, E. 1986. *Physical Fundamentals of Remote Sensing.* Springer-Verlag, Berlin.

Schanda, E. 1987. Microwave modelling of snow and soil. *Journal of Electromagnetic Waves and Applications* 1, 1: 1–24.

Scharfen, G., Barry, R., Robinson, D., Kukla, G. and Serreze, M. 1987. Large-scale patterns of snow melt on Arctic sea ice mapped from meteorological satellite imagery. *Annals of Glaciology* 9: 200–5.

Schiffer, R. and Rossow, W. 1983. The International Satellite Cloud Climatology Project (ISCCP): the first project of the World Climate Research Program. *Bulletin of the American Meteorological Society* 64: 779–84.

Schneider, S. 1977. Operational satellite assessment of snow cover and river ice in the Saint John Basin. World Meteorological Organisation, World Weather Watch, *Saint John Basin Pilot Project Task Force Report Number 6–2*.

Schwaller, M., Olson, C., Benninghoff, W. and Wehnes, K. 1986. Preliminary report on satellite remote sensing of Adélie penguin rookeries. *Antarctic Journal of the United States* 21, 5: 205–6.

Shine, K., Henderson-Sellers, A. and Barry, R. 1984. Albedo-climate feedback: the importance of cloud and cryosphere variability. In Berger, A. and Nicolis, C. (eds), *New perspectives in climate modelling (Developments in Atmospheric Science 16)*: 135–55. Elsevier, Amsterdam.

Simonett, D. and Davis, R. 1983. Image analysis – active microwave. In Colwell, R. (ed.), *Manual of Remote Sensing*, Volume 1: 1,125–81. American Society of Photogrammetry.

Sissala, J. 1969. Observations of an Antarctic Ocean tabular iceberg from the NIMBUS 2 satellite. *Nature* 224: 1,285.

Slater, P. 1980. *Remote sensing: optics and optical systems*. Addison-Wesley Publishing Company, Reading, Massachusetts.

Sloggett, D. 1989. *Satellite data: processing, archiving and dissemination. Volume 1: Applications and infrastructure. Volume 2: Functions, operational principles and design.* John Wiley and Sons, Chichester, UK.

Smith, W. 1989. Satellite soundings – current status and future prospects. *Remote sensing of atmosphere and oceans. Proceedings of COSPAR 27th Plenary Meeting*, Espoo, Finland, 18–29 July 1988: 363–72.

Staelin, D. 1980. Progress in passive microwave remote sensing: nonlinear retrieval techniques. In A. Deepak (ed.), *Remote sensing of atmospheres and oceans*: 259–75. Academic Press, Orlando, FL.

Star, J. and Estes, J. 1990. *Geographic information systems: an introduction*. Prentice Hall, Englewood Cliffs, New Jersey.

Stearns, C. and Wendler, G. 1988. Research results from Antarctic automatic weather stations. *Reviews of Geophysics* 26, 1: 45–61.

Steffen, K., Barry, R. and Schweiger, A. 1989. DMSP SSM/I NASA algorithm validation using primarily LANDSAT and secondarily DMSP and/or AVHRR visible and thermal infrared satellite imagery. *NASA Report NASA-CR-182979*. NASA, Washington DC.

Stephenson, S. and Bindschadler, R. 1990. Is ice-stream evolution revealed by satellite data? *Annals of Glaciology* 14: 273–77.

Stewart, R. 1985. *Methods of Satellite Oceanography*. University of California Press, Berkeley, California.

Stiles, W., Ulaby, F., Fung, A. and Aslam, A. 1981. Radar spectral observations of snow. *Proceedings of IGARSS '81*, Washington DC: 854–68.

Stiles, W. and Ulaby, F. 1982. Dielectric properties of snow. *CRREL Special Report 82–18*: 91–103.

Swift, C. 1980. Passive microwave remote sensing of the ocean – a review. *Boundary-Layer Meteorology* 18: 25–54.

Swift, C., Hayes, P., Herd, J., Jones, W. and Delnore, V. 1985. Airborne microwave measurements of the Southern Greenland ice sheet. *Journal of Geophysical Research* 90: 1,983.

Swithinbank, C. 1973. Higher resolution satellite pictures. *Polar Record* 16, 104: 739–41.

Swithinbank, C. 1985. A distance look at the cryosphere. *Advances in Space Research* 5, 6: 263–74.

Swithinbank, C. 1988. Antarctica. In Williams, R. and Ferrigno, J. (eds), *Satellite Image Atlas of Glaciers of the World (Volume B)*. *USGS Professional Paper 1386-B*. Books and Open-File Reports Section, US Geological Survey, Federal Center, Box 25425, Denver, CO 80225.

Thomas, R., Sanderson, T. and Rose, K. 1979. Effect of climatic warming on the West Antarctic ice sheet. *Nature* 277: 355–8.

Thomas, R., Martin, T. and Zwally, H. 1983. Mapping ice-sheet margins from radar altimeter data. *Annals of Glaciology* 4: 283–8.

Thomas, R., Bindschadler, R., Cameron, R., Carsey, F., Holt, B., Hughes, T., Swithinbank, C., Whillans, I. and Zwally, H. 1985. Satellite remote sensing for ice sheet research. *NASA Technical Memo 86233*. NASA, Washington DC.

Thomson, R. (ed.). 1989. Space and airborne technology applications to Antarctic operations. *SCAR Working Group on Logistics Symposium*, Hobart, Australia, 1–2 September 1988. Department of Scientific and Industrial Research, Antarctic Division, Christchurch, New Zealand.

Troy, B., Hollinger, R., Lerner, R. and Wisler, 1981. Measurement of the microwave properties of sea ice at 90GHz and lower frequencies. *Journal of Geophysical Research* 86: 4,283–9.

Tsang, L., Kong, J. and Shin, R. 1985. *Theory of microwave remote sensing.* John Wiley and Sons, New York.

Ulaby, F., Moore, R. and Fung, A. 1981. *Microwave remote sensing, volume 1. Microwave remote sensing: fundamentals and radiometry.* Addison-Wesley Publishing Company, Reading, Massachusetts.

Ulaby, F., Moore, R. and Fung, A. 1982. *Microwave remote sensing, Volume 2: Scattering and emission theory.* Addison-Wesley Publishing Company, Reading, Massachusetts.

Ulaby, F., Moore, R. and Fung, A. 1985. *Microwave remote sensing, Volume 3: From theory to applications.* Artech House, Dedham, Massachusetts.

Ulander, L. 1988. Observations of ice types in satellite altimetric data. *Proceedings of IGARSS '88,* Edinburgh, 13–16 September 1988. *ESA SP-284 (IEEE 88CH2497–6),* Volume 2: 655–8.

United Nations General Assembly, 1987. Principles relating to remote sensing of the Earth from space. *A/RES/41/65.* United Nations, New York.

Untersteiner, N. (ed.). 1983. Air-sea-ice: research program for the 1980s. *Report APL-UW 8306.* Applied Physics Laboratory, University of Washington, Seattle.

Untersteiner, N. (ed.). 1984. Passive microwave remote sensing for sea ice research. *Report of the NASA Science Working Group for the Special Sensor Microwave Imager (SSM/I).* Oceanic Processes Branch, NASA HQ, Washington DC.

Untersteiner, N. (ed.). 1986. *The Geophysics of Sea Ice.* Plenum Press, New York.

Untersteiner, N. and Thorndike, A. 1982. Arctic data buoy program. *Polar Record* 21, 131: 127–35.

Vesecky, J., Samadini, R., Daida, J., Smith, M. and Bracewell, R. 1988. Observations of sea ice motion and deformation using synthetic aperture radar: automated analysis algorithms. In *Instrumentation and Measurements in the Polar Regions. Proceedings of the Workshop, January 1988:* 47–65. Marine Technology Society, San Francisco Bay Region Section, 2411 Valley Street, Berkeley, CA 94702, USA.

Viereck, L., Slaughter, C., Way, J., Dobson, C. and Christensen, N. 1990. Taiga forest stands and SAR monitoring for sub-Arctic global change. *International Conference on the Role of the Polar Regions in Global Change,* 11–15 June 1990, University of Alaska, Fairbanks: 27.

Vonder Haar, T. 1974. Measurement of albedo over polar snow and ice fields using NIMBUS-3 satellite data. In *Advanced concepts and techniques in the study of snow and ice resources,* National Academy of Sciences: 161–8.

Vornberger, P. and Bindschadler, R. In press. Multispectral analysis of ice sheets using coregistered SAR and LANDSAT imagery. *International Journal of Remote Sensing.*

Wackerman, C., Jentz, R. and Shuchman, R. 1988. Sea ice type classification of SAR imagery. *Proceedings of IGARSS '88,* Edinburgh, 13–16 September 1988. *ESA SP-284 (IEEE 88CH2497–6),* Volume 1: 425–8.

Wadhams, P. 1980. An analysis of ice profiles obtained by submarine sonar in the Beaufort Sea. *Journal of Physical Oceanography* 25, 93: 401–24.

Wadhams, P. 1986. The seasonal ice zone. In Untersteiner, N. (ed.), *The geophysics of sea ice:* 825–991. Plenum Publishing Corporation.

Wadhams, P. and Holt, B. 1990. Waves in frazil and pancake ice and their detection on SEASAT SAR imagery. Submitted to *Journal of Geophysical Research (Oceans).*

Wadhams, P., Squire, V., Goodman, D., Cowan, A. and Moore, S. 1988. The attenuation rates of ocean waves in the marginal ice zone. *Journal of Geophysical Research* 93, C6: 6,799–818.

Wark, D. and Popham, R. 1962. Ice photography from the meteorological satellites TIROS I and TIROS II. *Report No. 8,* US Department of Commerce Weather Bureau, Washington DC.

Warren, B. 1981. Deep circulation of the world ocean. In Warren, B. and Wunsch, C. (eds), *Evolution of physical oceanography,* 6–41. Massachusetts Institute of Technology Press, Cambridge, Massachusetts.

Warren, D. and Turner, J. 1988. Cloud track winds from polar orbiting satellites. *Proceedings of IGARSS '88,* Edinburgh, 13–16 September 1988. *ESA SP-284 (IEEE 88CH2497–6),* Volume 1: 549–50.

Warren, S. 1982. Optical properties of snow. *Reviews of Geophysics and Space Physics* 20, 1: 67–89.

Weeks, W. and Ackley, S. 1982. The growth, properties and structure of sea ice. *CRREL Monograph 82–1*. US Army CRREL, Hanover NH.

Weinreb, M., Hamilton, G., Brown, S. and Koczor, R. 1990. Nonlinearity corrections in calibration of Advanced Very High Resolution Radiometer infrared channels. *Journal of Geophysical Research* 95, C5: 7,381–8.

Wendler, G. 1973. Sea ice observations by means of satellite. *Journal of Geophysical Research* 78, 9: 1,427–48.

Wiesnet, D. 1979. Satellite studies of fresh-water ice movements on Lake Erie. *Journal of Glaciology* 24, 90: 415–26.

Wiesnet, D. 1980. A satellite mosaic of the Greenland ice sheet. *Proceedings of the Riederalp Workshop, September 1978, IASH-AISH Publication Number 126*: 343–48.

Wiesnet, D. and Matson, M. (eds). 1983. Remote sensing of weather and climate. In Colwell, R. (ed.), *Manual of Remote Sensing*, Volume 2: 1,305–69. American Society of Photogrammetry.

Wilheit, T. 1979. A model for the microwave emissivity of the ocean's surface as a function of wind speed. IEEE *Transactions of Geoscience and Electronics* GE-17: 244.

Wilheit, T., Chang, A. and Milman, A. 1980. Atmospheric corrections of passive microwave observations of the ocean. *Boundary-Layer Meteorology* 18: 65–77.

Williams, R. 1976. Vatnajökull Icecap, Iceland. In Williams, R. and Carter, W. (eds), ERTS-1, a new window on our planet, *US Geological Survey Professional Paper 929*: 188–93.

Williams, R. and Ferrigno, J. 1988. LANDSAT Images of Antarctica. In Swithinbank, C. Antarctica. Volume B of *Satellite Image Atlas of Glaciers of the World*, edited by Williams, R and Ferrigno, *US Geological Survey Professional Paper 1386–B*: 139–278. Books and Open-File Reports Section, USGS, Federal Center, Box 25425, Denver, CO 80225, USA.

Winebrenner, D. and Tsang, L. 1988. Sea Ice characterization measurements needed for testing of microwave remote sensing models. *Instrumentation and Measurements in the Polar Regions. Proceedings of the Workshop, January 1988*: 1–13. Marine Technology Society, San Francisco Bay Region Section, 2411 Valley Street, Berkely, CA 94702, USA.

WMO. 1970. WMO Sea-ice nomenclature, terminology, codes, and illustrated glossary. *WMO/OMM/BMO No. 259, TP145*. Secretariat of the World Meteorological Organization, Geneva.

WMO. 1986. Scientific plan for the World Ocean Circulation Experiment. *World Climate Research Programme Publications Series No. 6, WMO/TD-No. 122*. World Meteorological Organization, Geneva, Switzerland.

WMO. 1987a. Report of the second session of the Working Group on Sea Ice and Climate, Seattle, 27–31 October 1987. *World Climate Research Programme Report WCP-128*, World Meteorological Organization, Geneva.

WMO. 1987b. The Southern Ocean. Report of the WOCE Core Project 2 Planning Meeting, Bremerhaven, 20–3 May 1986. *World Climate Research Programme Report WCP-138*, World Meteorological Organization, Geneva.

WMO. 1989. Information on meteorological and other environmental satellites. *WMO-411*. World Meteorological Organization, Geneva.

Wolfe, W.L. and Zissis, G.J. 1989. *The infrared handbook*. The Infrared Information Analysis Center, Environmental Research Institute of Michigan.

Wunsch, C. and Gaposhkin, E. 1980. On using satellite altimetry to determine the general circulation of the oceans with application to geoid improvement. *Reviews of Geophysics and Space Physics* 18: 725–45.

Zwally, H. 1977. Microwave emissivity and accumulation rate of polar firn. *Journal of Glaciology* 18, 79: 195–215.

Zwally, H., Comiso, J., Parkinson, C., Campbell, W., Carsey, F. and Gloersen, P. 1983a. Antarctic sea ice, 1973–1976: satellite passive-microwave observations. *NASA Special Publication SP-459*. NASA, Washington DC.

Zwally, H., Parkinson, C. and Comiso, J. 1983b. Variability of Antarctic sea ice and CO_2 change. *Science* 220: 1,005–12.

Zwally, H. and Gloersen, P. 1977. Passive microwave images of polar regions and research applications. *Polar Record* 18, 116: 431–50.

Zwally, H., Major J., Brenner, A. and Bindschadler, R. 1987. Ice Measurement by GEOSAT Radar Altimetry. In *Navy GEOSAT Mission. Special Issue, Johns Hopkins University Applied Physics Laboratory Technical Digest* 8, 2: 251–4.

Zwally, H., Thomas, R. and Bindschadler, R. 1981. Ice-sheet dynamics by satellite laser altimetry. *NASA Technical Memorandum 82128*. NASA GSFC, Greenbelt, Maryland.

PART 2

Details of
individual satellites

Part 2 of this monograph lists the most important remote sensing satellites whose data are relatively easily accessible, together with the most immediately forthcoming satellites. The information is arranged in chronological order of launch of satellite or of the first satellite of a series, and is presented in as uniform a format as possible. For each satellite (or series), information on the useful lifetime and orbital parameters is given first, followed by the characteristics of the sensors carried, the particular applications and limitations of the data to polar research, the format and the availability of the data. Each section concludes with a short list of references and further reading.

1. Television Infrared Observation Satellites (TIROS) I–X

Operational lifespan and orbital characteristics

Dates: 1 April 1960–2 April 1966.
Semimajor axis: 7,010–7,159km (8,022 for TIROS IX, 7,172 for TIROS X).
Orbital height: 614–974km. (705–2,582 for TIROS IX, 751–837 for TIROS X).
Inclination: 47.8°–58.5° (TIROS I–VIII), 96.43°–98.7° (TIROS IX and X).
Period: 97.34–100.44 minutes (119.23 for TIROS IX, 100.76 for TIROS IX).
Notes: First generation TIROS/NOAA, first series. Research and development meteorological satellites. TIROS IX and X were the first sun-synchronous satellites.

Sensor characteristics

Vidicon:
   VIS (wavelength 0.45–0.65μm). Ground resolution (GR) 2.0–3.0km. Half inch television cameras. Swath width 800–1,200km.
Automatic Picture Transmission (APT) (on TIROS VIII):
   VIS (wavelength 0.45–0.65μm). GR 7.5km. Swath width 1,200km.
Scanning Radiometer (SR) (on TIROS II, III, IV, and VII):
   5 bands: 2 VIS (wavelength 0.5–0.75 and 0.2–0.6μm); 3 TIR (6.0–6.5, 8.0–12.0 and 8.0–13.0μm; 14.0–16.0μm on TIROS III).
An Earth radiation budget experiment was included on some of the missions.

Polar applications

Measurement of the seasonal and/or annual gross variability of sea ice distribution and deformation, and cloud pattern recognition.

Limitations

Coarse geometric resolution. Obscured by cloud/fog. Inoperative in darkness. Difficult to distinguish ice from cloud. Panoramic and Earth curvature distortion. Data are difficult/impossible to use in a quantitative fashion. The orbits of this series mainly precluded extensive coverage of the polar regions. This was especially true for TIROS I–VIII, the sea ice coverage of which was effectively limited to regions such as the Gulf of St Lawrence. Geolocation of APT imagery was difficult, as it contained no latitude or longitude grid. No station received APT pictures of Antarctica. Each satellite had a very short lifetime, which precluded continuity of data both spatially and temporally. A major limitation of the early SR experiment was the uncertainty in the absolute value of the data.

Data format

Individual picture frames: 35mm microfilm in catalogue format, 2 or 3 days per reel.

Data availability and archiving dates

Data from 1 April 1960–20 April 1966 are archived at: the Satellite Data Services Division (SDSD), NOAA/NESDIS, Princeton Executive Centre, 5627 Allentown Road, Camp Springs, MD 20746, USA, telephone (301)–763–8399, telex 248376 OBSWUR, FAX 7638443. Note: Pre-1970 data are in semi-permanent storage.

Final meteorological radiation CCTs (products) from TIROS II, III, IV, VII are archived at: National Space Science Data Center (NSSDC), Code 633.4, NASA Goddard Space Flight Center, Greenbelt, MD 20771, USA, telephone (301)–286–6695, telex 89675 NASCOM GBLT, TWX 7108289716. NSSDC is a node on NASA's Space Physics Analysis Network (SPAN), and can be accessed in two ways. Electronic mail can be directed to (SPAN) NSSDC::REQUEST. For access to a menu of information, limited data directory and limited data display, requesters may log on to the NSSDC node with NSSDC as the user ID. No password is required. Telemail addresses are JLGREEN, JVETTE or JKING (check these addresses with NSSDC, as they may be outdated). The Internet address is request @ nssdc.gsfc.nasa.gov.

Note: data requests from researchers outside the USA must be addressed to the Director, World Data Center A for Rockets and Satellites, NSSDC, Code 630.2, NASA Goddard Space Flight Center, Greenbelt, MD 20771, USA, telephone (301)–286–2864, telex 89675 NASCOM GBLT, TWX 7108289716, network address (SPAN) NSSDC::REQUEST, Internet address request@nssdc.gsfc.nasa.gov.

For access to a menu of information, the Master Directory (MD), which is a limited data directory, and limited data displays, requesters may use SPAN to log onto the NSSDC node, with NSSDC as Username. No password is required. NSSDC may also be reached by Telenet; current procedures are available from the NSSDC Network Hotline (301)–286–7251. The MD is continually being expanded and developed, and allows a user to search for useful data sets by several methods. For a full list and description of TIROS data and data products held at the NSSDC, consult Ng and Stonesifer (1989). Note: no historical APT data are available.

References and important further reading

Astheimer, R. 1961. Infrared radiometric instruments on TIROS II. *Journal of the Optical Society* 51: 1,386–93.

Bandeen, W., Halev, M. and Strange, I. 1965. A radiation climatology in the visible and infrared from the TIROS meteorological satellites. *NASA Technical Note D–2534*. NASA, Washington DC.

Bird, J., Morrison, A. and Chown, M. 1964. World atlas of photography from TIROS Satellites I to IV. *NASA Contractor Report CR–98*. NASA, Washington DC.

Ng. C. and Stonesifer, G. 1989. Data catalog series for space science and applications flight missions. Volume 4B: descriptions of data sets from meteorological and terrestrial applications spacecraft and investigations. *NSSDC/WDC–A–R&S 89–10*. Available from NSSDC.

Schnapf, A. 1985. The TIROS meteorological satellites – twenty-five years: 1960–1985. In Schnapf, A. (ed.), *Monitoring Earth's ocean, land and atmosphere from space – sensors, systems and applications*. American Institute of Aeronautics and Astronautics, New York: 606–20.

2. Nimbus 1, 2, 3 and 4

2nd generation NASA research and development meteorological satellites.

Nimbus 1

Operational lifespan and orbital characteristics

Dates: 28 August 1964–22 September 1964.
Semimajor axis: 6,968km.
Orbital height: 412–768km.
Inclination: 98.7°.
Period: 96.5 minutes.
Notes: Sun-synchronous, near-polar orbit.

Sensor characteristics

Automatic Picture Transmission (APT):
    VIS (wavelength 0.45–0.65µm). GR 4.6km. Swath width 1,660km.
Advanced Vidicon Camera Subsystem (AVCS):
    VIS (wavelength 0.45–0.65µm). GR 2km. Swath width 830km. Up to 2 full orbits of data
    (192 pictures) were recorded on board for subsequent playback.
High Resolution Infrared Radiometer (HRIR):
    IR (wavelength 3.4–4.2µm). GR 3–8km. Swath width 2,700km.

Nimbus 2

Operational lifespan and orbital characteristics

Dates: 15 May 1966–17 January 1969.
Semimajor axis: 7,519km.
Orbital height: 1,103–1,179km.
Inclination: 100.4°.
Period: 108.15 minutes.
Notes: Sun-synchronous orbit.

Sensor characteristics

APT, AVCS and HRIR:
    Spectral bands as for Nimbus–1. GR 4.6km for APT, 1.85km for AVCS and 8km for
    HRIR (swath width 2,700km).
Medium Resolution Infrared Radiometer (MRIR):
    5 bands in the wavelength range of 0.2–30.0µm. Nadir-viewing. GR 55km.

Nimbus 3

Operational lifespan and orbital characteristics

Dates: 14 April 1969–April 1972.
Semimajor axis: 7,483km.
Orbital height: 1,075–1,135km.
Inclination: 99.9°.
Period: 107.40 minutes.
Notes: Sun-synchronous orbit.

Sensor characteristics

HRIR:
    i)   Band 1: VIS and NrIR (wavelength 0.7–1.3μm). GR 8.5km (both bands). Swath width 2,700km.
    ii) Band 2: IR (wavelength 3.4–4.2μm).
Image Dissector Camera System (IDCS):
    VIS (wavelength 0.45–0.65μm). GR 2.2–8.5km. Swath width 1,400km.
Medium Resolution Infrared Radiometer (MRIR):
    As for Nimbus–2.
Satellite Infrared Spectrometer (SIRS):
    7 bands in the range of 11.0–15.0μm. Swath width 220km.
Infrared Interferometer Spectrometer (IRIS):
    operated in the range of 5.0–20μm. Surface IFOV a 144km diameter circle.
Interrogation, Recording and Location System (IRLS):
    a data collection system for interrogating ground-based instruments.

Nimbus 4

Operational lifespan and orbital parameters

Dates: 8 April 1970–30 September 1980.
Semimajor axis: 7,476km.
Orbital height: 1,095–1100km.
Inclination: 99.9°.
Period: 107.29 minutes.
Notes: Sun-synchronous orbit.

Sensor characteristics

IDCS:
    VIS (wavelength 0.45–0.65μm). GR 2.2km. Swath width 1,400km.
Temperature-Humidity Infrared Radiometer (THIR):
    i)   Band 1: IR (wavelength 6.5–7.0μm). GR 22.6km. Swath width 3,000km.
    ii) Band 2: TIR (wavelength 10.5–12.5μm). GR 7.7km. Swath width 3,000km.

IRLS:

see Nimbus–3

Other sensors carried include a SIRS (14 bands in range of 11–36μm), IRIS (6.25–25μm range), a Backscatter Ultraviolet Spectrometer (BUV) (12 bands in range of 2,500–3,400Å) and a Selective Chopper Radiometer (SCR) (6 bands around 15μm). For a more detailed description of these sensors, see the *Nimbus 4 User's Guide*, available from NSSDC.

Polar applications (Nimbus 1–4)

APT and AVCS:

as for the same instruments flown on the earlier TIROS, ESSA and NOAA satellites. The AVCS designed to record remote cloud cover pictures for subsequent playback; APT gave real-time images. Low sun elevation angles in the polar regions (particularly in the range of 1–10°) highlight much topographic detail. Delineation of leads/polynyas. Detection and tracking of large tabular icebergs. Good imagery of ice extent, particularly in the Antarctic (Predoehl 1966); sea ice-open ocean boundaries plotted to an accuracy of 3–10km (Popham and Samuelson 1965). The spatial resolutions are sufficient for synoptic-scale studies of ice-open water regimes.

HRIR:

measured radiance in the range of 210–330K. TIR night-time surface temperature, cloud, and large-scale sea ice–open water or sea ice–land ice boundary mapping. Monitoring of leads and polynyas in winter. VIS/NrIR: delineation of areas of surface melt on sea ice and ice sheets – monitored local changes attributable to the progress of the austral spring thaw – the reflectivity of ice and snow diminishes sharply in the wavelengths from 0.7–1.3μm but less at VIS wavelengths when the surface becomes wet. Large iceberg detection and tracking.

MRIR:

measurement of albedo over polar snow and ice fields. A radiation budget sensor.

IDCS:

daytime cloud mapping and large-scale monitoring of sea ice extent and snowcover. Delineation of leads and polynyas. Cartographic delineation of the Antarctic coastline.

THIR:

large-scale monitoring of ice boundaries during polar darkness. Band 1 provides information on the moisture and cirrus cloud content of the upper troposphere and stratosphere, and the location of jet streams and frontal systems. Band 2 was designed to measure cloud cover, and temperatures of cloud tops, land and ocean surfaces.

IRLS:

data (eg air temperature and surface pressure) relay from remote instrument packages on the surface, recorded onboard the satellite for subsequent transmission to receiving stations when within line-of-sight. Sea ice deformation and drift – used in AIDJEX in 1972 to monitor remote surface stations, and tracked the Arctic ice island T3. A forerunner of the Random Access Measurement System (RAMS) of later NIMBUS satellites and the current NOAA ARGOS system.

SIRS, IRIS and BUV:

measured vertical profiles of atmospheric properties and temperatures. BUV was a prototype for the Nimbus–7 Solar Backscatter Ultraviolet and Total Ozone Spectrometer (SBUV/TOMS); also provided albedo data.

Limitations (Nimbus 1–4)

APT and AVCS:

as for the same instruments flown on the earlier/contemporary TIROS, ESSA and NOAA satellites.

HRIR:

limited by cloud and haze. Ground resolution too poor for most ice and snow studies. Diurnal temperature variations may cause ambiguous interpretations of TIR data. Use limited under summer melt conditions, when thin ice in particular may appear as open water. Lack of an unambiguous signature for discriminating clouds from ice. Film strips are difficult to use in a quantitative fashion. Data require enhancement. Sensor saturation effects in high latitudes at certain times.

MRIR:

observations of albedo during austral summer only. Surface measurements limited by cloud.

IDCS:

limited by cloud, darkness and sensor saturation problems. Image distortion and problems of geolocation. Poor ground resolution. Of little use for the detailed quantitative study of snow/ice.

THIR:

limited by cloud and poor ground resolution. Great care must be taken in relating TIR data to surface conditions, especially under spring/summer melt conditions.

IRLS:

relatively crude positional accuracy.

SIRS, IRIS, SCR and BUV:

not designed to measure the Earth's surface (apart from surface temperature, eg IRIS provided SST data). Coverage limited by cloud cover.

NIMBUS was esentially a research and development series. Intermittent coverage of higher latitudes.

Data format (Nimbus 1–4)

AVCS:

4 × 6 inch black and white (B&W) microfiche (world montage catalogue) for NIMBUS–2 only. Useful for browse purposes.

HRIR:

i) Meteorological radiation data on CCTs.
ii) Photofacsimile film strips (night-time and daytime for Nimbus–3), 70mm B&W negatives.
iii) Film and CCT data catalogued on 4 × 6 inch B&W microfiche.

MRIR:

i) Meteorological radiation data on CCTs.
ii) Photographic display on 5 × 4 inch B&W negatives.
iii) Pictorial data catalogue on 4 × 6 inch B&W microfiche.

THIR:

i) 70mm black and white negatives of data from both bands. Day- and night-time orbital swaths displayed in strips, pole to pole.
ii) CCTs containing data from both bands.

iii) Data catalogue on 4 × 6 inch B&W microfiche.

IDCS:

4 × 6 B&W microfiche (world montage catalogue).

SIRS, IRIS, SCR and BUV:

all on CCT only, EBCDIC. Format details on request from NSSDC.

Note: data can be provided in a format other than standard, but these products are likely to be more expensive.

Data availability and archiving dates

Data are archived at: National Space Science Data Center. For full address, see TIROS I–X. Data from the following dates are archived:

Nimbus–1: 28 August 1964–22 September 1964, for both AVCS and HRIR.

Nimbus–2: 15 May 1966–31 August 1966 for AVCS; 15 May 1966–15 November 1966 for HRIR; 15 May 1966–28 July 1966 for MRIR.

Nimbus–3: 14 April 1969–31 May 1970 for IDCS; 14 April 1969–31 May 1970 for HRIR; 14 April 1969–31 May 1970 for MRIR.

Nimbus–4: 8 April 1970–8 April 1971, for both IDCS and THIR. BUV 10 April 1970–6 May 1977.

Data from the other sensors are also available from NSSDC.

References and important further reading

Barnes, J., Chang, D. and Willand, J. 1972. Application of ITOS and NIMBUS infrared measurements to mapping sea ice. *Final Report Contract 1–36025, Publication 8G93–F.* Department of Commerce, NOAA NESS.

Haas, I. and Shapiro, R. 1985. The NIMBUS satellite system: research and development platforms for the 1970s. In Schnapf, A. (ed.), *Monitoring Earth's ocean, land and atmosphere from space – sensors, systems and applications.* American Institute of Aeronautics and Astronautics, New York: 71–95.

Heath, D. 1979. Comparisons of global ozone trends inferred from the BUV experiment on NIMBUS–4 and the ground-based network. In Bolle, H. (ed.), *Remote sounding of the atmosphere from space*: 113–16. *Proceedings of the 21st Plenary Meeting of COSPAR,* Innsbruck, Austria 29 May–10 June 1978. Pergamon Press, Oxford, New York.

Popham, R. and Samuelson, R. 1965. Polar exploration with NIMBUS meteorological satellites. *Arctic* 18, 4: 246–55.

Predoehl, M. 1966. Antarctic pack ice: boundaries established from NIMBUS–1 pictures. *Science* 153, 3738: 861–3.

Rashke, E. and Bandeen, W. 1970. The radiation balance of the planet Earth from radiation measurements of the satellite NIMBUS–II. *Journal of Applied Meteorology* 9: 215–38.

Shenk, W. and Salomonson, V. 1972. A multispectral technique to determine sea surface temperatures using NIMBUS–II data. *Journal of Physical Oceanography* 2, 2: 157–67.

3. Environmental Science Services Administration (ESSA) 1, 3, 5, 7, 9 and ESSA 2, 4, 6, 8

Operational lifespan and orbital characteristics

Dates: 3 February 1966–15 November 1972.
Semimajor axis: 7,152–7,846km.
Orbital height: 702–1,508km.
Inclination: 97.9–102.12°.
Period: 100.35–115.28 minutes.
Notes: Second series TIROS/NOAA. The first operational meteorological satellites. Sun-synchronous orbits.

Sensor characteristics

Advanced Vidicon Camera Subsystem (AVCS) – on ESSA 1, 3, 5, 7, 9:
  VIS (wavelength 0.45–0.65µm). GR 2.2–3.0km. Swath width 2,300km. Up to 36 pictures could be stored onboard for subsequent playback to command and data acquisition stations at Wallops Island (Virginia) and Gilmore Creek (Alaska).
APT – on ESSA 2, 4, 6, 8:
  VIS (wavelength 0.45–0.65µm). GR 3.7km. Swath width 3,100km.
IR Flat Plate Radiometer (FPR) – on ESSA 1, 3, 5, 7, 9:
  operated in the range of 7.0–30.0µm.

Polar applications

Large-scale sea ice boundaries, extent and deformation. Could distinguish major lead systems and polynyas, broad categories of sea ice type and snow cover. Cloud distributions and cyclogenesis. Large-scale ice sheet boundaries. Composite Minimum Brightness (CMB) images were generated from AVCS data at NOAA/NESDIS on an experimental basis as an aid to research into large-scale snow and ice mapping in both polar regions. Using CMB data from the AVCS of ESSA 9, Wendler (1973) could distinguish five categories of sea ice, ranging from open water to very compact ice with a dry snow cover. AVCS data have roughly the same spatial resolution as those from the APT, but have a wider tonal range, thereby facilitating the distinction between cloud cover and the ice surface. Aided ship routing in ice-infested waters. FPR monitored the intensity of emitted and reflected radiation.

Limitations

Coarse ground resolution. Obscured by cloud and fog. Inoperative in darkness. Often difficult to distinguish ice from cloud. Imagery from the early meteorological satellites was not coupled to detailed ground observations. A major problem in using APT imagery from the ESSA satellites involves gridding and location of individual frames (the AVCS is far superior to the APT in this respect, as coastlines and graticules are superimposed by computer onto the photographs). Panoramic and Earth curvature distortion. Low solar elevation reduces image quality. The data are difficult to use in a quantitative fashion. Many important features are not resolved. Areas of low ice concentration and/or young ice are difficult to detect and interpret. Inability to deal with transient ice features and short term melting and freezing. Cannot accommodate subtle tonal and pattern variations over large areas. FPR low resolution, cloud-limited.

Data format

AVCS:
   i) Individual picture frames:
       a) 35mm microfilm, 2 days per reel;
       b) 25 × 25cm negatives, 1 image per negative, ESSA 9 only (Note: these data are not as readily available as the earlier ESSA data);
       c) 15 × 20cm negatives, pole to pole strips, 2 negatives per day.
   ii) Mosaics:
       a) 25 × 25cm negatives, 3 per day (polar stereographic projection);
       b) 35mm microfilm, 1 reel per month;
       c) 7 track/556bpi CCT, 3 days per CCT (polar mosaics only).
   iii) Global atlas of relative cloud cover, 1967–1970:
       a) 7 track/556bpi CCT, daily relative cloud cover;
       b) 7 track/556bpi CCT, monthly cloud frequency amounts;
       c) 9 track/1600bpi CCT, monthly cloud summaries;
       d) 25 × 25cm negatives;
       e) 35mm microfilms.
   iv) 5-day Composite Minimum Brightness (CMB) images generated from AVCS data and covering both hemispheres in polar stereographic projections are available as: 35mm negatives for 20 November 1968–14 December 1970 (5-day, and a few 10-day, 'augmented resolution' composites were produced from November 1969–June 1970 – archived as 25.4cm negatives); and 25.4cm negatives for 3 February 1970–6 November 1972.
APT:
   i) Mosaics:
       a) 15 × 20cm negatives, 1 per day.

Data availability and archiving dates

Data for 4 February 1966–15 November 1972 archived at the Satellite Data Services Division, NOAA/NESDIS, Princeton Executive Center, 5627 Allentown Road, Camp Springs, MD 20746, USA, telephone (301)–763–8399, telex 248376 OBSWUR, FAX 7638443. Note: Pre-1970 data are in semi-permanent storage.

References and important further reading

Booth, A. and Taylor, V. 1969. Meso-scale archive and computer products of digitized video data from ESSA satellites. *Bulletin of the American Meteorological Society* 50: 431–8.

Streten, N. 1968. Some aspects of high latitude southern hemisphere circulation as viewed by ESSA–3. *Journal of Applied Meteorology* 7, 3: 324.

Streten, N. 1973. Satellite observations of the summer decay of the Antarctic sea ice. *Archiv für Meteorologie, Geophysik und Bioklimatologie Series A*, 22, 1: 119–34.

Swithinbank, C. 1971. Composite satellite pictures of the polar regions. *Polar Record* 15, 98: 743–4.

Taylor, V. and Winston, J. 1968. Monthly and seasonal mean global charts of brightness from ESSA 3 and ESSA 5 digitized pictures, February 1967–February 1968. *ESSA Technical Report NESC 46*, Washington DC.

Wendler, G. 1973. Sea ice observations by means of satellite. *Journal of Geophysical Research* 78, 9: 1427–48.

4. Improved TIROS Operational Satellites (ITOS) 1/National Oceanic and Atmospheric Administration (NOAA) 1–5

Operational lifespan and orbital characteristics

Dates: 23 January 1970–1 March 1979.
Semimajor axis: 7,575–7,894km.
Orbital height: 918–1,522km.
Inclination: 101.74–102.80°.
Period: 109.36–116.34 minutes.
Notes: Second generation operational meteorological satellites. Sun-synchronous orbits. Near-polar coverage. Typical equatorial crossings 09:30 and 21:30 LST. Two-day repeat coverage interval for AVCS, daily for SR (VIS), twice daily for SR (TIR), daily for VHRR (VIS), and twice daily for VHRR (TIR).

Sensor characteristics

AVCS (on ITOS I):
    VIS (wavelength 0.45–0.65μm). GR 2.2km. Swath width 2,300 km.
APT:
    see the ESSA series.
SR:
    i) Band 1: VIS (wavelength 0.50–0.70μm). Daytime. GR 3.2km. Swath width 2,900km.
    ii) Band 2: TIR (wavelength 10.50–12.50μm). Night-time only. GR 8.0km. Swath width 2,900km.

Very High Resolution Radiometer (VHRR) – on NOAA 2–5 only, 15 October 1972–1 March 1979:

  i) Band 1: VIS (wavelength 0.60–0.70μm). Day and night. GR 1–1.9km, swath width 2,580km (for both bands).

 ii) Band 2: TIR (wavelength 10.50–12.50μm). Night-time only.

Vertical Temperature Profile Radiometer (VTPR) – on NOAA 2–5:

  7 bands in the range of 13.4–18.7μm. Surface IFOV 60 × 60km.

Polar applications

AVCS:

  see the ESSA series.

APT:

  see the ESSA series.

SR:

  i) VIS: mapping of gross sea ice-open water boundaries, ice disbribution and deformation. Ice sheet boundaries.

 ii) TIR: mapping gross sea ice distribution and conditions, such as the ice edge or the shape and size of large polynyas, during periods of darkness (although the imagery is generally inferior to VIS imagery); cloud patterns and SSTs (to an accuracy of 1–4K over a range of 185–330K).

iii) 10-day CMB and composite maximum temperature (CMT) images, generated from NOAA–3, –4 and –5 SR data (both VIS and TIR). With a resolution of 55km, these experimental images aided large-scale mapping of average snow and ice cover extent. By this method, the albedo of each pixel is stored and the mimimum value of each is read out at the end of the 10-day collection period. Thus, cloud-free periods for each pixel were retained and the presence of snow and ice detected.

VHRR:

  Direct readout to the two command and data acquisition stations, at Wallops Island, Virginia and Fairbanks, Alaska, allowed coverage of the Arctic from Greenland and Labrador to beyond the Chukchi and Bering Seas.

  i) VIS: large-scale mapping of sea ice extent, structure and deformation; lead and polynya detection; cloud patterns; snow cover; delineation of ice sheet margins; depiction of oceanic eddies, fronts and upwelling.

 ii   TIR: large-scale mapping of night-time or winter sea ice extent and structure; delineation of large leads and polynyas; cloud patterns; SSTs and ice temperatures; delineation of ice sheet margins.

iii) Although the VHRR was designed primarily as a direct readout sensor, there was an onboard data storage capacity of about 8% of the daily data load. Special stored VHRR data coverage (high resolution picture transmission, or HRPT) was obtained of the Ross Sea from November 1972–January 1973 to assist US Navy forecasters in their base resupply missions. Similar coverage took place of the Antarctic Peninsula and the Weddell Sea. Special coverage was also arranged for areas of the Arctic, including the North Water area, Foxe Basin, East Greenland current, Iceland and northern Scandinavia (McClain 1974). DeRycke (1973) used VHRR imagery of the Ross Sea to track large ice floes and icebergs (and calving events). Detailed experimental ice analyses based on VHRR imagery were produced by NOAA/NESS and transmitted to the Alaska area for guidance to the National Weather Service there.

VTPR:

vertical profiles of atmospheric temperature and moisture every 12 hours – atmospheric corrections for VHRR SST computations. Also thermal emission from the Earth's surface and cloud tops.

Limitations

AVCS:

see the ESSA series (page 159).

SR:

coverage limited by tape recording capacity when out of line of site of a receiving station.
  i)   VIS: comments as for AVCS.
  ii) TIR: restricted by cloud/fog; poor resolution; imagery often require enhancement, especially in summer when melt conditions make interpretation difficult. SSTs measured with errors of about 1–4°C rms.
  iii) If there is cloud cover over a region throughout the 10-day data collection period, the CMB technique gives an erroneous result. Even if cloud cover is not present, a great deal can happen to a sea ice cover in 10 days.

VHRR:

as for SR above.
  i)   VIS: restricted by cloud/fog/darkness.
  ii) TIR: restricted by cloud/fog; newly formed ice is difficult to distinguish from open water; unreliable in summer, when wet ice may appear as open water; cannot distinguish thick FY ice from MY ice; TIR imagery usually requires enhancement; many important ice features are unresolved on VHRR imagery.
  iii) Projections of early VHRR images were such that transfers to more standard map projections were difficult without an appreciable loss of information and quality. Geometric distortion and intensification of atmospheric effects towards swath edge.

VTPR:

poor spatial resolution, and cloud-limited.

Data format

AVCS:

see the ESSA series (page 159).

SR:
  i) Mapped Mosaics:
      a) 25 × 25cm negatives, polar-stereographic projections, both hemispheres. On 15 September 1976, the format for hemispheric mosaics was changed from a 2048- to a 1024-grid structure, with both hemispheric mosiacs on one 25 × 25cm negative;
      b) 35mm microfilm, 2 or 3 months per reel;
      c) 7 track/556bpi CCT (VIS data only); 9 track/1600bpi CCT (VIS and TIR data).
  ii) Orbital Swath Images:
      a) 25 × 25cm negatives, one per track, VIS and TIR;
      b) 35mm microfilm, 1–2 months of data, up to 6 tracks.

iii) 10-day CMB and CMT images generated from the VIS and TIR data. 25.4cm negatives, both hemispheres, polar stereographic projection mosaics, one negative per day. Scale 1:110 million, resolution 55km.

iv) Sea Surface Temperature (SST) Products: digital VHRR and SR TIR data used to compute SST values for grid points at roughly 100km intervals. Where the sea surface was obscured by clouds, values from the most recent day without cloud interference were retained to complete the daily data field. One CCT containing SST values, their age in days, and other related statistical parameters were prepared each day. Data reproduced as:

  a) CCTs containing global SSTs for each day and a summary of the orbital passes processed. Each CCT contains 1 month's data;

  b) Negatives (25 × 25cm) displaying SST observations, and the age of the observation at each data point for both hemispheres;

  c) SST 10-Day Analysed Field Tapes: once per day, all satellite SST observations were merged into a polar stereographic projection field, which also contains land-sea tags, climatology temperatures, SST gradient and data age information. Ten analysed fields from ten successive days are written to CCT;

  d) Global Operational Sea Surface Temperature Computation (GOSSTCOMP) isoline charts: contour displays of the analysed fields in paper form. Arctic and Antarctic regions are enlarged in polar-stereographic projection.

VHRR:

  i) Individual frames:

  a) 25 × 25cm negatives, HRPT and VRec (recorded swaths), VIS and TIR. The area covered in direct readout mode is 5,000 × 2,200km for the satellite pass most nearly overhead. Coverage when out of line-of-sight of receiving stations limited to one 30° geocentric arc along the orbital track per orbit, and had to be specially requested. Each picture shows an area of about 2,200 × 2,200km, scale about 1:10 million.

  b) CCT: a limited amount of digital high resolution data retained, ie 1 reflectance value per day from the Arctic. Note: VHRR data are not geometrically rectified to a base map, and image registration can be a problem. See above for SST products. From March 1973–May 1974, the CCT format was 7-track 556bpi, this changing to 9-track 1600bpi from May 1974 to present.

VTPR:

  on CCT only.

Data availability and archiving dates

Data from 26 April 1970–16 March 1978 are archived at the Satellite Data Services Division (SDSD), NOAA/NESDIS, Princeton Executive Center, 5627 Allentown Road, Camp Springs, MD 20746, USA, telephone (301)–763–8399, telex 248376 OBSWUR, FAX 7638443.

  i) NOAA 1: data not retained in digital format.

  ii) SR Orbital Swath Images, November 1972–15 March 1978. SR Mapped Mosaics on 35mm microfilm or photographic products, November 1972–March 1978 (polar stereographic mosaics for both hemispheres are available on CCT for data from May 1974–February 1979). SR SST CCTs are available from December 1972–February 1979; SST negatives from November 1972–March 1979; SST 10-Day Analysed Field Tapes from 10 May 1973–February 1979; and GOSSTCOMP SST isoline charts from April 1976–February 1979 (with June 1976 missing). SR CMB and CMT images for both polar sectors from 1 December 1974–15 March 1978.

iii) VHRR: HRPT and VRec imagery is available from 16 November 1972–7 February 1979. A 90-day rotating file of VHRR VIS and IR data CCTs began on 1 January 1977. Copies of the full CCTs may be obtained, or copies of particular segments from several CCTs may be requested and combined onto output CCTs. Data products only were archived from April 1970–15 September 1976, during which time no data were retained in full resolution digital format. Some digital data for the Arctic, covering 15 December 1977–16 March 1978, were archived.

iv) Only since 1969 has coverage of Antarctica been consistent enough to be of operational value. SR VIS imagery data (daily mosaics with a 3km resolution) are archived at SDSD on 35mm negative film (from 1969) and on 25 × 25cm negatives from 1972 to present (except for March–December 1978). Since March 1973, SR IR night-time and IR daytime data (since May 1974) mosaics have been archived in digital format. Only a limited amount of VHRR imagery is available for 1972 and 1973.

Data are also available from other sources, eg Department of Electrical Engineering, University of Dundee, Dundee DDI 4HN, Scotland, telephone 0382–23181, telex 76293 ULDUND G: since 23 August 1976, hardcopy only.

DERIVED PRODUCTS

Legeckis (1978) has published a survey product of global SST fronts detected by VHRR. The following oceanic fronts have been seen in polar/sub-polar oceans:

a)   Southern Greenland, 13 September 1977 – VHRR image number 5086;
b)   Labrador Sea, 25 August 1975 – VHRR image number 3542;
c)   Newfoundland, 8 October 1977 – VHRR image number 5395;
d)   Bering Sea, 25 March 1975 – VHRR image number 2120;
e)   Macquarie Ridge, 21 September 1976 – VHRR image number 660*;
f)   Ross Sea, 6 December 1976 – VHRR image number 1613;
g)   Weddell Sea, 22 March 1977 – VHRR image number 2917; and
h)   Drake Passage, 7 October 1976 – VHRR image number 859*.

Images marked * have been computer reprocessed, ie geometrically corrected and enhanced for the temperature range at the SST front. These images (in original format) are available from SDSD. See also NOAA AVHRR.

References and important further reading

Aber, R. and Vowinckel, E. 1972. Evaluation of North Water spring sea ice cover from satellite photographs. *Arctic* 25: 263–71.

Ahlnäs, K. 1979. IR enhancement techniques to delineate surface temperature and sea ice distributions. *Proceedings of the 13th International Symposium on Remote Sensing of Environment*, 23–27 April 1979, Volume 1: 1,067–76. ERIM, Ann Arbor, Michigan.

Conlan, E. 1973. Operational products from the ITOS Scanning Radiometer data. *NOAA Technical Memorandun NESS 52*. US Department of Commerce, Washington DC.

DeRycke, R. 1973. Sea ice motions off Antarctica in the vicinity of the eastern Ross Sea as observed by satellite. *Journal of Geophysical Research* 73, 36: 8,873–9.

Gruber, A. 1977. Determination of the Earth-atmosphere radiation budget from NOAA satellite data. *NOAA Technical Report NESS 76*. NOAA, Department of Commerce, Washington DC.

Legeckis, R. 1978. A survey of worldwide sea surface temperature fronts detected by environmental satellites. *Journal of Geophysical Research* 83, 9: 4,501–22.

McClain, E. 1974. Some new satellite measurements and their application to sea ice analysis in the Arctic and Antarctic. In *Advanced concepts and techniques in the study of snow and ice resources*: 457–66. US National Academy of Sciences, Washington DC.

McGinnis, D., Pritchard, J. and Wiesnet, D. 1975. Determination of snow depth and snow extent from NOAA–2 satellite very high resolution radiometer. *Water Resources Research* 11: 897.

Streten, N. 1974. Large-scale sea ice features in the western Arctic basin and the Bering Sea as viewed by the NOAA–2 satellite. *Journal of Arctic and Alpine Research* 6, 4: 333–45.

Wiesnet, D. 1980. A satellite mosaic of the Greenland ice sheet. *World glacier inventory, Proceedings of the Riederalp Workshop*, September 1978. *International Association of Hydrological Sciences Publication* 126: 343.

5. Landsat 1 and Landsat 2

Operational lifespan and orbital characteristics

Dates: Landsat 1 (formerly ERTS–A) 23 July 1972–6 January 1978. Landsat 2 (formerly ERTS–B) 22 January 1975–27 July 1983, interrupted 1979–80.

Semi-major axis: 7,290km.

Orbital height: 903–921km.

Inclination: 99.1°.

Period: 103.27–103.28 minutes. (Exactly repeating orbit with 18 day repeat period for each satellite).

Notes: circular, sun-synchronous orbits. Equatorial crossing time 09:42 local.

Sensor characteristics

Multispectral Scanner (MSS):
   i) Band 4: VIS (wavelength 0.5–0.6μm). GR 79m (all bands). Swath width 185km (all bands), with 10% forward lap and 14% sidelap at the equator, increasing towards the poles. Data rate 15Mbps (for all 4 bands). All bands operated simultaneously. Each scene provided an image of 34,225km². Radiometric resolution 6 bits. Sampling rate 9.958ms, corresponding to 56m on the ground. MSS data are digitised to 128 grey tones for bands 4–6, and 64 for band 7.
   ii) Band 5: VIS (wavelength 0.6–0.7μm).
   iii) Band 6: NrIR (wavelength 0.7–0.8μm).
   iv) Band 7: NrIR (wavelength 0.8–1.1μm).
Return Beam Vidicon (RBV):
   3 cameras focused on the same ground area:
   i) Band 1: VIS (wavelength 0.475–0.575μm). GR 80m (all bands). Swath width 185km (all bands), providing a simultaneous view, with a 14% overlap and a 10% forward lap at the equator, increasing towards the poles.
   ii) Band 2: VIS (wavelength 0.58–0.68μm).

iii) Band 3: NrIR (wavelength 0.69–0.83 μm).
Note: imagery is normally only acquired on the descending portion of the Landsat orbit; the
    ascending part is normally reserved for 'housekeeping' tasks.
Data Collection System (DCS):
    interrogation, storage and relay of data from remote ground experiments for subsequent
    transmission. Similar to ARGOS on current NOAA satellites.

Polar applications

MSS:

i) *Land ice*: accurate mapping of glacier extent (Figure II.1) and ice sheet margins.
    Calving/tracking of tabular icebergs. Icebergs embedded within an icefield are more
    easily detectable using NrIR (due to their low surface temperatures). Planimetric
    mapping of floating ice fronts, under-ice coastlines, grounded ice walls, the direction
    of ice motion, ice divides as defined by ridges, superimposed ice, ice rises and rumples,
    ice rifts, grounding lines, surface equilibrium zones, and glacier surges. Rates of ice
    stream and ice front movement from sequential imagery. Mapping of ice-free areas.
    Good delineation of flow and stream lines, moraines and crevasses (indicating areas of
    strong deformation). Areas of ablation, onset of melt and albedo changes. Low sun
    elevation angle highlights topographic features that may not be visible on the ground.
    Extent of debris overlying the ice. Snow cover mapping in lower latitude regions.
    Detection of areas of blue (bare) ice, which may contain meteorites. Mapping of
    extra-glacial geomorphology. Bands can be digitally combined/enhanced to highlight
    subtle features.

ii) *Sea ice*: accurate ice/water boundaries. Floe size and shape distributions. Distribution
    and dynamics of leads and polynyas. Short-term floe dynamics or velocity vectors by
    the tracking of recognisable ice features in time on sequential images; this technique
    was used during AIDJEX during the 1970s to calculate ice stress, deformation, strain,
    strain rate and vorticity (Hall 1980). Combination with, and verification of, data
    from coarser resolution sensors, eg Nimbus–5 ESMR (Comiso and Zwally 1982).
    Differences in reflectance between bands 4 and 5 and the near-infrared band (7) have
    been found to be useful for differentiating ice types and discriminating characteristics
    of the ice or snow surface (Barnes and others 1974). Ice concentration, particularly in
    complex MIZs. Mapping of fast ice extent. Large pressure ridge extent. The detection
    of onset of melt and melt ponding. Features can be mapped accurately to a scale of
    least 1:250,000. Ocean sediment patterns delineating currents and eddies, etc outside
    ice edge.

    a) Band 4: detects all ice types. Very responsive to thin ice. Senses lower ice
        concentrations better than the other bands. Detection of initial stages of ice
        growth.

    b) Band 5: detects most ice types. Moderately responsive to thin ice. Along with
        band 4, the most useful for ice edge feature mapping.

    c) Band 6: detects dry, thin ice. With band 7, most information on sea ice distri-
        bution and lead.

    d) Band 7: detects dry, thin ice, and changes in surface characteristics. Best delinea-
        tion of small leads and floes in areas of high concentration. Wet floes can be
        delineated by comparing bands 5 and 7. Most useful for identifying features
        revealed by changes in ice surface characteristics. NrIR bands are the best for ice

00752-058

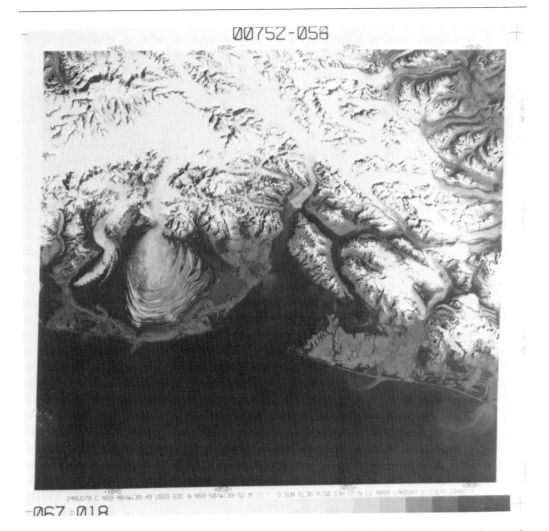

-067 ⌐018

Figure II.1 Landsat 2 false colour MSS image (I.D. 21675-19482) of the St Elias Mountains on the Alaska–Canada border, 24 August 1979. The Malaspina Glacier, left centre, covers an area of more than 2,000km. Hubbard Glacier can be seen in the centre of the image; see also Figure II.16. Courtesy of Jane Ferrigno, USGS.

sheet surface detail. Best for distinguishing cloud and snow cover (especially when combined with band 4). Fewer detector saturation problems than the other bands. Distinguishes grey, grey-white and older ice.

RBV:

Did not produce imagery suitable for polar studies from LANDSATs 1 and 2.

Limitations

Both RBV and MSS restricted by cloud, darkness and sensor saturation over snow ice masses (at certain times of the year, ie from November to January in the Antarctic). MSS band 5 worst affected by saturation, with band 7 unaffected. Maximum latitudinal coverage to 82° –

over Antarctica, this covers Ice Streams E and D but not A, B and C. Narrow swath width. Coverage overlap good in high latitudes, but followed by a 'holiday' away from the area. Features must be resolved to be identifiable; estimates of open water area derived from VIS, NrIR, IR and TIR data tend to be too high, and care must be taken in estimating ice concentration from these data due to the presence of unresolved leads, small floes and thin ice. Estimates in the amounts of thin ice can also be complicated by the masking effect of a snow cover. Absolute image registration and geolocation remains a problem, particularly when no land features are present. Often difficult to identify thickly debris-covered glaciers. At the launch of Landsat 1, only four ground stations existed which were capable of receiving real-time data (3 in USA and 1 in Canada). Restricted duty cycle over polar regions and limited onboard recording capacity; latter affected by recorder technical problems. Old ice, smaller pressure ridges, hummocks and individual melt features cannot be distinguished.

i) MSS Band 4: tends not to differentiate thicker sea ice forms. Bands 4 and 5 may not detect icebergs embedded in an icefield unless a low sun angle shows them up as topographic detail.

ii) Band 5: poor response to newest sea ice forms. Difficult to distinguish individual floes in areas of high concentration with bands 4 and 5. Over ice sheets, MSS bands 4 and 5 give little information on surface properties, such as snow grain-size variations. Bands 4 and 5 do not discriminate well between snow/ice and clouds, as all three reflect highly at these wavelengths (in many MSS scenes of the Antarctic, the cloud-free ice sheet surface has often been misinterpreted as cloud-covered during quality-control analysis).

iii) Bands 6 and 7: wet snow cover reduces albedo and causes interpretational difficulties.

RBV:

this original system was not a success, and RBV operated for only 45 hours with LANDSATs 1 and 2, compared to 7,462 hours of MSS operation.

DCS:

few data collected from buoys etc in polar regions using this system.

Note: Landsat 1 direct readout only was possible after March 1973, due to onboard tape recorder failure, therefore further limiting coverage of Antarctica. Many of the CCTs holding early Landsat images have deteriorated; a large number of these have not been transferred to the current computer format and are now unusable. Many images are, however, usable.

Data format

i) CCTs:
    a) 9-track, 1600/6250bpi, Tape Set format. MSS all bands (band sequential [BSQ] or interleaved by line [BIL]), RBV single subscene or set of 4 subscenes. Radiometric and geometric corrections included;
    b) 7-track, 800bpi (from EROS Data Centre).

ii) Standard Photographic Data Products:
    a) 55.8mm (2.2 inch) film, positive and negative, scale 1:3,369,000;
    b) 18.5cm (7.3 inch) film negative (black and white [B&W] only), film positive or paper print (B&W or colour composite for MSS, B&W only for RBV), scale 1:1,000,000;
    c) 37.1cm (14.6 inch) paper print, B&W or colour composite for MSS, B&W only for RBV, scale 1:500,000;

d) 74.2cm (29.2 inch) paper print, B&W or colour composite for MSS, B&W only for RBV, scale 1:250,000;

e) Note: unless otherwise requested by the customer, standard colour composites will be generated using MSS bands 4, 5 and 7;

f) Each SDP displays geographical coordinates, a grey scale, centre and satellite subpoint coordinates, orbit number, date, solar elevation, etc;

g) 70mm quick-look imagery, based upon band 6 and distributed as microfiche.

Note: several correction and enhancement procedures can be applied by EOSAT on request.

Data availability and archiving dates

i) The main receiving, archive and processing centre (containing data from June 1972 to the present): The Earth Observation Satellite Company (EOSAT) Customer Services, EROS Data Centre (EDC), Sioux Falls, SD 57198, USA, telephone (605)–594–6950, FAX (605)–594–6589, TWX 9106680310. Alternatively, Landsat data can be ordered from the EOSAT Customer Services, 4,300 Forbes Boulevard, Lanham, MD 20706, USA, telephone (301)–552–0537 or 1–800–344–9933 (within the USA.), telex 277685 LSAT UR, FAX 301–552–0507. Order forms and current price lists available on request.

Listings of available MSS and TM data for a particular area of interest are available at/from EOSAT; this service is provided free of charge. The maximum acceptable cloud cover, minimum image quality standards and the time of year preferred can also be specified. At the EDC, Landsat scenes are corrected to 'system level' to compensate for spacecraft altitude variations. Scenes contain minimal geometric distortion. *LANDSAT Data Users Guides* are available from EOSAT.

Landsat 1 was turned off on 20 October 1976, but turned on again on 1 June 1977. MSS band 4 failed on 3 March 1977, and was finally turned off on 27 October 1977. Payload operations terminated on 6 January 1978. Landsat 2 data transmission from 9 April 1975–25 January 1982.

Note: all data produced by EDC before the EOSAT contract was signed (27 September 1985) remain in the public domain. All data products produced by the EDC after this date fall under strict copyright restrictions, regardless of when the datasets were originally recorded by the satellites.

ii) Some E Arctic data collected at Kiruna in Sweden, and available from: Eurimage Operations Office – ESRIN–CP64, Via Galileo Galilei, 00044 Frascati, Italy, telephone 39–6–9426285 or 39–6–9401218, telex 610637 ESRIN I EURIMAGE. The UK EARTHNET National Point of Contact is: Mr M. Hammond, National Remote Sensing Centre, Royal Aircraft Establishment, Farnborough, Hampshire GU14 6TD, telephone 0252 24461, telex 858442. For details of all data and facilities available at the UK NRSC, see the *National Remote Sensing Centre Data Users' Guide.*

iii) The US Geological Survey (USGS) Image Processing Facility in Flagstaff, Arizona has been collecting and enhancing MSS images of Antarctica. For further information, contact USGS Headquarters at 12201 Sunrise Valley Drive, National Center, Reston, VA 22092, USA.

iv) Canadian receiving stations/national and regional archives include:

a) Shoe Cove, Newfoundland: operational since mid-1977, covering S Greenland, S Baffin Island and Labrador up to 71°N;

b) Prince Albert, Saskatchewan: has covered much of the Canadian Arctic (up to approximately 74°N) since July 1972. For information regarding the acquisition of data in Canada, contact:

   i) CCRS User Assistance Group, 717 Belfast Rd., Ottawa, Ontario, K1A 0Y7, telephone (613)–952–0500, telex 0533777.

   ii) Integrated Satellite Information Service Ltd., PO Box 1630, Prince Albert, Saskatchewan, S6V 5S7, Canada, telephone (306)–764–3602, telex 074 29242.

Data can be ordered from:

a) Data Acquisition Division, Canada Centre for Remote Sensing (CCRS), 2464 Sheffield Rd., Ottawa, Ontario, K1A 0Y7, Canada, telephone (613)–952–2717, telex 0533777 CA. Catalogues of LANDSAT data coverage and availability, including current price lists and order forms, are available from this source.

b) The CCRS LANDSAT Order Desk, Prince Albert Receiving Station, Prince Albert, Saskatchewan, S6V 5T2, Canada.

v) Hatoyama receiving station in Japan has received MSS data since 1978, the station mask covering much of the seasonally sea ice-covered Sea of Okhotsk. CCTs and photoproducts are available from: The Data Service Department, Remote Sensing Technology Centre of Japan (RESTEC), Uni-Roppongi Building 7–15–17, Roppongi, Minato-ku, Tokyo 106, Japan, telephone 03–403–1761, telex 2426780 RESTEC J, FAX (81)–3–403–1766.

vi) The University of Alaska-Fairbanks holds a regional archive of Landsat data, offering coverage of the Bering, Chuckchi and Beaufort Seas (W Arctic). Initially (1972), data were acquired directly from NASA, and formed a subset of the EDC holdings. Starting in 1978, additional images acquired from CCRS. Since 1982, has acquired data through the Quick Look Project from NOAA Gilmore Creek station. Many of these images are not in the EDC data base. On-line database/inventory – accessed via SPAN or the NASA Master Directory (NMD). SPAN access by SET HOST FRED and username ALI; the LINK command from the NMD. Contact GeoData Center, Geophysical Institute, University of Alaska-Fairbanks, AK 99775–0800, telephone (907)–474–7487.

References and important further reading

Barnes, J. and Bowley, C. 1973. Use of ERTS data for mapping snow cover in the western United States. *Proceedings of the Symposium on Significant Results obtained from ERTS–1. NASA Report SP–327*: 855–62. NASA GSFC, Greenbelt, MD 20771.

Barnes, J., Bowley, C., Chang, D. and Willand, J. 1974. Application of satellite visible and infrared data to mapping sea ice. In *Advanced concepts and techniques in the study of snow and ice resources*: 467–76. US National Academy of Sciences, Washington DC.

Colvill, A. 1977. Movement of Antarctic ice fronts measured from satellite imagery. *Polar Record* 18: 390–4.

Comiso, J. and Zwally, H. 1982. Antarctic sea ice concentrations inferred from NIMBUS–5 ESMR and LANDSAT imagery. *Journal of Geophysical Research* 87: 5836–44.

Dowdeswell, J. and McIntyre, N. 1986. The saturation of LANDSAT MSS detectors over large ice masses. *International Journal of Remote Sensing* 7, 1: 151–65.

Hall, R. 1980. A test of the AIDJEX ice model using LANDSAT images. In Pritchard, R. (ed.), *Sea ice processes and models*: 89–101. University of Washington Press, Seattle, Washington.

Krimmel, R. and Meier, M. 1975. Glacier applications of ERTS images. *Journal of Glaciology* 15, 73: 391–402.

Maul, G. and Gordon, H. 1975. On the use of the Earth Resources Technology Satellite (ERTS–1) in optical oceanography. *Remote Sensing of Environment* 4: 95–128.

Southard, R. and MacDonald, W. 1974. ERTS–1 imagery applications in polar regions. *Antarctic Journal of the United States* 9, 3: 61–7.

US Geological Survey. 1979. *LANDSAT Data User's Handbook*. US Geological Survey, EROS Data Center, Sioux Falls, South Dakota, USA.

Williams, R., and Carter, W. (eds). 1976. ERTS–1: a new window on our planet. *Geological Survey Professional Paper 929*. US Government Printing Office, Washington DC, USA.

Williams, R., Meunier, T. and Ferrigno, J. 1983. Blue ice, meteorites, and satellite imagery in Antarctica. *Polar Record* 21: 493–6.

6. Nimbus 5

Operational lifespan and orbital characteristics

Dates: 11 December 1972–November 1982.
Semimajor axis: 7,474km.
Orbital height: 1,089–1,102km.
Inclination: 100.0°.
Period: 107.25 minutes.
Notes: Sun-synchronous orbit.

Sensor characteristics

THIR:
    As for Nimbus–4, except that GRs are now 22.5km for band 1 and 8.2km for band 2.
Surface Composition Mapping Radiometer (SCMR):
    i) Band 1: VIS (wavelength 0.8–1μm). GR 0.8km (all bands). Swath width 800km (all bands).
    ii) Band 2: IR (wavelength 8.3–9.3μm).
    iii) Band 3: TIR (wavelength 10.5–11.3μm).
Nimbus–E Microwave Spectrometer (NEMS):
    Passive microwave. 5 channels: 22.234, 31.4, 53.65, 54.9 and 58.8GHz (wavelengths 1.35, 0.96, 0.56, 0.55 and 0.51cm). Antenna beamwidth 10°. Spatial resolution at nadir 180km. RF bandwidth 250MHz (all channels). Integration time 2s (all channels). Temperature sensitivity 0.1–0.29K rms. Dynamic range 100–325K (all channels). Long term absolute accuracy 2K (all channels).
Electrically Scanning Microwave Radiometer (ESMR):
    Passive microwave (centre frequency 19.35GHz, wavelength 1.55cm). H polarisation. RF bandwidth 250MHz. Integration time 47μs. Absolute accuracy 2K. GR 25 × 25km (160 × 45km at the end of the scan). Swath width 3,000km. Data rate 200bps. Dynamic range 50–330K. 78 scan positions varying from 50° to the left to 50° to the right of the

subsatellite track (nadir). If all 78 beam positions were used, full coverage of almost the entire polar regions could be obtained from a sequence of 6 satellite orbits, or one half day of good data. However, with restrictions on beam positions and with additional rejection of data because of occasional instrument problems, about three days of data were generally required to provide near-complete polar coverage.

IRLS:

tracking and data relay.

Other sensors carried include an Infrared Temperature Profiling Radiometer (ITPR) (7 bands in the range of 3.7–20µm, 32km GR); and a SCR (16 bands in the range of 2.08–133.3µm, GR 25km). See Sabatini (1972) for further details.

Polar applications

THIR:

Band 1: moisture and cirrus cloud content of the upper troposphere and stratosphere, and the location of jet streams and frontal systems. Band 2: cloud cover and cloud top and ocean surface temperatures. Provided continuous reference data for use by the other instruments. Can combine with ESMR data to compute ice sheet emissivity contours, and sea ice emissivity. Mapping of gross ice boundaries during darkness. Can infer sea and land ice/snow surface temperatures (under cloud-free conditions), if surface emissivity known. See NOAA AVHRR and ERS–1 ATSR-M.

SCMR:

useful for the medium-scale delineation of ice sheet edge, ice/water boundaries and the detection of bare glaciers. Thermal properties of the surface.

NEMS:

an experimental microwave sounder designed to obtain vertical profiles of atmospheric temperature and water vapour/liquid water content. Channels 1 and 2 sensitive over oceans to water vapour and liquid water, and indicated surface temperature and emissivity variations over land. The combination of the average of the radiometric temperatures measured by channels 1 and 2 and their frequency gradient showed the presence of well defined microwave signatures of snow and ice surfaces over large areas (Kunzi and others 1976). These two frequencies are similar to those used in channels 4 and 5 of the Nimbus–7 SMMR.

ESMR:

i) *Sea ice*: at 19.35GHz, there is a sharp contrast in the emissivities of open water and sea ice. New ice (<0.1m thick) has an emissivity of 0.45–0.92; FY ice (≥0.1m thick) 0.92; for MY ice it is 0.84; for 'summer ice' it is 0.45–0.95; and that of open water is about 0.44. Can estimate sea ice concentrations and extent to an accuracy of approximately ±15% in regions of predominantly FY ice under freezing conditions (Zwally and others 1983). Linear interpolation algorithm. Also, temporal and regional large-scale variations in ice boundaries, compactness, progression of freeze-melt (central Arctic), and large polynya growth and decay (the famous Weddell Polynya was discovered on ESMR imagery).

The amount of ice to survive the summer (ie MY fraction) can be estimated in certain seasons and locations by analysing the increase in brightness temperature (TB) when the mixture of MY ice and open water is approximately replaced by a mixture of MY ice and new ice (Carsey 1982). Particularly useful in the study of Antarctic sea ice, which is more homogeneous (less emissivity variability) as a microwave target than Arctic sea ice. The 3,000km swath width and coarse resolution allowed the creation of maps of both polar regions (poleward of latitude 55°) almost in their entirety daily.

ii) *Land ice*: onset of melt for seasonal snow cover. Monitoring of ice sheet regional ablation. Detection of ice streams and drainage basins. Potential estimation of snow accumulation rates to an accuracy of ±20%; the emissivity varies from 0.65–0.75 in regions of Antarctica known to have low accumulation rates, and from 0.8–0.9 where accumulation rates are high (Zwally 1977).

iii) *Over open ocean*: wind speed (but not direction) and atmospheric water vapour and liquid water measurements.

See Nimbus–7 SMMR and DMSP SSM/I.

IRLS:

used for tracking and interrogating surface experiment packages and ice-strengthened buoys. Data received from the remote sites were subsequently relayed to the NASA GSFC in Maryland. Iceberg drift. See ARGOS.

ITPR and SCR:

vertical profiles of atmospheric temperature and water vapour.

Limitations

SCMR:

limited by cloud and poor detail of surface features, especially at higher elevations. Relatively poor spatial resolution. Few high quality images were obtained before it became inoperative a few weeks after launch.

NEMS:

only channels 1 and 2 could be employed for surface observations, and then only of large-scale phenomena. Limited by a very poor spatial resolution.

ESMR:

single channel and polarisation – only limited success in distinguishing different ice types within IFOV, especially in the Arctic and under melt conditions. The ESMR ice concentration algorithm assumes a unique value of ice emissivity, which is approximately true only if one ice type predominates. Carsey (1982) sets the lower limit of ice concentration that can be detected accurately from ESMR data at 15% when weather conditions are unknown. Ice concentration retrievals in areas containing a complex mixture of new, FY and MY ice may have an accuracy of only 25%. Melting during the austral spring and summer causes serious ambiguities in the interpretation and extraction of ice concentrations; at this time, the microwave signature of MY ice approaches that of FY ice. Effective TB of open water is about 130K, sea ice about 210–250K, and a free water content of only 1% in the snowcover may increase the TB by about 49K. Real-time algorithms using ESMR data were generally unable to discriminate between FY and MY sea ice. Inability to accommodate variations in the physical temperature of the radiating portion of the ice. Not possible to remove atmospheric contribution using single channel data.

Poor resolution of new/thin ice (<5cm ie new ice, dark nilas, unconsolidated frazil and grease ice); it may interpret thin/wet ice as open ocean (particularly during summer). Conversely, ice >5cm thick (ie light nilas, grey ice and grey-white ice) has emissivities similar to those of FY ice. The vast continuous pancake ice covers of Antarctica, when forming, are resolved as lower concentration covers of FY ice. Melt ponds (in Arctic in particular) can reduce its emissivity to about 0.44 (at 18GHz). Further ambiguities may be caused by the flooding of the snow-ice interface with sea water by ice depression under the weight of a snowcover or by wave overwashing; this greatly increases the salinity of the surface layer. Moreover, highly saline frost flowers formed on refrozen lead ice may persist to present a very bright passive microwave target. Detailed mapping/monitoring not possible, due to wide angle effects and very coarse resolution. Individual floes and leads not resolved.

   Use limited in the complex MIZs and close to coastlines, where the overlap of some ocean or land pixels may contaminate sea ice retrievals; also antenna pattern and sidelobe effects. Freeze-thaw cycles (on diurnal, semi-diurnal and seasonal time scales) tend to produce a more granular, layered snow cover, which may suppress the effective emissivity. Due to the very coarse resolution, it is difficult to relate the data to *in situ* measurements. The microwave emissivity of land ice is poorly understood. For land ice, V polarisation is desirable to minimise the effect of surface reflection and maximise the ability to observe radiation from the volume layer of the snow. See also Nimbus–7 SMMR and DMSP SSM/I.

THIR:
limited by cloud and poor resolution. Difficulty in relating data to surface conditions, especially under summer melt conditions. See NOAA AVHRR and ERS–1 ATSR–M.

IRLS:
coverage was not as frequent as that of the ARGOS system on the NOAA satellites. Larger positional errors.

ITPR and SCR:
not designed to measure surface features. Poor spatial resolution.

Data format

SCMR:
calibrated and Earth-located radiance data in photographic and CCT format (EBCDIC). Data are grouped into 7-minute sections, and mapped on a Mercator projection covering 80°N to 80°S.

NEMS:
CCTs or 4 × 6 inch black and white (B&W) microfiche of TBs. Each 9-track, 1600bpi CCT contains 2–6 days of data.

THIR:
all data available as photofacsimile film, 70 × 70mm B&W negatives; some data on CCTs.

ESMR:
   i)   4 × 6 inch B&W microfiche of derived oceanic rainfall.
   ii) 70mm B&W photofacsimile film.
   iii) selected colour images, 10 × 8 inch positive or negative transparencies (15 December 1972–10 February 1973).
   iv) CCTs containing Earth-located calibrated brightness temperatures (CBTs) and sea ice concentrations and MY fraction. The CBT tapes were the primary source of calibrated radiometer data used for sea ice analysis in the magnificent Antarctic and Arctic sea ice atlases produced at the NASA GSFC Oceans and Ice Branch by Zwally and others (1983) and Parkinson and others (1987). The CBT tapes contain the time, the calibration parameters, the measured TBs, and the corresponding geographical

coordinates. Each 6250bpi CBT tape holds about 180 files, each of which contains one orbit of data. Mapped onto a polar stereographic projection, grid size 293 × 293; each map cell represents a surface area varying from about 32 × 32km near the poles to 28 × 28km near 50° latitude. Three day average maps provide almost complete coverage of polar regions; use only the centre of the observation swath (1,360km as opposed to 2,500km). These secondary mapped products are available for either TB or sea ice concentration data.

ITPR and SCR:

CCT only.

Data availability and archiving dates

NEMS, SCMR, ITPR, THIR and ESMR data are archived at National Space Science Data Center (for full address, see TIROS I–X). Note: requests to NSSDC from researchers outside USA must be addressed to the Director, World Data Center A for Rockets and Satellites, Code 630.2, NASA GSFC, Greenbelt, MD 20771, USA.

The following products are also available from National Snow and Ice Data Center (NSIDC), WDC–A for Glaciology, Campus Box 449, University of Colorado, Boulder, CO 80309, telephone (303)–492–5171, telex 257673 (WDCA UR), telemail MAIL/USA [NSIDC/OMNET], VAX mail via SPAN: KRYOS::NSIDC:

i)   ESMR: December 1972–May 1977, although good data exist only up to late 1976, when the instrument lost calibration and the data became streaked with lost beams. Four years of 3-day averaged TBs, Antarctic and Arctic (2 CCTs each); 1 year (1974) of 1-day averaged TBs, Antarctic (1 CCT); monthly averaged TBs and sea ice concentrations, Arctic and Antarctic (1 CCT each). 9-track CCTs, 1600/6250bpi, EBCDIC or ASCII, 80 characters/record, block length variable. See Parkinson and others (1987) and Zwally and others (1983). Ancillary meteorological data (surface air temperature and pressure) also available in the same grid format from NSIDC.

ii) THIR: 19 December 1972–12 March 1975.

iii) NEMS: 17 December 1972–31 October 1973.

iv) SCMR: 13 December 1972–26 December 1974.

References and important further reading

Carsey, F. 1982. Arctic sea ice distribution at end of summer 1973–1976 from satellite microwave data. *Journal of Geophysical Research* 87, C8: 5,809–35.

Cavalieri, D. and Parkinson, C. 1981. Large-scale variation in observed Antarctic sea ice extent and associated atmospheric circulation. *Monthly Weather Review* 109: 2,323–36.

Chang, A. and Wilheit, T. 1980. Remote sensing of water vapor, liquid water, and wind speed at the ocean surface by passive microwave techniques from the NIMBUS–5 satellite. *Radio Science* 14: 793–802.

Comiso, J. and Zwally, H. 1980. Corrections for anomalous time-dependent shifts in brightness temperature from the NIMBUS–5 ESMR. *NASA Technical Memorandum TM–82055*. NASA GSFC, Greenbelt, Maryland.

Kunzi, K., Fisher, A. and Staelin, D. 1976. Snow and ice surfaces measured by the NIMBUS–5 Microwave Spectrometer. *Journal of Geophysical Research* 81, 27: 4,965–80.

Parkinson, C., Comiso, J., Zwally, H., Cavalieri, D., Gloersen, P. and Campbell, W. 1987. Arctic sea ice, 1973–1976: satellite passive-microwave observations. *NASA Report SP–489*. NASA, Washington DC.

Sabatini, R. (ed.). 1972. *NIMBUS–5 Users Guide*. NSSDC, NASA GSFC, Greenbelt, MD 20771, USA.

Zwally, H. 1977. Microwave emission and accumulation rate of polar firn. *Journal of Glaciology* 18, 79: 195–215.

Zwally, H., Comiso, J., Parkinson, C., Campbell, W., Carsey, F. and Gloersen, P. 1983. Antarctic sea ice, 1973–1976: satellite passive-microwave observations. *NASA Report SP–459*. NASA, Washington DC.

7. The Defense Meteorological Satellite Program (DMSP)

Block 5B/C

Operational lifespan and orbital characteristics

Dates: 17 August 1973–14 July 1977.
Semimajor axis: 7,210km.
Orbital height: approx. 811–852km.
Inclination: 98.9°.
Period: 101.6 minutes.
Notes: Formerly called the Data Acquisition and Processing Program (DAPP). USAF operational, military meteorological satellites. Two satellites in operation at any one time. Sun-synchronous orbits.

Sensor characteristics

Sensor AVE Package (SAP):
  i) High resolution (HR) mode band 1: VIS/NrIR (wavelength 0.4–1.1μm). GR 3.7km. Swath width 3,025km (for all bands).
  ii) HR mode band 2: TIR (wavelength 8–13μm). GR 4.4km.
  iii) Very high resolution (VHR) mode band 1: VIS/NrIR (wavelength 0.4–1.1μm). GR 0.62km.
  iv) VHR mode band 2: TIR (wavelength 8–13μm). GR 0.62km.
Vertical Temperature Profile Radiometer (VTPR, or Supplementary Sensor E [SSE]) on block 5B (satellites F2, F3 and F5) and 5C (satellites F1 and F2):
  8 bands in range of 11.0–15.0μm. GR 37km. Swath width 185km.

Block 5D–1

Operational lifespan and orbital parameters

Dates: 11 September 1976 to present.
Semimajor axis: 7,231km.

Orbital height: 813–892km.
Inclination: 98.9°.
Period: 102.0 minutes.
Notes: Sun-synchronous orbits. Two satellites in operation at any one time.

Sensor characteristics

Operational Linescan System (OLS):
  i) Band 1: VIS (wavelength 0.4–1.1μm). GR 0.62km for fine resolution data (2.40km for night-time mode, smoothed and stored data). Swath width 3,012km.
  ii) Band 2: TIR (wavelength 8.0–13.0μm. Note: changed to 10.5–12.6μm from 6 June 1979 with the launch of spacecraft F–4). GR 0.56km for fine resolution data (stored data is smoothed to 2.8km resolution). Swath width 3,012km. Continuous data collection. Rectified, polar stereographic image products have a GR of 5.4km.
Special Sensor Microwave/Temperature (SSM/T), in operation since 1985:
  Scans across-track through nadir. Frequencies 50.5, 53.2, 54.35, 54.9, 58.4, 58.825 and 59.4GHz in the oxygen resonance band. NEΔT 0.3K for bands 1–5, 0.4K for band 6 and 0.5K for band 7. Beam width 14.4°. Footprint at nadir 175–200km. Provides temperatures at 15 levels in the atmosphere to 10mb (about 30km). Scan across-track through nadir to ±36°. Scans in synchronisation with the Special Sensor H.
Special Sensor C (SSC):
  a pushbroom scan radiometer on spacecraft F–4. Wavelength 1.51–1.63μm (IR). The along-track scan was provided by the forward motion of the spacecraft, while the 40.2° cross-track scan was provided by a linear array of 48 detector elements at the image plane of a wide-angle lens.
Special Sensor H (SSH), or VTPR:
  16 bands in the 9.6–30μm range. GR 39km. Swath width 2,200km.

Block 5D–2

Operational lifespan and orbital characteristics

Dates: Spacecraft F–8 launched on 19 June 1987.
Orbital height: approx 833km.
Inclination: 98.8°.
Period: 102 minutes.
Notes: Circular sun-synchronous orbits. Two satellites in orbit at any one time. Coverage up to a maximum latitude of 87°. Ascending node equatorial crossing time 06:12. Satellites are launched 'on demand', ie not until essential instruments begin to show signs of failure – a demand is issued 90 days before the launch of a replacement satellite.

Sensor characteristics

Special Sensor Microwave Imager (SSM/I):
  All channels passive microwave. Conically scanning (Figure II.2).

i) Channel 1: frequency 19.35GHz, wavelength 1.55cm. Antenna beamwidth (AB) 1.98°. Geometric footprint dimension (GFD) 70 × 45km. Specified sensitivity (temperature resolution) 0.8K. Earth incidence angle 53.1°. Data swath width 1,394km for all channels, which covers an active scan of 102.4°, compared to 50° for the Nimbus–7 SMMR. The swath is organised into 64 pixels for the 5 lower frequency channels, and 128 pixels for the 85.5GHz channels.

ii) Channel 2: frequency 22.235GHz, wavelength 1.35cm. Sensitivity 0.8K. AB 1.72°. GFD 60 × 40km.

iii) Channel 3: frequency 37.00GHz, wavelength 0.81cm. Sensitivity 0.6K. AB 1.06°. GFD 38 × 30km.

iv) Channel 4: frequency 85.50GHz, wavelength 0.35cm. Sensitivity 1.1K. AB 0.45°. GFD 16 × 14km. Note: the H channel failed on 26 February 1990.

v) All channels both V and H polarisations, apart from channel 2 (V only). RF bandwidth is 500MHz for channels 1 and 2; 2,000MHz for channels 3 and 4; and 3,000MHz for channel 5. Radiometric accuracy 1.5K for all channels. Dynamic temperature range 375K (all channels). A total system calibration carried out every scan period (1.9s).

The next SSM/I will be launched with a 16:30 ascending node. There will be six more SSM/I launches up to 2000, and 5 SSM/IS (Sounder) launches up to 2010. The SSM/IS has same channels, apart from changing 85.5GHZ to 92GHz and the addition of two sounder channels. All will be on Block 5D–3 satellites.

OLS:

see Block 5D–1. The 5D–2 OLS system has the same performance requirements as the 5D–1 system, although certain improvements have been made, eg an increased IR digitisation (from 7 to 8 bit), and an improved sensitivity to low temperature values (190–210K). The TIR band changed to 10.2–12.8μm. Also included a third band of wavelength 0.5–0.9μm; this Photomultiplier Tube (PMT) is a low light level VIS sensor that measures reflected moonlight. Three digital tape recorders on board; each can store 20 minutes of interleaved VIS and TIR fine resolution (0.6 × 0.6km) data. Analogue filtering and digital averaging are used to smooth data to 2.4 × 2.4km for onboard global storage. Telemetry and special meteorological sensor data are included within the primary smoothed data stream. Real-time encrypted transmission of 0.6km and 2.4km data.

Special Sensor Microwave/Temperature–2 (SSM/T–2):

measures radiation at the water vapour resonance line (183GHz) with ancillary channels at 91.5 and 150GHz. Spatial resolution at the surface 50km.

Infrared Temperature and Profile Sounder (SSH–2) – on spacecraft F8:

upgraded version of SSH on Block 5D–1, similar to HIRS/2 on the NOAA TIROS–N series. 16 bands in range of 3.7–30μm. GR 60km. Swath width 2,204km. NEΔT 2.5–3.0K rms.

Polar applications

SAP:

2 scanning radiometers, mechanically coupled. Similar applications to NOAA VHRR and AVHRR. Mapping of cloud and atmospheric circulation conditions in relation to snow cover and sea ice boundaries. SSTs. Monitoring the aurora borealis.

SSE:

vertical temperature and water vapour profiles in the atmosphere.

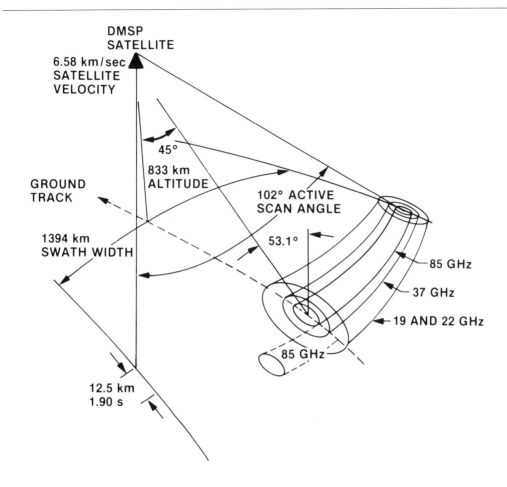

Figure II.2 The scan geometry of the DMSP SSM/I. The rotating antenna sweeps the surface in two alternating modes, one in which all four frequencies are recorded and another in which only 85GHz data are recorded. The use of a single antenna results in different ground resolutions for each frequency. From Untersteiner and others (1984).

SSM/T and SSM/T–2:

   vertical atmospheric temperature profiles. Temperature sounding over previously in-accessible cloudy regions of the globe, and to higher latitudes than was previously possible using IR sensors alone.

SSH and SSH–2:

   humidity, temperature (both sensors) and ozone (SSH only) profiles.

SSC:

   an experimental unit used in conjunction with the VIS band of the OLS. simultaneous in-orbit use of these two sensors proved that snow/cloud scene discrimination can be obtained through the combination of NrIR and VIS data. The success rate in this discrimination during the lifetime of F–4 was estimated at 90%.

OLS:

detects changes in snow and ice extent (boundaries) on regional to global scales (Figure II.3). Cyclogenesis associated with sea ice extent. Fast ice/pack ice boundaries. Detection and delineation of polynyas and large lead patterns (also in winter); leads >300m wide (±50m) detected under cloud-free conditions. VIS band: daytime water-ice boundaries and clouds. TIR band: day/night SSTs, winter ice/water gross boundaries and cloud. First estimates of interannual variations of summer snow melt and associated albedo changes. Iceberg detection, motion vectors and trajectories. PMT mode helpful in detecting clouds, snow and ice at night (Figure II.4). See also NOAA AVHRR.

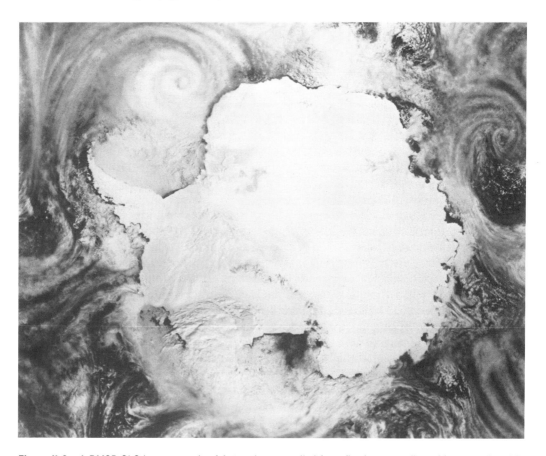

Figure II.3 A DMSP OLS image mosaic of Antarctica, compiled from five images collected between 4 and 25 November 1986, VIS band, 2.7km resolution. Courtesy of NSIDC, Boulder, Colorado.

SSM/I:

i) *Sea ice*: similar to NIMBUS–7 SMMR, but no SST capability, and optimised for the detection and measurement of atmospheric water vapour and sea ice. Improved estimates of ice extent, concentration, MY fraction, ice type and ice edge delineation (Figure II.5). Geolocation accuracy to one third of the spatial resolution. TB differences are greatest at higher frequencies for ice types, but lowest for ice/water contrast. The emissivity of new ice is almost the same as that of FY ice at 37GHz, but lower at

Figure II.4 DMSP PMT low light level imagery over Greenland and Scandinavia, 8 January 1987, at about 23:00 LST, two days after full moon. The image shows reflected moonlight from clouds, snow and ice. The city lights of N Europe are also visible. PMT imagery can be processed in a similar fashion to daytime VIS imagery, and can help distinguish low clouds from a cloud-free surface at night (when VIS data are unavailable). Courtesy of Robert d'Entremont, USAF.

19.35 GHz. For frequencies below about 40 GHz, the emissivity of FY ice is almost constant, the emissivity of MY ice decreases with frequency, and the emissivity of open water increases with frequency. At 85.5 GHz, there is a large difference in the emissivities of new and older sea ice. At this frequency, very strong returns have been recorded from saline frost flowers on the surface of newly refrozen Antarctic lead ice (Comiso and others 1989). Preliminary work suggests that improvements in the retrieval of snow and ice parameters may result from the use of the 85.5 GHz channel data over the entire ice pack. These data, in combination with those of lower frequency channels, appear to have the capability to detect thin ice (newly refrozen leads and polynyas), and may be particularly useful in detecting Antarctic second year/MY ice on its distinctive snow cover signature. The TB at 85.5 GHz decreases as the snow thickness increases. Channel 2 provides atmospheric water vapour and liquid water data to correct for the effects of these on the other channels.

ii) *Land ice*: extent, water equivalent and onset of snow melt for seasonal snow cover (particularly channels 1 and 3). Detection of ice sheet regions where summer melting

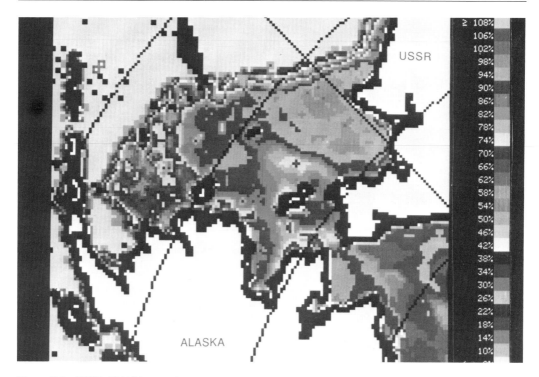

Figure II.5   DMSP SSM/I image of sea ice concentrations in the Bering Sea, 4 April 1988, obtained by applying the Comiso algorithm (Comiso and Sullivan 1986).

occurs (increasing the emissivity). Potential estimate of ice sheet accumulation rates. Channel 4: snowcover mapping.

iii) Wind speed (but not direction) over open ocean to an accuracy of $\pm 5.3 \text{ms}^{-1}$ in the range of $3-25 \text{ms}^{-1}$ at a 25km resolution.

iv) Area and intensity of precipitation over open ocean. Used for correcting GEOSAT ALT data for water vapour effects (Emery and others 1989).

See also Nimbus–5 and –6 ESMR, and Nimbus–7 SMMR (for a more detailed discussion of multi-frequency passive microwave remote sensing).

Limitations

SAP and OLS:

VIS bands limited by cloud and darkness. TIR limited by cloud. Lack of an unambiguous signature for the discrimination between clouds and ice, both in the VIS and TIR. Block 5 B/C HR data difficult to interpret due to gain variations along the scan line. Resolution for both SAP radiometers deteriorated with off-nadir angle. Imagery has not been readily available, and gaps often occur in polar coverage; this is true of the earlier data in particular. Digital data would be more useful. TIR data collected over sea ice are unreliable under summer melt conditions and when the ice temperature is similar to that of the surrounding water. Also, TIR data are not useful in distinguishing thick FY and MY ice. Occasional appearance of sun glint and cloud shadows on the daytime, ascending node filmstrips. See also NOAA AVHRR.

SSE, SSH, SSH–2 and SSC:

not designed to measure surface features. Coarse ground resolution.

SSM/T and SSM/T–2:

no use for surface observations, as the frequencies correspond to the oxygen and water vapour resonance bands.

SSM/I:

possesses no SST capability. Limitations in sea ice retrievals occur at frequencies of ≥40GHz, due to a loss in ice/water contrast caused by heavy cloud cover (especially in summer), and a drop in TB relative to lower frequencies in shear zones with heavy snow accumulation. Pixels containing coastlines may show erroneous values. Although the ground resolution is an improvement over SMMR, the difficulty remains in relating coarse resolution data to *in situ* measurements. Small, unmeasured sectors of 2.4° latitude occur at the North and South Poles. Snow cover variation effects on observed emissivity are poorly understood. No wind speed measurements possible over ice. Summer melt conditions can cause serious problems in the interpretation of microwave data. Tends to underestimate concentration of pancake/thin ice covers.

  i) Channel 1 (19.35GHz): can be used to extract surface information only if atmospheric effects are corrected (ie if used in conjunction with channel 3). Some melt ambiguities remain.

 ii) Channel 2 (22.23GHz) and channel 3 (37GHz): little surface information in summer, except for a wet snow signal. Channel 2 sensitive to atmospheric water vapour and liquid water. Channel 3 sensitive to ocean effects, which can give spurious values around the ice edge. Melt at the snow/ice interface may give areas of spuriously high MY fraction within areas of predominantly FY ice. The presence of a thick or very granular snow cover causes an unusually large variability in the emissivity of each ice type at frequencies of ≥37GHz.

iii) Channel 4 (85.5GHz): sensitive to atmospheric attenuation and snow cover variations, especially refrozen granular snow. Snow cover appears to suppress the TB at this frequency. Discrimination between thick FY ice and new/thin FY ice ambiguous using this frequency alone. Use of a high frequency channel in conjunction with a lower frequency channel is necessary to overcome this ambiguity and compute multitype ice concentrations.

See also Nimbus–5 and –6 ESMR, and Nimbus–7 SMMR (for a more detailed discussion of multi-frequency passive microwave remote sensing).

Data format

SAP and OLS:

few digital data currently available (see next section). Data consist of 3 positive transparency products produced by USAF and Scripps Institution of Oceanography at various print sizes (4 × 5, 8 × 10, 11 × 14 and 16 × 20 inch), 35mm slides and negatives, and 8 × 10 inch transparancies:

  i) limited coverage at a GR of 0.6km for areas around direct readout receiving stations;

 ii) global coverage on single orbit strips, GR 2.7km;

iii) global mosaics compiled from several orbits with latitudes and longitudes, GR 5.4km. Mosaics are the only pre-gridded product available. Available since December 1975 (the other products have been available since 1973).

The dynamic range (16 grey shades) can be subjected to a number of enhancement techniques. All DMSP filmstrips are treated to a cylindrical rectification to reduce image foreshortening. The map format, although not as convenient as that of Landsat, is simpler to use than that of the NOAA satellite data. Can be geometrically corrected and displayed on standard map projections, with coastlines added. Note: early archived data may have been processed in any one of the optional formats.

SSM/I:

various CCT and optical disc formats. The NSIDC archive (see below) contains:

i) Calibrated Brightness Temperatures:
    a) global swath-oriented SDRs for all 7 channels (accuracy 0.10K);
    b) daily gridded averages at 25km GR for the 5 lower frequency channels (304 × 448 cells for N hemisphere, 316 × 332 for S hemisphere), and at 12.5km GR for the 85.5GHz channels (608 × 896 cells for N, 632 × 664 for S). Both polar regions. The SSM/I grids have a standard latitude at 70°, chosen to minimise geometric distortion in the MIZs. Population counts (ie the number of swath scene stations averaged into the individual cells per day) are not included.
    c) 100km Browse Imagery (6-day metafile images) for the 19.35GHz channels, both polar regions;
    d) daily Monitor Areas Histograms, all 7 channels, covering the Antarctic. To monitor long-term sensor performance.

ii) Sea Ice Concentration products generated using the NASA GSFC Team Algorithm described by Cavalieri and others (1984) (NSIDC 1990):
    a) 3-day gridded averaged estimates of total ice concentration, FY ice concentration and MY fraction on a 50km grid, both polar regions;
    b) daily ice edge maps (>15% concentration), in the form of sea ice distribution contours, on a 12.5km grid, both polar regions. These data are not stored on CD–ROM. Instead, software is provided to enable the user to compute the ice edge for any specified percentage of ice concentration from the 85GHz data;
    c) 100km Browse Images (6-day averages), giving total ice concentration, FY ice concentration and MY ice concentration, both polar regions. Allow rapid visual scanning of the SSM/I gridded products. Stored as rasterised arrays which can be quickly converted to display code for the Tektronix 4107 format.
    d) research is continuing to improve ice concentration retrievals in problem areas, eg thin ice, melt conditions, etc. Such improvements will subsequently be integrated into the CDMS.

iii) An alternative algorithm is offered by NSIDC, the cluster analysis technique of Comiso (1985), to extract ice concentrations and MY fraction from the TB data.

iv) The CD-ROMs are compatible with the MS-DOS operating system. The data are stored in 2-byte integer format. To support compatibility with the VMS operating system, extended attribute records (XABs) are supplied on the discs (these are transparent in the MS-DOS environment). Software for the VMS and Macintosh operating systems has been developed; IMAGIC allows the correct interpretation of the data values in the SSM/I (and SMMR) CD-ROM grids on Macintosh computers. This software is available from JPL/NODS, Mail Stop 300/323, 4800 Oak Grove Drive, Pasadena, CA 91109, USA, telephone (818)–354–6980. Hardware requirements are a 286 processor, 640 kilobytes of memory, a hard disc (at least 20 megabytes), a CD-ROM drive and file software manager software supporting the ISO 9660 standard, and a monitor with an EGA or VGA graphics adaptor. A math(s) coprocessor is recommended for faster data extraction. See Nimbus–7 SMMR for further details of hardware requirements.

SSH and SSH–2:

6250bpi, 9-track binary CCT for total calibrated radiance and total ozone data.

SSM/T, SSM/T–2, SSC and SSE:

CCT, various formats.

Data availability and archiving dates

SAP and OLS:

OLS VIS and IR hardcopy imagery are available, covering the period from 1973 onwards. Complete basic data available from late 1978. A dataset exists from 1973 with daily global IR and VIS coverage at a spatial resolution of 150–250km (only a few CCTs available). Data prepared from a USAF global digital intensity file are archived in hardcopy form at National Snow and Ice Data Center (NSIDC), WDC–A for Glaciology, Campus Box 449, University of Colorado, Boulder, CO 80309, USA, telephone: (303)–492–1834, telex: 257673 (WDCA UR), telemail: MAIL/USA [NSIDC/OMNET], VAX mail via SPAN: KRYOS::NSIDC, NSIDC has served as the main archive of DMSP data since March 1983.

NSIDC can offer custom searches, custom photographic products and on-site use. DMSP imagery is archived only after operational military use (usually 45–60 days), and is therefore only useful for research purposes. Note: no archive of digital data exists, only hard-copy photographic products. Negotiations are underway with the US Department of Defense to allow the archiving of digital data at NSIDC (possibly from mid-1991 onwards). The data, which will be subdivided into auroral, polar and meteorological data, will be available retrospectively two or three weeks after collection. Routine products will include HDDTs and CCTs, and a digital browse file; hard-copy images will be prepared upon request.

Imagery containing aurora borealis available from NSSDC (for details of address, see TIROS I–X). See Ng and Stonesifer (1989).

The Antarctic Research Center at Scripps Institution of Oceanography (SIO) receives digital DMSP (and NOAA AVHRR) data from McMurdo and Palmer Stations (the latter since January 1990) in Antarctica. Several high resolution (0.55km) images are collected each day and night, year round, and covering 85% of Antarctica (up to 90°). The system can differentiate between clouds and ice in two ways: visually, by using the low sun angle; or, by using the IR bands, on the basis of temperature, as low clouds are generally warmer than the ice surface. Time-sequencing of images can be performed. Previously at McMurdo, data were often only retained for two–seven days. Scientists wishing to obtain data or work interactively with the system should contact Robert Whritner, Antarctic Research Center, Ocean Research Division, A–014, Scripps Institution of Oceanography, La Jolla, CA 92093, USA, telephone (619)–534–3785.

Note: gaps often occur in the coverage of the higher latitudes. VIS and TIR data are freely available to scientists with suitable receiving facilities for regions south of 60° latitude; the finer resolution data have only recently been de-classified. All other data, including those collected over the Arctic, remain encrypted, and cannot therefore be collected in real-time outside the military community, although hardcopy data are available from NSIDC (see above).

SSH, SSH–2, SSC, SSE and SSM/T:

data archived at Satellite Data Services Division (SDSD), NOAA/NESDIS, Room 100, Princeton Executive Center, 5627 Allentown Road, Camp Springs, MD 20746, USA, telephone (301)–763–8399, telex 248376 OBSWUR, FAX 7638443. SSM/T level IB data (from 18 August 1987 to present) are also available from the National Climatic Data Center (NCDC), Federal Building, Asheville, NC 28801-2696, USA, telephone (704)–259–0682, telex 6502643731, FAX 704–259–0876. Some SSH total ozone and calibrated radiance data archived at NSSDC. See Ng and Stonesifer (1989).

SSM/I:

i)   NSIDC (see above for address) receives the data on CCTs from Fleet Numerical Oceanography Center and Air Force Global Weather Central through NOAA's SDSD, and corrects the swath antenna temperature data for antenna pattern effects. They are then reformatted into a rapid access archive stored on optical disc. The corrected Sensor Data Record (SDR) TB data are binned into polar grids, both N and S hemispheres. NSIDC has been funded by NASA to develop a computer-based Cryospheric Data Management System (CDMS) at NSIDC to extract polar SSM/I data and make it available to non-operational users within two or three weeks of collection. CDMS is the first operational remote node on the NASA Ocean Data System (NODS), and is an enhanced version of the JPL Pilot Ocean Data System (PODS).

The CDMS archive maintains a data catalogue and inventory, browse image collection, and TB monitor areas. The software utilises a menu-driven user interface. The user can first query the data catalogue to determine what data sets are available within selected temporal and geographical limits. The NODS software automatically extracts data selected by the user. Data may be displayed or ordered interactively. Note: there are practical limits to on-line data extraction from the archive, and it is not possible to extract and plot the SSM/I locations for an entire day for the whole world. There are fewer restrictions on the amount of data that can be ordered on CCT/CD–ROM – the SSM/I produces about 76Mbytes of swath data per day. The archival system is evolving as requirements and technology change.

Users can access CDMS via commercial long-distance dial-up services, NASA's Space Physics Analysis Network (SPAN), or the TELENET packet-switched network (via SPAN). Users may access the CDMS via telephone through two principal routes:

a)   a 1200 baud modem directly to the CDMS VAX–11–750. Dial (303) 492–2364; or

b)   a 1200 or 2400 baud modem to the University of Colorado's AT&T ISN. Dial (303) 492–1900 to reach the ISN modem pool. Enter ACS/KRYOS at the 'DIAL:' prompt. The most direct way to access the VAX–11/750 is to perform a remote logon from another VAX computer via the DECnet command, $SET HOST KRYOS. Remote users may also access CDMS through the SPAN network with 300 and 1200 baud modems. These users are required to dial a TELENET PAD number to access the NASA Data Access Facility (DAF) security computer. Once connected, access to KRYOS is transparent. Users requiring this type of access should contact: Valerie Thomas, Code 633.0, NASA Goddard Space Flight Center, Greenbelt, MD 20771, USA, telephone (301)–286–4740.

Data can be ordered and tabular output obtained by using almost any terminal (eg VT100, PC with communications package, etc). To receive plots at the terminal requires a VT100 with Retrographics VT640, HP2647a, Tektronix 4014, Ramtek 6211, Tektronix 4107 or VT125. A Tektronix 4107 terminal is required to view the gridded image data loaded into the system, including the browse images. The investigator can access the data via an inexpensive workstation ('286, '386 AT, Macintosh–II,

VAXstation, SUN, etc). For further practical information, contact the CDMS Manager at NSIDC. *CDMS Notes* newsletter is published free of charge on a regular basis by NSIDC.

Data from 27 June–8 July 1987 were not released for processing by the US Navy. Moreover, the sensor was turned off 3 December 1987–12 January 1988 to protect the instrument components from overheating due to solar radiation. The solar array panel could not be positioned to intercept the sun's rays during the winter solstice (20 December) and to shield the sensor. SSM/I temperatures were observed to be sharply increasing since 18 November 1987. The gridding of SSM/I TBs has revealed geolocation problems. CDMS is implementing a scheme developed by the University of Massachussetts/University of Colorado to minimise this error. Routine checks in Earth location and TB statistics are carried out.

ii) Ice products are also generated for the US Fleet Numerical Oceanography Center (FNOC), using algorithms developed by the Naval Research Laboratory (Hollinger and others 1987) and implemented by Hughes Aircraft Corporation. These products, which are used operationally by the US Navy Polar Oceanography Center in Suitland, MD, are available through the SDSD (from early 1988); see TIROS I–X for address. They are unsuitable for detailed ice analyses.

iii) The following 'open ocean' SSM/I data are available (on CCTs in binary integer format) from the JPL/NODS archive:

a) semi-monthly, 25×25km cells of wind speed over open ocean;

b) semi-monthly, 25×25km cells of vertically integrated water vapour content of the atmosphere over open oceans;

c) semi-monthly, 25×25km cells of cloud/rain/vertically integrated liquid water content of the atmosphere over open ocean. Global data July 1987–December 1988. Users are provided by NODS with a FORTRAN subroutine to unpack a logical record and place the properly scaled parameters into a common area;

d) $1° \times 1°$ gridded cells of vertically integrated atmospheric water vapour. This data set covers 15 July–16 August 1987. It is global in coverage, and is contained on one CCT. For further details, see Emery and others (1989).

JPL/NODS, M/S 300–323, Jet Propulsion Laboratory, California Institute of Technology, Pasadena, CA 91109, USA, telephone (819)–354–6980 (Elizabeth Smith), telex 675429 (attention NODS), FAX (818)–393–6720 (attention NODS), STANS::EAS on SPAN, and NODS.JPL on OMNET.

iv) Level Ib, Temperature Data Records and Sensor Data Records are available from NCDC (see above for address).

References and important further reading

Barry, R. and Miles, M. 1988. Lead patterns in Arctic sea ice from remote sensing data: characteristics, controls and atmospheric interactions. *Preprints of the Second Conference on Polar Meteorology and Oceanography, 29–31 March 1988*: 40–43. American Meteorological Society, Boston, Massachussetts.

Cavalieri, D., Gloersen, P. and Campbell, W. 1984. Determination of sea ice parameters with the NIMBUS-7 SMMR. *Journal of Geophysical Research* 89, D4: 5,355–69.

Comiso, J. 1985. Remote sensing of sea ice using multispectral microwave satellite data. In Deepak, A., Fleming, H. and Chahine, M. (eds), *Advances in remote sensing methods*: 349–70. A. Deepak Publishing, Hampton, Virginia.

Comiso, J., Grenfell, T., Bell, D., Lange, M. and Ackley, S. 1989. Passive microwave *in situ* observations of winter Weddell Sea ice. *Journal of Geophysical Research* 94: 10,891–905.

d'Entremont, R., Bunting, J., Felde, G. and Thomason, L. 1987. Improved cloud analysis using visible, near-infrared, infrared and microwave imagery. *Proceedings of IGARSS '87*, University of Michigan, Ann Arbor, 18–21 May 1987, Volume 1: 51–6.

Dickinson, L., Boselly, S. and Burgmann, W. 1987. Defense Meteorological Satellite Program (DMSP) user's guide. *Report AWS TR–74–250*. Air Weather Service, Scott Air Force Base, Illinois.

Emery, W., Born, G., Baldwin, D. and Norris, C. 1989. Satellite derived water vapor corrections for GEOSAT altimetry. *Journal of Geophysical Research* 95, C3: 2,953–64.

Gloersen, P. and Hubanks, P. 1989. Scalar winds from SSM/I in the Norwegian and Greenland Sea during NORSEX. *Proceedings of IGARSS '89*, Vancouver, Canada, 10–14 July 1989: 1,090–4.

Grody, N., Gray, D., Novak, C., Prasad, J., Piepgrass, M. and Dean, C. 1985. Temperature soundings from the DMSP microwave sounder. In Deepak, A., Fleming, H. and Chahine, M. (eds), *Advances in remote sensing retrieval methods*: 249–68. A. Deepak Publishing.

Hollinger, J., Lo, R., Poe, G., Savage, R. and Pierce, J. 1987. *Special Sensor Microwave/Imager User's Guide*. Space Sensing Branch, Naval Research Laboratory, Washington DC.

Meyer, W. 1985. The Defense Meteorological Satellite Program: a review of its impact. In Schnapf, A. (ed.), *Monitoring Earth's ocean, land, and atmosphere from space – sensors, systems, and applications*. New York, American Institute of Aeronautics and Astronautics: 129–49.

Mizera, N., Gorney, D. and Roeder, J. 1984. Auroral X-ray images from DMSP F–6. *Geophysical Research Letters*, 2: 255–8.

Ng, C. and Stonesifer, G. 1989. Data catalog series for space science and applications flight missions. Volume 4B: descriptions of data sets from meteorological and terrestrial applications spacecraft and investigations. *NSSDC/WDC–A–R&S 89–10*. Available from NSSDC.

NSIDC. 1990. *DMSP SSM/I Brightness temperature grids for the polar regions on CD-ROM, User's Guide*. NSIDC, Colorado.

Steffen, K., Barry, R. and Schweiger, A. 1989. DMSP SSM/I NASA algorithm validation using primarily LANDSAT and secondarily DMSP and/or AVHRR visible and thermal infrared satellite imagery. *NASA Report NASA–CR–182979*. NASA, Washington DC.

Svendsen, E., Mätzler, C. and Grenfell, T. 1987. A model for retrieving total sea ice concentration from a spaceborne dual-polarized passive microwave instrument operating near 90GHz. *International Journal of Remote Sensing* 8, 10: 1,479–87.

Untersteiner, N. (ed.). 1984. *Passive microwave remote sensing for sea ice research*. NASA, Washington DC.

8. Skylab

Operational lifespan and orbital characteristics

Dates: 14 January–23 January 1974 (The Earth Resources Experiment Programme [EREP]).
Semimajor axis: 6,808km.
Orbital height: 422–437km.
Inclination: 50.04°.
Period: 93.11 minutes.

Sensor characteristics

EREP consists of RADSCAT S–193 (a combined active/passive sensor):

i) A passive microwave radiometer and a radar scatterometer (SCATT). Frequency 13.9GHz. RF bandwidth 200MHz. Dynamic range 50–350K. Scatterometer incidence angle 15–50°. GR 10–15km. Radiometer swath width 11–170km; SCATT swath width 800km.

ii) Radar Altimeter (ALT): frequency 13.9GHz. RF bandwidth 100MHz. Antenna beamwidth 1.5°. Pulsewidth (uncompressed) 130ns. Had an experimental pulse compression mode. PRF 250Hz. ALT precision accuracy ±1–2m over open ocean.

iii) Hand-held camera shots were taken of the Gulf of St Lawrence whenever cloud conditions permitted. Multispectral imagery were collected.

Polar applications

Sea ice studies were limited to hand-held photography of the Gulf of St Lawrence, James Bay and the Sea of Okhotsk. The main contribution of SKYLAB was as a testbed for future polar applications sensors. It provided the first detailed maps of sea surface wind fields, and gave encouraging profiles of the geoid to an accuracy of ±5m.

Limitations

RADSCAT was an experimental sensor, with a limited spatial and temporal coverage (up to a maximum latitude of 50°). During the experiment, the SCATT was set to record with the transmitter switched off; the radiometer data were collected with a damaged antenna, the effect of which was to increase the effective bandwidth and contaminate the data. Few proper surface experiments took place to verify the data. The only remote sensing experiment in conjunction with the Skylab–4 overpass related to ice was conducted in the Gulf of St Lawrence and Atlantic Ocean, January–February 1974. Each photograph was taken at a different and unspecified angle relative to nadir. Data collection was constrained by the amount of film/CCT that the astronauts could carry. Radar altimeter: tracking data were too coarse to enable the detection of any dynamic signals of the ocean surface.

Data format

Standard photographic products. Digital waveform data only for ALT. The active and passive microwave data collected from airborne sensors in association with the Skylab overpass are described by Dunbar and Weeks (1975) and Campbell and others (1975).

Data availability and archiving dates

Photographs from Skylab (and other manned space flights) are archived at and available from: Customer Services, EROS Data Centre (EDC), Sioux Falls, SD 57198, USA, telephone (605)–594–6950 or 1–800–367–2801, TWX 9106680310. Other data are available from the Technical Applications Center, University of New Mexico, Albuqurque, NM 87131, USA, telephone (505)–277–3662.

References and important further reading

Campbell, W., Ramseier, R., Weaver, R., and Weeks, W. 1975. SKYLAB Floating Ice Experiment. *Final Report E75–10161, NASA–CR–147446*, NASA, Space and Applications Directorate, Lyndon B. Johnson Space Center, Houston, TX 77058, USA.

Dunbar, M. and Weeks, W. 1975. Interpretation of young ice forms in the Gulf of St. Lawrence using side-looking airborne radar and infrared imagery. *Cold Regions Research and Engineering Laboratory Report 337*.

McGoogan, J. 1975. Satellite altimeter applications. *IEEE Transactions on Microwave Theory and Technology* MTT–23: 970.

NASA. 1977. Skylab explores the Earth. *NASA Special Report SP–380*. NASA, Washington DC.

NASA. 1978. SKYLAB EREP investigations summary. *NASA Special Report SP–399*. NASA, Washington DC.

9. Geodetic Earth Observation Satellite-3 (GEOS-3)

Operational lifespan and orbital characteristics

Dates: 10 April 1974–1 December 1978.
Semimajor axis: 7,224km.
Orbital height: 839–853km.
Inclination: 115.0°.
Period: 101.82 minutes.
Notes: circular orbit.

Sensor characteristics

Radar Altimeter (ALT):
Active microwave: frequency 13.9 GHz (K_u-band). RF bandwidth 80MHz. Antenna beamwidth 2.6°. Pulse width (uncompressed) 1μs, 12.5ns (compressed). PRF 100Hz. Precision <50cm rms. Significant wave height (SWH) accuracy ±10% (range of 1–20m). Nominal pulse limited footprint diameter: 3.8km in high intensity mode, 14.3km in global mode.

Polar applications

GEOS–3 provided the first accurate ice sheet surface elevation measurements (over S Greenland), to an estimated overall accuracy of about ±50cm (over the smoother regions of the central ice sheet). Potential accurate delineation of the sea ice edge regardless of season, weather conditions and ice surface wetness. Ice surface roughness characteristics. Oceanic conditions adjacent to ice edge, including SWH and wind speed (but not direction). Accurate delineation of ice sheet and shelf margins. See also Skylab, Seasat, ERS–1, GEOSAT and TOPEX/Poseidon.

Limitations

Coverage limited to a maximum latitude of 65°, and by a narrow footprint at nadir. Principal error sources for GEOS–3 data include attitude error/orbit ephemeris and unmeasured geophysical effects (eg wet and dry tropospheric effects, ionospheric electron content). The orbital characteristics of GEOS–3 led to a variable repeat pattern, whereby its ground coverage was not easy to predict, and data could not be acquired exactly where required. High SNR and tracking errors often occurred over rough or undulating ice surfaces (the ALT was designed to operate over open ocean). Attempts to relate profiles at the Bering Sea ice edge to the measurement of sea ice freeboard provided spuriously high values; these are thought to result from the combined effects of freeboard plus an ALT bias induced by the specular nature of the reflecting ice surface as compared with the open water return (Stanley and others 1979). Surface ocean current measurements are limited to large-scale features. See also Skylab, Seasat, ERS–1, GEOSAT and TOPEX/Poseidon. GEOS-3 data are not sufficiently accurate to warrant reprocessing.

Data format

i) Hardcopy format: a computer printout listing a computed ice index in terms of latitude and longitude.
ii) CCTs:
   a) I–Tapes: contain raw data, sample rate 10 per second.
   b) G–Tape: contains only products, ie smoothed sea surface height, once per 1.024s, and all derived oceanographic parameters.

Data availability and archiving dates

Data from 14 April 1975–1 December 1978 (the G– and I–Tapes) are archived at and available from:
i)   Satellite Data Services Division (SDSD), NOAA/NESDIS, Princeton Executive Center, 5627 Allentown Road, Camp Springs, MD 20746, USA, telephone (301)–763–8399, telex 248376 OBSWUR, FAX 7638443 (the G–Tapes and I–Tapes).
ii) NSSDC. for full address, see TIROS I–X. See Agreen (1982) for further information.
iii) National Climatic Data Center, Federal Building, Asheville, NC 28801–2696, USA, telephone (704)–259–0682, telex 6502643731, FAX 704–259–0876.

Greenland Ice Sheet data processed by the Oceans and Ice Branch of NASA GSFC are available from National Snow and Ice Data Center (NSIDC), WDC-A for Glaciology, Campus Box 449, University of Colorado, Boulder, CO 80309, USA, telephone 303–492–5171, telex 257673 (WDCA UR), telemail MAIL/USA [NSIDC/OMNET], VAX mail via SPAN:KRYOS::NSIDC.

The geophysical data records, and derived global surface wind speed data, are available from JPL/NODS, Mail Stop 300–323, Jet Propulsion Laboratory, California Institute of Technology, Pasadena, CA 91109, USA, telephone (819)–354–6980 (Elizabeth Smith), telex 675429 (attention NODS), and FAX (819)–393–6720 (attention NODS). The NODS data system contains, in machine readable form, not only the ALT data but also extensive sets of surface observations used to calibrate the instruments, and a bibliography of descriptions of the satellite, the instruments, the calibrations and the application of the satellite data to oceanography.

Note: the ALT performed very well from 21 April 1975 to July 1979. Operation terminated on 1 June 1979 as a result of the degradation of the high intensity mode in late 1978. From late 1978–July 1979, it operated primarily in global mode.

References and important further reading

Agreen, R. 1982. The 3.5 year GEOS–3 data set. *NOAA Technical Memo, NOS NGS 33.* NGS, Rockville, Maryland, USA. Also available from JPL/NODS.

Brooks, R., Campbell, W., Ramseier, R., Stanley, H. and Zwally, H. 1978a. Ice sheet topography by satellite altimetry. *Nature* 274, 5,671: 539–43.

Brooks, R., Roy, N. and Stanley, H. 1978b. Sea ice boundary determination from satellite radar altimetry. *EOS Transactions of the AGU* 260.

Dwyer, R. and Godin, R. 1980. Determining sea ice boundaries and ice roughness using GEOS–3 altimeter data. *NASA Contractor Report 156862.* National Technical Information Service, Springfield, Virginia.

Stanley, H., Brooks, R. and Brown, G. 1979. Ice freeboard determination by satellite altimetry. *Proceedings of the International Workshop on Remote Estimation of Sea Ice Thickness.* St Johns, Newfoundland, 25–6 September 1977: 295–304.

10. Synchronous Meteorological (SMS)/ Geostationary Operational Environmental Satellite (GOES) series

Operational lifespan and orbital characteristics

Dates: 17 May 1974 to present.
Semimajor axis: 42,048–47,257km.
Orbital height: 33,367–48,390km.
Inclination: 1.9–0.2°.
Period: 1,430–1,436 minutes (geostationary).
Longitude: Normally 75°W (GOES–East) and 135°W (GOES–West). Only one satellite operational 29 July 1984–26 February 1987, at a longitude of 98°W.
Notes: Operational meteorological satellites. Coverage up to a maximum latitude of 55–60°. Repeat coverage twice per hour.

Sensor characteristics

Visible Infrared Spin Scan Radiometer (VISSR) – on SMS 1 and 2, GOES 1–3 (1975–1978):
    i)   Band 1: VIS (wavelength 0.55–0.70μm). GR 14km. Swath width horizon to horizon.
    ii) Band 2: TIR (wavelength 10.5–12.6μm). GR 8km. Temperature sensitivity 0.5°C.
VISSR Atmospheric Sounder (VAS) – on GOES 4 (September 1980) onwards:
    a spin-scan imaging device designed for day and night two-dimensional cloud cover imaging.
    i)   Band 1: VIS (wavelength 0.55–0.70μm). GR 14km. Swath width horizon to horizon.
    ii) Band 2: TIR (12 bands in the range of 4.496–14.81μm). GR 16km.
Improved VAS – on GOES 'Next' series satellites (1992 onwards):
    similar to VAS, but will be a stable (not spin) axis sensor. Will add an IR band (3.7μm) and a modified TIR band, 4km resolution.
Passive Microwave Radiometer (PMR) – on Goes 'Next' series:
    5 channels: 92, 118, 150, 183 and 230GHz. GR 35–15km. 500 × 500km area scanned in 10 minutes.
Data Collection System (DCS):
    frequency range 136.77 and 137.77MHz. Surface platforms may either transmit on a timed basis or be commanded to transmit the data to the satellite. 261 radio channels available.
WEFAX:
    a communications service transmitting meteorological data from GOES and NOAA satellites to users, ie forecasting services.

Polar applications

VISSR and VAS:
    designed for day and night two-dimensional cloud cover imaging. The complex array of 12 TIR detectors in the VAS provide data from which it is possible to infer the three-dimensional structure of atmospheric temperature and water vapour distribution. Occasionally, sea ice can be seen at the southern/northern extremes of the image, but this product is not a primary source of sea ice data. Monitoring of air mass movements to and from polar regions. Band 2 useful for night-time operation. Operational, time-lapse snow cover mapping in the USA. SST products from the TIR data are useful for ocean front analysis, and cover a scene 50° in latitude and longitude around the subpoint of each satellite. Earth radiation budget.
DCS:
    has been used by NOAA scientists to collect meteorological data from Bering Sea.
PMR:
    temperature and pressure soundings in presence of clouds, and precipitation estimation.

Limitations

VISSR and VAS:
    Band 1 limited by cloud/fog/darkness. Band 2 limited by cloud/fog. Relatively low spatial resolution (especially for the TIR band), becoming even poorer at high latitudes. Poor monitoring of air masses within polar regions. Although geostationary satellites can 'see' to a maximum latitude of about 81°, image foreshortening and the thickness of atmospheric slant become severe after about 60°.

DCS:

poor coverage of the high latitudes (above approximately 75°). Cannot determine location of drifting buoys (this technique relies on Doppler shift, ie the motion of the satellite relative to the buoy).

PMR:

not suitable for surface measurements.

Data format

VISSR and VAS:

  i) Hardcopy or digital formats: 8 × 8km or 4 × 4km data in TIR, 2, 4 or 8km resolution VIS data.
 ii) Photographic products:
    a) Full disc and sector (25 × 25cm negatives), VIS and TIR, every half hour; or
    b) Full disc VIS and TIR microfilm (35mm), 15 days/reel (recorded every hour).
iii) Digital CCTs: 800, 1600/6250bpi. The data are archived in the form of calibrated counts which can be converted to geophysical units by means of a fixed look-up table, and are supplied in digital arrays with navigation information appended and a given centre latitude and longitude. Software is supplied to use the navigation data to calculate the latitude and longitude of any pixel in the array.

Data availability and archiving dates

Data from 27 June 1974 to the present are archived at SDSD, NOAA/NESDIS, Princeton Executive Center, 5627 Allentown Road, Camp Springs, MD 20746, telephone (301)–763–8399, telex 248376 OBSWUR, FAX 7638443. Only selected reduced resolution data are retained on CCT. Full resolution digital data are archived (since 1978) at, and reproduced for SDSD by, the Space Sciences Engineering Center, University of Wisconsin, 1225 West Dayton Street, Madison, WI 53706, telephone (608)–262–3762. Products from NESS Control Centre, Suitland, MD, which receives the data via DOMSAT, are turned over to SDSD within one day. Some historical data are also available from NSSDC (for full address, see TIROS I–X). Various data and data products (including Earth imagery from 10 March 1975 to present, cloud motion vectors from 1 October 1974 to present, and International Satellite Cloud Climatology Project data from 1 July 1983 to present) are also available from the National Climatic Data Center (NCDC), Federal Building, Asheville, NC 28801–2696, USA, telephone (704)–259–0682, telex 6502643731, FAX 704–259–0876. For further information, see WMO (1989).

References and important further reading

Gibson, J. (ed.). 1984. *GOES Data User's Guide*. NOAA/NESDIS, Washington DC.

Greaves, J. and Shenk, W. 1985. The development of the geosynchronous weather satellite system. In Schnapf, A. (ed.), *Monitoring Earth's ocean, land, and atmosphere from space – sensors, systems, and applications*. American Institute of Aeronautics and Astronautics, New York: 150–81.

Kerut, E. and Haas, G. 1979. Geostationary and orbiting satellites applied to remote ocean buoy data acquisition. *Proceedings of the 13th International Sumposium on Remote Sensing of Environment, 23–7 April, 1979*, Volume 1: 519–34. ERIM, Ann Arbor, Michigan.

Minnis, P. and Wielicki, B. 1989. Comparison of cloud amounts derived using GOES and LANDSAT data. *Journal of Geophysical Research* 93, D8: 9,385–9,403.

Ng, C. and Stonesifer, G. 1989. Data catalog series for space science and applications flight missions. Volume 4B: descriptions of data sets from meteorological and terrestrial applications spacecraft and investigations. *NSSDC/WDC–A–R&S 89–10*. Available from NSSDC.

WMO. 1989. Information on meteorological and other environmental satellites. *WMO–411*, World Meteorological Organization, Geneva.

11. Nimbus 6

Operational lifespan and orbital characteristics

Dates: 12 June 1975–April 1977.
Semimajor axis: 7,476km.
Orbital height: 1,092–1,104km.
Inclination: 100.0°.
Period: 107.30 minutes.
Notes: Sun-synchronous orbit.

Sensor characteristics

THIR:
  i) Band 1: IR (wavelength 6.5–7.0µm). GR 22.6km. Swath width 3,000km.
  ii) Band 2: TIR (wavelength 10.5–12.5µm). GR 7.7km. Swath width 3,000km.

Scanning Microwave Spectrometer (SCAMS):
  Passive microwave: 5 bands at frequencies 22.235, 31.65, 52.85, 53.85 and 55.45GHz (wavelengths 1.35, 0.95, 0.58, 0.56 and 0.54cm). RF bandwidth 220MHz (all channels). Integration time 0.95s (all channels). NEΔT 0.2–0.6K rms. Absolute radiometric accuracy 1.5K. GR 145km at nadir, 330km at the scan limit (all channels). Swath width 2,618km (all channels).

ESMR:
  Passive microwave: 2 channels, centre frequency 37GHz, wavelength 0.81cm. H and V polarisation. RF bandwidth 250MHz. Absolute accuracy 2K. GR 20km. Swath width 1,270km. Data rate 300bps. Incidence angle from 50° to the left to 50° to the right of the subsatellite track.

Random Access Measurement System (RAMS):
  data collection and relay system, similar to that flown on earlier satellites in the series. Carrier frequency 401.2MHz. Number of platforms (capacity): 1,000 global, 200 in view. Satellite period 108 minutes. Data capacity 8–8 bit words. Modulation PSK. Modulation Index ±60°. Modulation rate 100Hz. Bit rate 100bps. Frame length 1s. Transmitter repetition rate 1 min. Short term stability 1:10^9 (100µs). Medium term drift 0.28Hz min^{-1} (15min). Transmit power 2.4 watts. Spurious emission with ±15kHz: –50dbc; beyond

±15kHz: −20dbc. Position-fix accuracy approximately 5km rms. The NOAA Data Buoy Office, NSTL Station, MS 39529, USA, developed an air-droppable RAMS (ADRAMS) for deployment from aircraft in remote areas inaccessible by ship.

Other sensors carried include an Earth Radiation Budget (ERB) sensor (12 bands in range of 0.2–50.0µm), a High Resolution Infrared Spectrometer (HIRS) (17 bands in the range of 0.69–17µm), a Pressure Modulated Radiometer (PMR) and a Limb Radiance and Inversion Radiometer (LRIR) (4 bands in the range of 9.6–37.0µm).

Polar applications

THIR:
  as for Nimbus–5.
SCAMS:
  potential large-scale distinction of new and old sea ice, and ice from open water. Over ice sheets, the data have been used to identify areas of dry firn by volume scattering characteristics (Rotman and others 1981). Possible inference of accumulation rates. Snow cover mapping/onset of melt. Applications similar to the Nimbus–5 NEMS. Observations at only three oxygen-band frequencies (52.85, 53.85 and 55.45GHz) were sufficient to retrieve temperature profiles of the atmosphere for the range of 100–1000mbar.
ESMR:
  as for Nimbus–5, except that the change in frequency from 19.35 to 37GHz effectively doubled the contrast between first year (FY) and multi-year (MY) sea ice. It also tripled the sensitivity to atmospheric water droplets while maintaining its sensitivity to water vapour over open ocean. Rainfall rate over open ocean. The quality of data deteriorated after 1976, but was still used for operational ice forecasting by US Navy/NOAA Joint Ice Center until 1983. See also NIMBUS–5 ESMR.
RAMS:
  several commercial oil drilling companies used RAMS tracked buoys in the Arctic to measure currents bearing icebergs. The US Coastguard used drifting buoys to measure ocean surface current velocities in the Labrador current – the results could be used to model and predict iceberg drift behaviour (Cote and others 1982). The 1977 and 1978 International Ice Patrols used ADRAMS buoys deployed from C–130 aircraft onto icebergs calving from Greenland. Used to track ice-strengthened buoys in the E Arctic (Vinje 1978). ADRAMS buoys were deployed in 1978 to determine the forces impacting offshore structures close to the north coast of Alaska. The Alaska Department of Fish and Game tracked polar bears. CNES and NASA conducted an experiment to track drifting icebergs in the Antarctic – successful tracking took place of three icebergs for roughly one year over a range of 2,500km with a locational accuracy of 4–7km. Synoptic RAMS (SYNRAMS) buoys were used during AIDJEX in the early 1970s. The SYNRAMS instrument consisted of a hydrophone, a thermistor, and a barometer; positional fixes to an accuracy of 2–5km in radial position also transmitted.

  ADRAMS buoys were deployed on sea ice in the Weddell Sea in 1979–80 by Dr S. Ackley of the US Army Cold Regions Research and Engineering Laboratory (CRREL). They transmitted temperature, barometric pressure and positional data, thereby enabling an improved determination of the constitutive relationship governing the movement and deformation of sea ice in the region. Allow the study of drift/strain fields in relation to atmospheric forcing. Buoy data are essential in filling in the huge gaps in meteorological data collection between terrestrial stations. One of the buoys tracked the motion of the ice island station T–3. See also Nimbus–5 IRLS and NOAA ARGOS.

ERB:
  large-scale reflected and emitted radiation from the Earth's surface and atmosphere. Snow/ice albedo and BRDF.

HIRS, PMR and LRIR:
  vertical profiles of temperature, water vapour and ozone (LRIR) in the atmosphere.

Limitations

THIR:
  as for Nimbus–5.

SCAMS:
  very coarse resolution. Difficulty in relating measured brightness temperatures (TBs) to surface and subsurface characteristics. Most of the channels are severely affected by intervening atmospheric water vapour and liquid water effects.

ESMR:
  as for Nimbus–5, but more sensitive to water droplets in the atmosphere. Difficulty in relating measured TBs to surface and subsurface detail. Still only a single channel instrument, with the ambiguity in interpretation remaining. See also Nimbus–5 ESMR.

RAMS:
  poorer coverage than NOAA ARGOS system. Temporal resolution too coarse for many applications; fixes often limited to one per day. Gaps occur in the data (although fewer towards the poles). Positional errors tend to be larger, ie of the order of 5km rms. The highly stable oscillators required by the ground transmitters contribute substantially to the cost. The average velocity of a moving transmitter can only be estimated from frequency measurements acquired from two consecutive overpasses of the satellite; this requirement often not met.

ERB, HIRS, PMR and LRIR:
  not suitable for studying Earth's surface. Poor spatial resolution. LRIR horizontal coverage limited to the latitudinal band from 64°S to 84°N at 4° increments.

Data format

THIR:
  as for Nimbus–5.

SCAMS:
  CCTs and 70 × 70mm black and white (B&W) negative film.

ESMR:
  CCTs containing TBs, and 70 × 70mm B&W film.

RAMS:
  either CCTs or hardcopy computer printouts.

Data availability and archiving dates

Data archived at NSSDC (for full address, see TIROS I–X). All requests from non-US researchers must be addressed to the Director, WDC–A for Rockets and Satellites, at the same address. Data can be provided in a different format, but this is likely to be more expensive.

SCAMS:
> 15 June 1975–29 May 1976.

THIR:
> 18 June 1975–6 May 1977.

ESMR:
> 17 June 1975–11 August 1977.

RAMS:
> data are best acquired from the various principal investigators whose work is described in Cote and others (1982).

ERB, HIRS, PMR and LRIR:
> see Ng and Stonesifer (1989) for archive dates.

References and important further reading

Chang, A., Choudhury, B. and Gloersen, P. 1980. Microwave brightness of polar firn as measured by NIMBUS–5 and –6 ESMR. *Journal of Glaciology* 25: 85–91.

Cote, C., Taylor, R. and Gilbert, E. 1982. NIMBUS–6 Random Access Measurement System Applications Experiments. *NASA Report SP–457*. NASA Scientific and Technical Information Branch, Washington DC.

Liou, K. and Yeh, H. 1979. Remote sounding of the temperature profile and cloud thickness in cirrus cloudy atmospheres from NIMBUS–VI HIRS channels. In Bolle, H. (ed.), *Remote sounding of the atmosphere from space*: 157–60. Pergamon Press, Oxford and New York.

Ng, C. and Stonesifer, G. 1989. Data catalog series for space science and applications flight missions. Volume 4B: descriptions of data sets from meteorological and terrestrial applications spacecraft and investigations. *NSSDC/WDC–A–R&S 89–10*. Available from NSSDC.

Rotman, S., Fisher, A. and Staelin, D. 1981. Analysis of multiple-angle microwave observations of snow and ice using cluster analysis techniques. *Journal of Glaciology* 27, 95: 89–97.

Sissala, J. (ed.). 1975. *The NIMBUS–6 Users Guide*. NASA GSFC, Greenbelt, MD 20771.

Smith, G., Rutan, D. and Bess, T. 1990. Atlas of albedo and absorbed solar radiation from NIMBUS–6 Earth Radiation Budget data set – July 1975–May 1978. *NASA Reference Publication 1230*. NASA, Washington DC.

Vinje, T. 1978. Sea ice conditions and drift of NIMBUS–6 buoys in 1978. *Norsk Polarinstitutt Arbok 1978*. Oslo: 57–66.

12. GMS/Himawari

Operational lifespan and orbital characteristics

Dates: 14 July 1977–present.
Orbital height: 35,900km.
Inclination: 0°.
Period: 1,436 minutes (geostationary).
Longitude: 140°E.
Notes: Japanese operational meteorological programme.

Sensor characteristics

VISSR:
i)   VIS (wavelength 0.55–0.75μm). GR 1.25km.
ii) TIR (wavelength 10.5–12.5μm). GR 5km. Swath width limb to limb.
Also carries a Data Collection System (DCS) and WEFAX.

Polar applications

Operational information on sea ice conditions and weather conditions in the Sea of Okhotsk.
Cloud amount. Winds derived from cloud motion. Snow cover monitoring. SSTs. See also
GOES, Meteosat and Insat.

Limitations

See GOES.

Data format

i)   Image data (hard-copy): original film negative, printed positive, microfilm and animated
     16mm film.
ii) Digital data on CCT: original VISSR imagery and extracted meteorological parameters
     (cloud-motion wind, SST, etc).
iii) Contour charts.

Data availability and archiving dates

Details from: Meteorological Satellite Center, 3–235 Nakakiyoto, Kiyose, Tokyo 204,
Japan, telephone (0424)–93–1111, FAX (0424)–92–2433. Data can be purchased from:
International Meteorological Planning Division, Japan Weather Association, 2–9–2 Kanda
Nishikicho, Chiyodaku, Tokyo 101, Japan, telephone (03)–238–0480, FAX
(03)–262–9549. Data are retained in archive for 1–10 years. For further information, see
WMO (1989).

References and important further reading

Hirai, M., Watanabe, K., Tsuru, H., Miyaki, M. and Kimura, M. 1975. Development of the Geostationary Meteorological Satellite (GMS) of Japan. *Proceedings of the 11th International Symposium on Space Technology and Science*: 461–5, Tokyo, Japan.

WMO. 1989. Information on meteorological and other environmental satellites. *WMO–411*, World Meteorological Organization, Geneva.

13. Meteosat

Operational lifespan and orbital characteristics

Dates: 23 November 1977 to present
Semimajor axis: 41,681–42,288km.
Orbital height: 34,913–35,973km.
Inclination:0.7–1.0°
Period: 1,412–1,442 minutes (geostationary).
Longitude: 0°
Notes: Operational meteorological satellite programme, operated by Eumetsat from Darmstadt, GDR. Latitudinal coverage approximately 55°N to 55°S.

Sensor characteristics

i)   2 VIS bands (wavelength 0.4–1.1µm). GR 2.5–5km. 5,000 samples × 5,000 lines.
ii) 1 TIR band at 5.7–7.1µm and 2 at 10.5–12.5µm. GR 5km. Swath width limb to limb. Two images per hour. 2,500 samples × 2,500 lines.
Also carries a Data Collection System (DCS).

Polar applications and limitations

See GOES. Meteosat, GOES and GMS participated in the first Global Atmospheric Research Programme (GARP), which had an important Southern Oceans element.

Data format

9-track, 6250bpi CCTs. Images and operational products (cloud motion wind vectors, SST, cloud analysis, etc). WEFAX data.

Data availability and archiving dates

Meteosat Data Services, European Space Operations Centre, Robert Bosch Str. 5,6100 Darmstadt, Germany. For further information, see WMO (1989).

References and important further reading

Schmetz, J. 1989. Cloud observations from Meteosat and the inference of winds. *COSPAR Advances in Space Research 9*, 7: 91–9.
WMO. 1989. Information on meteorological and other enironmental satellites. *WMO–411*, World Meteorological Organisation, Geneva.

14. Landsat 3

Operational lifespan and orbital characteristics

Dates: 5 March 1978–31 March 1983.
Semimajor axis: 7,287km.
Orbital height: 900–918km.
Inclination: 99.14°
Period: 103.21 minutes. Exactly repeating orbit, repeat period 18 days.
Notes: Sun-synchronous orbit. Landsat 3 followed Landsat 2 by nine days, thereby providing 9-day repetitive coverage.

Sensor characteristics

MSS:
As for Landsats 1 and 2, but with the addition of Band 8: TIR (wavelength 10.4–12.6µm): GR 240m (all other bands 79m); swath width 185km.
RBV:
Band 1: Panchromatic, VIS/NrIR (wavelength 0.505–0.75µm), × 2. Geometric resolution 40m. Swath width 98km × 2, the two RBV images slightly overlapping. Four overlapping RBV scenes were registered to be approximately equivalent to one 185 × 185km MSS scene.
Data Collection System (DCS):
identical to Landsats 1 and 2.

Polar applications

MSS:
Bands 4–7 as for Landsats 1 and 2. Band 8 failed soon after launch.
RBV:
Only 11 excellent quality RBV subscenes of Antarctica exist in archive. These, however, reveal a wealth of glacial surface detail, especially with a sun eleveation angle of 5–7°.
DCS:
collection of data from remote platforms.

Limitations

MSS:

as for Landsats 1 and 2. Band 8 failed soon after launch (in July 1978). Some Landsat 3 (and 1 and 2) MSS images contain various internal distortions caused by spacecraft roll, pitch and yaw during picture acquisition; difficult to produce rectified maps from these data unless a large number of fixed ground-control points are available.

Cloud-cover assessment appearing on either computer or microfiche summaries of each archived image is often quite unreliable (Williams and Ferrigno 1988). Snow is often mistaken for clouds and vice versa. Moreover, the 1972–80 16mm microfilm cassettes (see below) contain only band 5 images, often over-exposed, which make it almost impossible to distinguish clouds from snow. Band 7 images on 70mm (1:3,369,000 scale) archival film rolls stored at NASA GSFC must be used to make a more definitive determination of cloud cover extent. Fortunately, Williams and Ferrigno (1988) have prepared invaluable tables of optimum Landsat 1, 2 and 3 images of Antarctica.

RBV:

severely restricted data availability over snow and ice due to detector saturation caused by the high reflectivity of the surface, combined with technical problems. Once blinded by snow/ice, the sensor took many seconds to recover (if at all). Also, system problems caused a data backlog at NASA GSFC, and a data gap at EDC. Coverage up to a maximum latitude of 82°.

DCS:

underused in polar regions.

Data format

MSS:

as for Landsats 1 and 2. 16mm microfilm cassettes for quick reference. Note: 1972–80 cassettes contain only band 5 images. Band 7 images are stored on 70mm (1:3,369,000 scale) archival film rolls at NASA GSFC.

RBV:

originally, the imagery was processed directly from wide-band video tape data to 70mm film. This changed after September 1980, when digital products of the RBV data were extracted at NASA GSFC. Wide-band video tape data were then subjected to analogue-to-digital preprocessing, and were corrected both radiometrically and geometrically to produce high density digital tapes (HDDTs). These data were then transmitted to the EDC via DOMSAT, where they were used to generate 241mm photographic images at a scale of 1:500,000. CCTs also generated, although not all RBV scenes from Landsat 3 are available in CCT format.

Note: several correction and enhancement procedures can be applied by EOSAT on request.

Data availability and archiving dates

Main archive: EOSAT (for full address, see Landsats 1 and 2). RBV imagery is available for inspection on 70mm film roll/18cm film clip at EDC. Some MSS band 8 data are available. Landsat 3 transmitted data 2 April 1978–November 1982.

The US Geological Survey (USGS) has been compiling detailed maps of parts of Antarctica using digitally-enhanced MSS imagery. Special purpose images have been prepared showing bedrock or ice features only, eg blue ice. For further information, contact the USGS Headquarters at 12201 Sunrise Valley Drive, National Center, Reston, VA 22092, USA.

Data are also available from European, Canadian, Japanese and Alaskan receiving stations: details (addresses, etc) as for Landsats 1 and 2.

References and important further reading

Ferrigno, J., Williams, R. and Kent, T. 1983. Evaluation of LANDSAT 3 RBV images. In Oliver, R., James, P. and Jago, J. (eds), *Antarctic Earth Science: Proceedings of the Fourth International Symposium on Antarctic Earth Sciences*, 16–20 August 1982, University of Adelaide: 446–9. Australian Academy of Science, Canberra/Cambridge University Press.

Rango, A. 1979. Remote sensing of snow and ice: a review of research in the United States 1975–1978. *NASA Technical Memo 79713*, GSFC, Greenbelt, MD 20771.

Swithinbank, C. 1988. Antarctica. Volume B of *Satellite image atlas of glaciers of the world. US Geological Survey Professional Paper 1386–B*. Books and Open-File Reports Section, USGS, Federal Center, Box 25425, Denver, CO 80225, USA.

Williams, R. and Ferrigno, J. 1988. LANDSAT Images of Antarctica. In Swithinbank, C., Antarctica. Volume B of *Satellite image atlas of glaciers of the world*, edited by Williams, R. and Ferrigno, J. *US Geological Survey Professional Paper 1386–B*. Books and Open-File Reports Section, USGS, Federal Center, Box 25425, Denver, CO 80225, USA.

15. The Heat Capacity Mapping Mission (HCMM)

Operational lifespan and orbital characteristics

Dates: 26 April 1978–31 August 1980.
Semimajor axis: 6,979km.
Orbital height: 560–641km.
Inclination: 97.6°
Period: 96.72 minutes (exactly repeating orbit). Repeat period 16 days.
Notes: Circular, sun-synchronous orbit. Crossing times 13:30 and 1:30 in mid-northern latitudes.

Sensor characteristics

Heat Capacity Mapping Radiometer (HCMR):
i) Band 1: VIS and NrIR (wavelength 0.55–1.1μm). GR 500m at nadir. Swath width 716km. Albedo range 0–100%. Noise equivalent radiance (NET) 0.2 milliwatt cm^{-1}.
ii) Band 2: TIR (wavelength 10.5–12.5μm). GR 600m at nadir. Swath width 716km. Range 260–340K. NEΔT 0.3K at 280K.

Polar applications

Band 1 provided measurements of the surface albedo in the sun illumination region. Precise day and night temperature difference measurements. Large-scale extent and dynamics of Arctic sea ice, particularly in winter. Detection and delineation of leads and polynyas. Monitoring and mapping of snow cover. Data can be related to melting/freezing conditions. Potential surface physical temperature. See also NOAA AVHRR and Landsat TM.

Limitations

Band 1 limited by cloud and darkness. Band 2 limited by cloud. Interpretation of band 2 data during melt periods and when the ice temperature is near to that of the surrounding water is difficult. Direct data readout only (no onboard recording facility): polar coverage limited to the Beaufort, Chukchi and Bering Seas (from the receiving station in Alaska); S Labrador and S Hudson Bay and from Iceland and the Greenland Sea up to about 73°N (from Lannion, France). The interpretation of thermal data is not a simple matter – affected by emissivity variations and cloud cover. Coverage up to a maximum latitude of 85°N. See also NOAA AVHRR and Landsat TM.

Data format

Day and night thermal data, geometrically corrected to allow digital overlay of the two datasets. Temperature difference and thermal inertia data then produced.
  i)   NSSDC/WDC–A:
       a)   9-track, 800bpi/1600bpi CCT, binary in BSQ format.
       b)   1:4,000,000 scale black and white images on 241mm print paper or positive/negative transparencies: daytime VIS and TIR, night-time TIR, temperature differences (night versus day), and thermal inertia.
  ii) ESA ESRIN:
       Standard data products similar to those from NSSDC/WDC–A.

Data availability and archiving dates

Data are available for May 1978–31 August 1980.
  i)   Alaskan data are archived at NSSDC/WDC–A (for details of addresses, etc, see TIROS I–X).
  ii) Lannion data are archived at ESA ESRIN/Information Retrieval Service, Online Services Division, Via Galileo Galilei, 00044 Frascati, Italy, telephone 39–6–9426285 or 39–6–9401218, telex 610637 ESRIN I EURIMAGE. The UK Earthnet National Point of Contact is: Mr M. Hammond, National Remote Sensing Centre, RAE, Farnborough, Hampshire GU14 6TD, telephone 0252 24461, telex 858442. For details of all data and facilities available at the UK NRSC, see the *National Remote Sensing Centre Data User's Guide*.
The *HCMM Data User's Guide for Applications Explorer Mission–A* (1980) is available from: Code 902, NASA GSFC, Greenbelt, MD 20771, USA.

References and important further reading

NASA. 1979. *The Heat Capacity Mapping Mission Users' Guide*. NASA GSFC, Greenbelt, MD 20771, USA.

Ng, C. and Stonesifer, G. 1989. Data catalog series for space science and applications flight missions. Volume 4B: descriptions of data sets from meteorological and terrestrial applications spacecraft and investigations. *NSSDC/WDC–A–R&S 89–10*. Available from NSSDC.

Short, N. and Stuart, L. 1982. The Heat Capacity Mapping Mission (HCMM) anthology. *NASA Report SP–465*. NASA GSFC, Greenbelt, MD 20771, USA.

16. TIROS-N, NOAA 6–11 and NOAA 'NEXT'

Operational lifespan and orbital characteristics

Dates: 19 October 1978 to the present. NOAA-10 and -11 currently active.

Semimajor axis: 7,195–7,235km.

Orbital height: 808–867km.

Inclination: 102.3° (TIROS-N), 98.75°–98.92° (NOAA-6 to -11).

Period: 101.29–102.09 minutes.

Notes: Near-polar, circular sun-synchronous orbits. Equator crossing times: a) even numbered satellites ascending node 1930, descending node 0730; b) odd numbered satellites ascending node 1430, descending node 0230. Starting with NOAA-8, satellites in the series are now designated Advanced TIROS-N (ATN). NOAA-D (-12) is scheduled for launch in May 1991.

Sensor characteristics

Advanced Very High Resolution Radiometer (AVHRR) – on TIROS-N, NOAA-6, -8, -10 and -12:

i) Band 1: VIS and NrIR (wavelength 0.55–0.90μm, 0.58–0.68μm for TIROS-N). 1.1km GR (at nadir; 2.5km resolution at the swath edge) High Resolution Picture Transmission (HRPT) data are available for all spectral bands, and 4km Global Area Coverage (GAC) for any two selected bands. Automatic Picture Transmission (APT) data are at 4km GR, Local Area Coverage (LAC) data at 1km (see data format). Swath width 2,580km (HRPT), 4,000km (APT). Samples per scan line 2048. Scanning rate 6Hz. 10-bit radiometric resolution (1024 digital levels). Data rate 665.4Kbps.

ii) Band 2: NrIR (wavelength 0.725–1.10μm).

iii) Band 3: IR (wavelength 3.55–3.93μm).

iv) Band 4: TIR (wavelength 10.05–11.05μm).

AVHRR/2 – on NOAA-7, 9 and 11 (23 June 1981 to present):

i) Bands 1, 2 and 3 as for AVHRR.

ii) Band 4: TIR (wavelength 10.50–11.50μm). NEΔT 0.12K at 300K (bands 3–5).

iii) Band 5: TIR (wavelength 11.50–12.50µm). GR, swath width and data rate of AVHRR/2 as for AVHRR.

AVHRR/3 – on NOAA 'Next' (-K, -L and -M, June 1993 onwards):

as for AVHRR/2, but with a sixth band, 1.58–1.64µm, allowing better discrimination between cloud and snow. Also, band 3 [3.63–3.9µm] will operate during night-time only, and band 2 will be 0.82–0.87µm.

TIROS Operational Vertical Spectrometer (TOVS), a currently operational 3-instrument system consisting of:

 i) High Resolution Infrared Spectrometer/2 (HIRS/2): measures incident radiation in 20 bands of the IR spectrum (from 0.660–14.98µm). Scan time 6.4s. Number of steps 56. Optical FOV 1.25°. Step angle 1.8°. Step time 100µs. Ground IFOV 17.4km at nadir; 58.5km cross-track × 29.9km along-track at end of scan. Distance between IFOVs: 42km along-track. Swath width 2,240km. Data rate 2,880bps.

 ii) Stratospheric Sounding Unit (SSU): a step-scanned far-IR spectrometer with 3 bands in the 15µm CO_2 absorption band. Uses pressure modulation technique to measure radiation emitted from CO_2 at the top of atmosphere. The 3 bands have the same frequency but different cell pressures. GR 147.3km at nadir. Angular FOV 10°. Number of Earth views/line 8. Time interval between steps 4s. Swath width 1,473km. Data rate 480bps.

iii) Microwave Sounding Unit (MSU): a passive scanning microwave spectrometer with 4 channels in the 5.5µm oxygen region (frequencies of 50.3, 53.74, 54.96 and 57.95GHz with channel RF bandwidths of 220MHz). NEΔT: 0.3K (all channels). Dynamic range: 0–350K (all channels). GR 109km. Swath width 2,320km. Data rate 320bps. TOVS data are transmitted as direct readout and are relayed via the communications satellite DOMSAT.

ARGOS Data Collection and Location System:

Carrier frequency 401.65MHz. Transmit power 3.0W. Ageing (during life) ±2kHz. Short-term stability (100ms): 1:10^9 (platform requiring location); 1:10^8 (platform not requiring location). Medium-term stability (20 min): drift 1:0.2Hz min^{-1} (15 min, requiring location). Long-term stability (2 hr) 1:±400Hz. Modulation Index ±1 radian. Modulation rate 400Hz. Bit rate 400bps. Frame length 390–950µs. Transmitter repitition rate for message: 40–60s (requiring location); 60–200s (not requiring location). Data sensors: 4–32 8-bit sensors for environmental data. Overpass time approximately 10 minutes. 4 messages can be received and processed simultaneously; with the launch of NOAA-K and subsequent spacecraft, an additional 4 data record units will be added, increasing the capacity fourfold. Total number of platforms (capacity): 4,000 global, 450 within view. Position-fix accuracy ±1km rms (see data format). Details of manufacturers of Platform Transmitter Terminals (PTTs) which are certified by ARGOS are available from Service ARGOS (see later for address).

Advanced Microwave Sounding Unit-A (AMSU-A) (on the NOAA 'Next' Series, beginning with NOAA-K, and EOS-A [in 1998]):

passive microwave across-track scanning device, 15 channels around the following frequencies: 23.8GHz (channel 1), 31.4GHz (channel 2), 50.8–58GHz (channels 3–14) and 89GHz (channel 15). GR 45km at nadir. Beam width 3.3° at each frequency. Swath width 2,300km. Data rate 3 Kbps. Global coverage every 12 hours poleward of latitude 30° with 2 satellites in orbit.

AMSU-B (on NOAA 'NEXT' Series and EOS-A):

passive microwave: 5 channels around 89GHz, 157GHz and 183.3GHz. Altitude 825km.

GR 15km. Swath width 2,300km. Data rate 6 Kbps.

Search and Rescue System (NOAA-8 onwards). NOAA-9 and -10 also carry an Earth Radiation Budget Experiment (ERBE) (wavelengths 0.2–50μm, limb to limb, 40km GR). Starting with NOAA-9, a Solar Backscatter Ultraviolet Radiometer (SBUV/2) is carried on afternoon ascending satellites – an improved version of the Nimbus-7 SBUV, operating at bands in range of 0.16–0.40, 160–400μm.

Polar applications

AVHRR and AVHRR/2:

routine monitoring of large-scale atmospheric, oceanic, sea ice and land ice features.

i) Bands 1 and 2 (VIS and NrIR): broad-scale daytime ice extent and cloud mapping. Delineation of sea ice/water boundary. Mapping fast ice. Leads ⩾1km wide. Polynyas. Concentration of sea ice types from nilas to thin FY ice (based on subtle tonal/textural variations). Ice edge features, including bands. Albedo changes of snow and ice, and the onset and progression of melt. Useful alternative to passive micro-wave sensors in giving large-scale sea ice concentrations during melt periods when the data from the latter are unreliable. Large-scale ice sheet/glacier extent and boundaries. Study of ice stream dynamics and features (boundaries, ridges, rises and grounding lines) (Figure II.6). Tabular iceberg calving and tracking (Figure II.7). The additional 1.57–1.78μm daytime band on NOAA 'Next' will provide better discrimination between snow and clouds – snow appears darker than cloud at this wavelength.

ii) Bands 3 (IR) and 4 (TIR): SSTs (accuracy 0.5–0.7K). Inferred ice and water physical temperatures (under cloud-free conditions). Winter/night-time cloud and ice distri-bution mapping (**Figure II.8**). TIR imagery is extensively used to study ocean features which cause observable variations in surface temperature, such as ice edge eddy features, currents, fronts, upwelling, and coastal mixing patterns. Identification of subtle ice sheet features based on thermal signature contrasts.

iii) Band 5 (TIR): as for bands 3 and 4. Added primarily to account for boundary layer water vapour effects in SST computations. Bands 4 and 5 are 'split-window'.

In practice, bands 1 and 4 are mainly used for ice analysis. Experimental CMB and CMT images (digitised VIS and TIR data) have been generated by NOAA to aid large-scale snow and ice mapping. Weekly N hemisphere snow cover maps are produced using VIS imagery from AVHRR (and GOES VISSR).

APT data of crucial importance to polar shipping. Cloud-tracking to estimate wind speed and direction. Study of polar lows. Poleward of 60° latitude, the distance between ground tracks is less than half the AVHRR's swath width, and any point on the Earth's surface is visible in two consecutive passes twice a day. Poleward of 70°, the overlap is large enough to guarantee coverage of a given point on at least three consecutive passes. The data are readily available, relatively cheap, and easy to display/manipulate (using software such as SEAPAK [McClain and others 1989; 1990]). The study of relatively short-lived features, such as sea ice fracture patterns. Automatic techniques have been developed to extract pack ice motion from sequential AVHRR imagery (Ninnis and others 1986); an alternative to the finer-scale tracking by buoys and SAR, filling in gaps on a regional scale. Intermediate resolution between SAR/high resolution VIS/TIR sensors (TM, MSS, HRV, etc) and passive microwave sensors – complementary. AVHRR data used with latter to produce operational ice charts. Used to verify concentration retrievals from passive microwave data. Also for

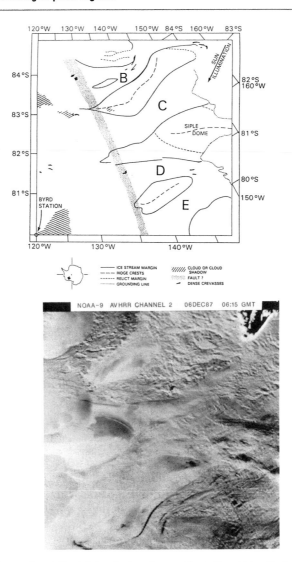

Figure II.6   (i) AVHRR image (band 2) of Antarctic ice streams B, C, D and E, collected at 06:15 GMT on 6 December 1987. (ii) Location map of the above image in West Antarctica. From Bindschadler and Vornberger (1990).

comparison with radar altimeter sea ice data, ie ice edge location. TIR data to monitor the emergence of katabatic winds from continental Antarctica onto the ice shelves – powerful when used with TOVS and automatic weather station data (Bromwich 1989). See also DMSP OLS, JERS-1 VTIR and ERS-1 ATSR/M. Note: digital data are more useful than hard copy.

TOVS:

four frequencies in the oxygen resonance band. Does not include the water sensitive channels present in its predecessors (the Nimbus NEMS and SCAMS). Designed so that the acquired data will permit calculation of:

i) vertical temperature profiles from the surface to 10mb;

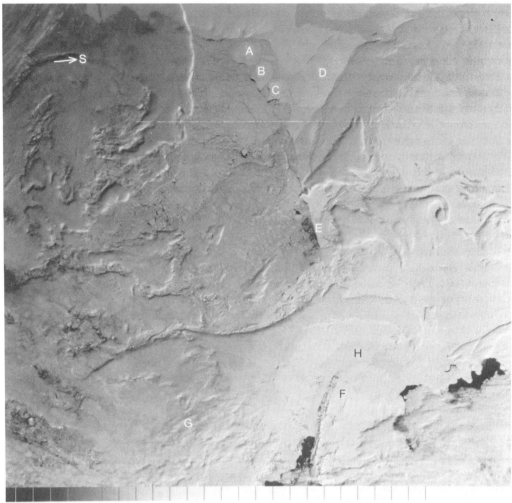

GIL 1 007:21:06:34 10671 1 91A N9 LVS 07JA87

Figure II.7 An AVHRR image (band 1) of the Weddell Sea, Antarctica, 7 January 1987. Note the calving of icebergs (A, B and C) from the Filchner Ice Shelf. In 1986 alone, more than 8,800 square miles of ice (approximate thickness 1,000 feet) detached from the Larsen and Filchner/Ronne Ice Shelves. D is Berkner Island, E a polynya off the Ronne Ice Shelf, F Alexander Island, G cloud-covered Antarctic Peninsula, and H King George VI Sound. Courtesy of NOAA NESDIS.

ii) water vapour content at 3 levels of the atmosphere; and

iii) total ozone content. Heat fluxes and temperature gradients in both polar regions. Monitoring the development of synoptic and smaller-scale weather systems, including polar lows (in conjunction with AVHRR). MSU was one of several pioneering instruments that demonstrated the value of cloud-penetrating microwaves for both soundings and surface observations. TOVS profiles available over the Global Telecommunications System are used extensively in operational numerical weather prediction systems. TOVS soundings can provide information on the structure of the

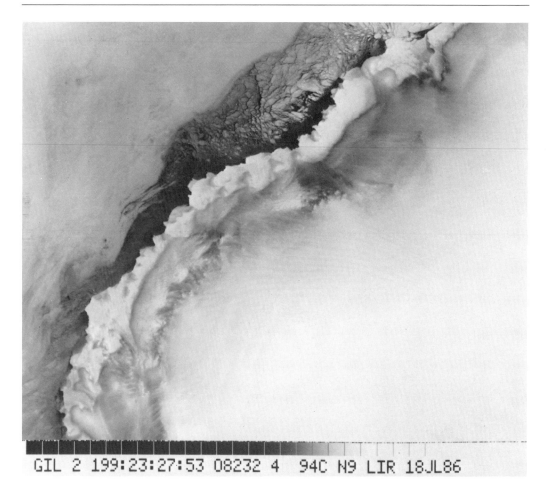

GIL 2 199:23:27:53 08232 4 94C N9 LIR 18JL86

Figure II.8 (i) An AVHRR TIR (band 4) image of the Antarctic coastline (Coats Land), 18 July 1986. Such images are of immense value in the monitoring of coastal polynyas and ice sheet margins during periods of polar darkness. Courtesy of Joey Comiso, NSA GSFC.

atmosphere at high horizontal resolution in areas where few radiosonde ascents are available. HIRS/2 data for atmospheric correction of AVHRR-derived SSTs.
ARGOS:

FROZEN OCEANS AND METEOROLOGY

A cost-effective and regular means of interrogating and relaying data from remote instrument packages on the surface, as well as tracking their position by measuring the Doppler shift in transmitter frequency (for accuracy, see *data format* section). The Cooperative Arctic Ocean Buoy Programme (CAOBP) has produced air pressure, air temperature and location data from the Arctic Basin since 1979 (Figure II.9) (Untersteiner and Thorndike 1982). More recent buoys can transmit wind speed and direction, SWH, SST, and even ocean conductivity, temperature and density data from under the ice (Burke and Anderson 1989). The International Ice Patrol operationally deploys transponders on icebergs to predict their drift in the Labrador Sea and NE Atlantic Ocean. Indeed, drifting

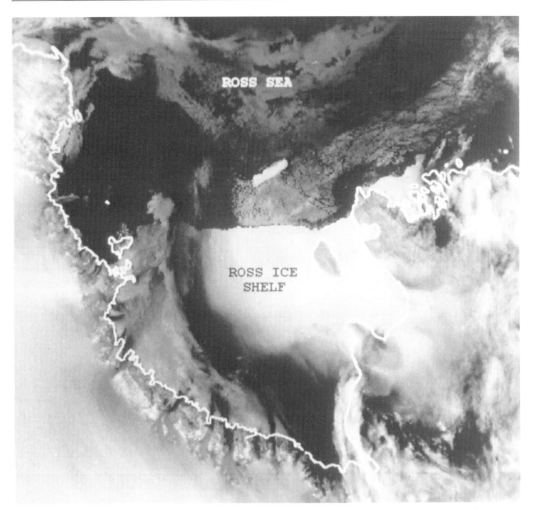

Figure II.8   (ii) NOAA 9 AVHRR thermal infrared (channel 4) satellite image at 1013 UTC 5 June 1988. Spatial resolution is 2km. The prominent dark (warm) signature is associated with katabatic winds from southern Marie Byrd Land crossing the Siple Coast and propagating horizontally all the way across the Ross Ice Shelf. Propagation distance across the shelf exceeds 1,000km. The satellite information was available on 8mm video cassette from the Antarctic Research Center of Scripps Institution of Oceanography, and processed on a SUN 4/110 workstation using the TeraScan software supplied by SeaSpace. Provided by David H. Bromwich and Jorge F. Carrasco, Byrd Polar Research Center, The Ohio State University, Columbus, Ohio.

buoys provide the only means of routinely measuring surface currents under, and winds over, ice-covered oceans, and as such, help to fill in the enormous data gaps typical of polar oceans. Similar gaps around the perimeter of the Antarctic continent have been filled in since 1980 by automatic weather stations (AWS), operated by the University of Wisconsin.

WILDLIFE TRACKING

Polar and sub-polar wildlife ecology. University of Alaska scientists are studying the dispersal and diving behaviour of Weddell seals in Antarctica by attaching satellite-linked

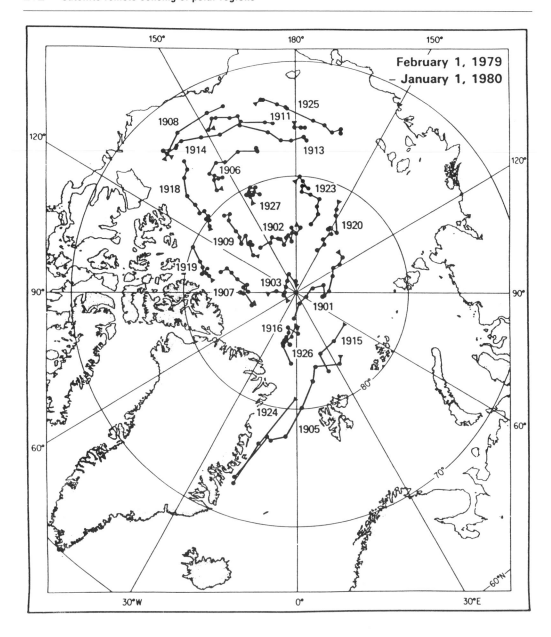

Figure II.9   The drift of ARGOS-tracked buoys forming the Arctic Ocean Buoy Network. From Parkinson and others (1987). The buoys also transmit surface air temperature and pressure data. Tracking the motion of buoys proviudes valuable information on both sea ice dynamics and kinematics. Inferences can be made from these data as to the behaviour of the underlying ocean. For further details, contact the Polar Science Center, University of Washington, 1013 NE 4th Street, Seattle, WA 98105, telephone (206)–543–1300. Frequent data reports are published.

time-depth recorders to the animals. When a seal leaves the water for the sea ice or a nearby beach, the recorder will relay information on its location and transmit summaries of the duration of dives and their depths via the satellite to the investigator. Early research on animal tracking was confined to very large animals, including moose and polar bears, as the PTTs were heavy. Smaller PTTs with a wider range of uses have recently been developed, including the tracking of birds; potentially useful in study of petrels (French 1988). The technological state of the art is described by Tomkiewicz and Beaty (1988).

FUTURE DEVELOPMENTS

ARGOS-II, to be included on the NOAA 'Next' satellite series, will offer improved locational accuracy, with an increase of 30% in the data rate. Buoys transmitting meteorological and oceanographic data via this system will play a central role in WOCE.

Over the next decade, more sophisticated buoys will provide additional data on such important variables as salinity, sea ice thickness/growth rate (which cannot otherwise be directly determined from space) and short-wave radiation. The Arctic Remote Autonomous Measurement Platform (ARAMP), currently under development, will be capable of measuring ocean current, wind speed and direction, surface waves, floe-floe jostling and acoustic noise (Prada and Baggeroer 1988). Improved estimates of ice growth rates, average salt fluxes to the ocean and heat and mass fluxes to the atmosphere will be possible by synthesising buoy data with SAR data. The development of a coordinated Antarctic buoy programme is a very high priority; buoy coverage there has tended to be somewhat fragmentary.

AMSU-A:

designed to measure a variety of meteorological and climatological parameters. 55GHz: atmospheric temperature profiles, liquid water abundance. 89GHz: similar frequency to channel 4 of the SSM/I (see DMSP) – atmospheric water vapour profile and sea surface wind speed. Potential polar applications: snow cover extent and type, sea ice concentration and type, land ice properties, eg firn distribution and state (wet versus dry). The 89GHz and 31.4GHz channels in particular are potentially useful in providing supplementary snow/ice data.

AMSU-B:

designed to measure atmospheric water vapour profiles. 90GHz and 166GHz channels in particular have potential in sea ice and land snow cover studies, and should provide excellent ice extent maps at 12-hour intervals for each satellite in orbit. The combined 176/190GHz channel may have some value in sea ice studies. At these higher frequencies, there is a larger difference in the emissivities of FY/young ice, second-year and MY ice (based upon aircraft and surface experiments), although the emissivity of water increases with frequency. AMSU-B could in theory be used to provide atmospheric corrections for all channels used in sea ice studies.

ERBE:

global radiation budget. Snow/ice albedo and BRDF.

SBUV/2:

total ozone content and vertical ozone distribution in atmosphere.

Limitations

AVHRR and AVHRR/2:
i) All bands cloud-limited. Bands 1 and 2 inoperative in darkness/winter.

ii) Band 1: no differentiation between thicker sea ice forms based on tonal/textural variations.

iii) Bands 1 and 2: snow cover can make thin/young ice types appear thicker/older. Cloud often difficult to distinguish from snow/ice, although research is underway to develop algorithms which are capable of automatically identifying cloud in polar regions, eg work is in progress at the British Antarctic Survey to develop a scheme using the five bands of AVHRR – based on multiple tests and using thresholds on the 3.7μm band to isolate fully and partly cloud-filled pixels. Although the system is capable of detecting most convective clouds and clouds composed of water droplets, problems are still experienced in identifying some uniform layers of cirrus which are featureless in both the VIS and IR bands.

iv) Band 2: does not respond well to thin sea ice, and wet ice can have similar signature to open water/new ice.

v) Bands 3, 4 and 5: will not distinguish ice when its temperature is close to that of surrounding water, particularly under summer melt conditions. New ice difficult to distinguish from open water. Cannot distinguish thick FY from MY sea ice. TIR imagery often requires enhancement. Noise rendered band 3 on the TIROS-N AVHRR almost useless; Fourier filters can remove much of the noise, although a degree of human intervention is still necessary (Warren 1989). Care must be taken using band 3 during daytime, when it measures both solar reflected and thermally emitted contributions.

Many important ice features are unresolved on AVHRR imagery. Tends to underestimate ice concentration, particularly in areas of low concentration (eg MIZs). Image registration is a problem, and the image is displayed on a coordinate system that is relatively awkward to use. Unrectified images appear foreshortened towards the horizon, although this can be geometrically rectified. At large scan angles, the atmospheric contamination is greater than at nadir.

Techniques developed to track clouds and sea ice from AVHRR data are hampered to a certain extent by the difficulty in matching features on sequential imagery. This problem occurs because AVHRR pixel size increases from 1km at the centre to 6km towards the outside of the swath, so that detail prominent in one image may be completely obscured on the next pass. Error sources in SST retrievals include residual cloud contamination, noise in band 3, calibration and navigation errors, skin/bulk temperature differences, and aerosols (the latter can be estimated using band 1 data). SST changes of up to a few degrees can arise in the course of the day, peaking in the late afternoon. Accurate SST values from the afternoon NOAA satellite (NOAA-9, etc) are known to suffer from this effect. See also DMSP OLS, JERS-1 VTIR and ERS-1 ATSR/M.

TOVS:

HIRS/2 and SSU are affected by cloud cover. The MSU is little affected by clouds. All three sensors have a very coarse horizontal spatial resolution. The temperature profiles are relatively smooth in the vertical and give poor resolution of the tropopause and boundary layer. Designed to operate over open ocean.

ARGOS:

the position-fix accuracy depends upon the number of satellite passes in which position fixes are made. Two consecutive passes are ideally required, but fixes are often only obtained on one pass, thereby decreasing the accuracy and reliability of the positional data. Gaps occur in the data, but these are fewer towards the poles where satellite orbit tracks converge. For those packages with anemometers, icing is a constant problem.

AMSU-A and AMSU-B:
> not designed to measure surface features. Coarse spatial resolution. All channels affected by atmospheric attenuation and emission, due at the higher frequencies mainly to water vapour. The 55GHz, 183GHz and 176/190GHz channels in particular can only be used under clear sky conditions. Atmospheric attenuation is particularly a problem in summer, when the water vapour content of the atmosphere is higher. Aircraft 92GHz microwave imagery has been effective in mapping characteristics of the Bering Sea MIZ – but limitations include a 10–20K loss in ice/water contrast imposed by heavy cloud cover, and a drop in TB of 25K relative to lower frequencies in shear zones with heavy snow accumulation. See DMSP SSM/I.

ERBE and SBUV/2:
> coarse spatial resolution. SBUV/2 not designed to measure surface features.

Data format

AVHRR and AVHRR/2:
i) Data are generated in and can be acquired in four different modes:
   a) HRPT: 1km resolution, direct readout, digitised to 10-bit precision. These data are the most frequently used and in the most useful format. HRPT is the successor to APT. HRPT data full resolution, ie 2048 pixels per scan line. SDSD stations at Gilmore Creek (Alaska), covering much of the W Arctic in direct readout, and Wallops Island, Virginia, covering SE Canadian Arctic. A satellite pass directly over an antenna site will be within view of that antenna (horizon to horizon) for about 15.5–16 minutes.
   b) Alternatively, full resolution (1km) LAC data may be recorded onboard the spacecraft (about 10 minutes per 102 minute orbit, equivalent to a ground strip 4,000km long) for subsequent dumping at one end of the NOAA/NESDIS Command and Data Acquisition (CDA) stations at Wallops Island, Virginia and Gilmore Creek, Alaska (current satellites carry five digital tape recorders). This is the only method of obtaining 1km resolution imagery from the large areas of the globe not covered by HRPT receiving stations. The data are digitised to 10-bit precision.
   c) The third source of digital data, GAC, are recorded for an entire orbit, then dumped at Wallops Island. By this method, 4km resolution data are derived from 1km data, ie four out of every five samples along the scan line are used to compute one average value, and that data from only every third scan line are processed. The 10-bit precision is retained. TIROS Information Processor (TIP) is the onboard computer system that formats the sensor data for transmission. TIP includes TOVS and auxiliary data. GAC data allow daily imaging of virtually the whole of the Earth's surface. Since a GAC pixel is not square, severe distortions of surface features are generated (these can be rectified).
   d) APT: a continuously transmitted signal which may be captured and displayed by relatively inexpensive omni-directional antennae and unsophisticated imaging equipment. APT data are a processed subset of two of the original AVHRR bands and contain analogue data with an effective resolution of about 4km.

Details of expected future direct broadcast service changes are shown in WMO (1989).
ii) Formats:
   a) Mosaics:

     i)   25 × 25cm negatives, N and S hemisphere polar stereographic projections (Note: the production of original negatives from NOAA polar orbiter satellite data ceased in April 1985. From this date, the SDSD has archived UNIFAX copies of these products in UNIFAX format only, some of which are in slightly different formats).

    ii) 35mm microfilm, 2–3 months per reel.

   iii) 9-track/1600bpi CCT. Stored as 2048 pixels per scan line, with 5 values (corresponding to each of the bands) per pixel.

b) Orbit swaths (GAC Pass-by-Pass):

    i)   25 × 25cm negatives, 1 negative per track.

    ii) CCT, Level 1b.

c) 7-day CMB and CMT images with a 55km resolution. 25 × 25cm negatives, 1 negative per day; and on 35mm microfilm.

d) Individual frames/datasets (HRPT/LAC):

    i)   25 × 25cm negatives.

    ii) CCT, Level 1b.

   iii) Data are available in two forms:

       a) Level 1b: calibrated radiance data.

       b) Level II: retrieval products data, February 1979 to present.

   iv) The following options are offered for CCT data:

       a) CCTs with or without terabit memory (TBM) header record.

       b) CBTs with or without IBM standard label.

       c) Two record sizes for full band LAC/HRPT data – 7400 or 14800 byte records, and

       d) Band data packed in 16 instead of 10 bit words.

    v) Weekly N hemisphere snow and ice charts have been produced by National Environmental Satellite Service (NESS), Suitland, MD since 1966. S hemisphere charts are also produced. Joint Ice Center charts have been digitised to give sea ice contentrations, ice types, surface features etc, and coded using SIGRID (sea ice grid) system – available from NSIDC on 9-track CCTs, 1600 or 6250bpi, EBCDIC or ASCII, block length 4000 bytes, dataset 1972 to present. The JIC charts are derived not only from NOAA satellite data but from other sources, eg passive microwave data. Software for reading data in SIGRID format are available for borrowing/copying from NSIDC, on 1 CCT, 9-track, 1600bpi, ASCII, containing seven files. Weekly N hemisphere snow charts are manually digitised on a subset of a 128 × 128 polar stereographic map; the resolution is 190km at 60°N (Matson and others 1986).

   vi) Data from Dundee: VHRR/AVHRR photofacsimiles are linearised to correct Earth curvature distortion, but are not Earth-rotation deskewed – this is less of a problem near the poles (linearisation – scale of image is very nearly constant in all directions and at all points on the image).

       a) Two types of imagery are available from Dundee:

          i)   Full pass with/without latitude, longitude and land outline grids.

          ii) Sectorised enlargements: by electronic stretching – no gridding, but linearised.

  vii) Note: on 11 April 1985, NOAA abandoned the TBM system as a means of storing ingested polar orbiter data. During the period of conversion from TBM tapes, SDSD gave priority to recovering the GAC datasets in the new archive format, and 88% of these were saved. At the time of going to press, it is understood that 49% of the TOVS and 6% of the LAC/HRPT data were saved. Users should contact SDSD regarding the availability of any Level 1b data ingested between October 1978 and April 1985.

viii) The imagery can be geometrically corrected by NOAA and displayed on standard map projections, with coastlines added. This permits accurate geographical registration of ice features and measurement of their movement by overlying frames. See McClain and others (1989 and 1990).

iii) Other data products:

a) Planetary albedo (PA) and outgoing longwave radiation (OLR). The algorithm used to obtain PA was changed in May 1988 (Wydick and others 1987). Daily values and monthly means in two formats: a 125×125 grid-per-hemisphere and a 2.5×2.5 grid. OLR products include daily day and night values, absorbed solar radiation, and available solar radiation.

b) Aerosol products from channel 1 data: weekly contour charts and CCTs of individual aerosol optical thickness observations.

Contact NOAA for information regarding new calibration coefficient values for VIS channel data.

TOVS:

HIRS/2 data are digitised to 13-bit precision; SSU and MSU data are digitised to 12-bit precision. The MSU data output represents an apparent TB after a 1.84s integration period per step. Two CCT products are available:

i) Raw TOVS (Level 1b) data.

ii) Sounding Product, 9-track/1600bpi CCT, one tape per week. The sounding products are archived as temperature profiles from the surface to 0.4mbar; precipitable water for three atmospheric layers; tropopause pressure and temperature; cloud amount and height; and total ozone amount estimates. Nominal resolution 250km.

ARGOS:

data are available either on CCTs or in hardcopy form on computer printouts. ARGOS can process the raw data if necessary. Transmission across global data communications networks ensures delivery of data in near real-time if required. Four levels of location accuracy are available:

i) Class 0 (special location): based on a minimum of two messages received by satellite. Both Doppler solutions are given.

ii) Class 1 (non-guaranteed location): 68% of results within 1km of true position.

iii) Class 2 (standard location): 68% of results within 350km of true position.

iv) Class 3 (quality location): 68% of results within 150m of true position.

AMSU-A and AMSU-B:

Digital data on CCTs, format to be determined.

ERBE and SBUV/2:

for details see Ng and Stonesifer (1989) and Oslik (1984). SBUV/2 data products are under technical evaluation by a NOAA/NASA science team. For further details regarding the format of all data from NOAA Polar Orbiter Satellites, see Kidwell (1988).

Data availability and archiving dates

Data from 30 October 1978 to present (TOVS from 1 January 1979) are archived at Satellite Data Services Division (SDSD), NOAA/NESDIS, Princeton Executive Center, 5627

Allentown Road, Camp Springs, MD 20746, USA, telephone (301)–763–8399, telex 2483760 BSWUR, FAX 7638443. NOAA/NESDIS operate a data retrieval service (Kidwell 1988). Users may also obtain one-to-two-day turnaround for requests for the current operational polar orbiter Level 1b data sets, and may request routine delivery of data sets with coverage of a specified area. Seven-day CMB and CMT images generated from AVHRR data are available from 1 January 1979.

Users can gain access to an on-line inventory of AVHRR and TOVS data, free of charge, via the SDSD Electronic Catalog System (ECS); requests can also be submitted in this way. The data range of ECS is from April 1985 to present. Contact Mary Hollinger or Kay Metcalf at SDSD for further information. NOAA also maintain an Earth System Data Directory; for details on how to access this using a PC, contact Gerald Barton at NOAA/NODC, 1825 Connecticut Avenue, NW, Washington DC 20235, USA. The NOAA directory uses software developed by NASA for the NASA Master Directory. Useful publications include *The Catalogue of Operational Satellite Products*, (NESS 109) and *The Satellite Data Users' Bulletin* (published on an irregular basis). Both are available from SDSD.

Note: SDSD will continue to archive data principally in the form of orbitally sequential radiances. This is inconvenient for many scientific users, who generally require geographically gridded products of geophysical parameters.

Various AVHRR digital and non-digital data and data products are also available from the National Climatic Data Center (NCDC), Federal Building, Asheville, NC 28801–2696, USA, telephone (704)–259–0682, telex 6502643731, FAX 704–259–0876.

Other stations receiving (and archiving) data:

 i)   Department of Electrical Engineering, University of Dundee, Dundee DD1 4HN, Scotland, telephone 0382–23181, telex 76293 ULDUND G: since 6 November 1978. Photographic or digital format. All data are stored in original unprocessed form, and are archived indefinitely. CCTs 9-track 1600bpi. Coverage of much of the European Arctic, including E Greenland and Svalbard.

 ii) Rude Skov, Danish Meteorological Institute: standard black and white 10 × 8 prints since 1972. Archives data from only one satellite pass per day.

 iii) Sondre Strømfjord, W. Greenland: joint Danish Meteorological Institute and Canadian Atmospheric Environment Service (365 Laurier Avenue West, Ottawa, Ontario K1A 0H3, Canada) receiving HRPT data.

 iv) Tromsø Telemetry Station, PO Box 387, Tromsø, Norway N–9001, telephone 083 84817, telex 64025: HRPT coverage of the Barents and Greenland Seas, the North Pole and much of the Canadian and Russian Arctic. It is run by a private foundation and acts as the Earthnet National Point of Contact for Norway. Tromsø is able to see 10 out of the 14 NOAA passes per satellite per day. At least one full pass is archived per day. Produces quick-look (QL) images; users can order processed pictures and/or CCTs. Since 1988, preprocessed AVHRR data have been distributed in addition to raw data; former include system correction, radiometric enhancements, Earth curvature/rotation, and image data in annotated and gridded polar stereographic projection. Imagery can be ordered prior to the time of interest. Future plans include the further development of processing algorithms, special processing, high-speed data transmissions to users, and further cooperation with ESA-EARTHNET.

 v)   CMS Lannion, France: archives data from all available satellite passes per day. Offers coverage of much of Greenland and the Labrador Sea, including Newfoundland. Lannion data are archived at: ESRIN/Information Retrieval Service, Online Services

Division, Via Galileo Galilei, 00044 Frascati, Italy, telephone 39–6–9426285 or 39–6–9401218, telex 610637 ESRIN I EURIMAGE. The UK EARTHNET National Point of Contact is: Mr M. Hammond, National Remote Sensing Centre, Royal Aircraft Establishment, Farnborough, Hampshire GU14 6TD, telephone 0252 24461, telex 858442.

vi)   Canada: reception of HRPT data from Edmonton and Toronto – Ice Information Centre, AES, 365 Laurier Ave West, Ottawa, Ontario K1A OH3. NOAA data are also received at i) Shoe Cove, which offers coverage of Greenland, Iceland, Labrador, Newfoundland and Baffin Island (picture centre at about 62°N), and ii) Prince Albert, Saskatchewan. For further information, contact: Data Acquisition Division, Canada Centre for Remote Sensing, 2464 Sheffield Road, Ottawa, Ontario, K1A OY7, Canada, telephone (613)–952–2717, telex 0533777 CA. CCRS has archived AVHRR data collected in Canada since 1983.

vii) GeoData Center, Geophysical Institute, University of Alaska-Fairbanks, AK 99775–0800, telephone (907)–474–7487, holds a regional archive of AVHRR data.

viii) The Naval Polar Oceanography Center, 4301 Suitland Road, Washington DC, USA, archives a large amount of NOAA polar imagery, hardcopy format, both polar regions. These are used operationally to produce ice charts (see above). NPOC is presently installing a digital ice forecasting and analysis system, thereby upgrading from manual to interactive interpretation of ice features. ERS-1 and Radarsat SAR data will be incorporated when available.

ix)   Japan: Meteorological Satellite Center, 3–235 Nakkiyoto, Kiyose, Tokyo 204, Japan, telephone (0424)–93–1111, FAX (0424)–92–2433. AVHRR and TOVS coverage of Sea of Okhotsk. Sea ice charts are published for shipping and fishing, based on these and GMS imagery.

x)   Antarctica: HRPT receiving stations at McMurdo Sound (since 30 October 1985) and Palmer station (since January 1990); Syowa (Japanese) has been receiving data since 1981 (Takabe and Yamanouchi 1989). Several images are collected a day, year round, by the US bases, covering 85% of Antarctica (to 90°). The software can differentiate between clouds and ice in two ways: visually, by using the low sun angle; or, by using the IR bands, on the basis of temperature, as low clouds are generally warmer than the ice surface. Time-sequencing of images can be performed. Previously at McMurdo, data were frequently only retained for two–seven days.

Raw HRPT data recorded on to HDDTs at McMurdo and Palmer are converted to 8mm video cassettes at Scripps Institution of Oceanography. Scientists wishing to obtain or work interactively with the system should contact Robert Whritner, Antarctic Research Center (ARC), Ocean Research Division, A-014, Scripps Institution of Oceanography, La Jolla, CA 92093, USA; telephone (619)–534–3785. Data are collected and stored on 8mm video cassettes. The user may select from a variety of media and formats.

Advance notification of data requirements by scientists could result in data being retained. Advanced processing and interpretation is carried out at SIO. It may be simpler to obtain Antarctic data via the normal LAC data route at SDSD, although SDSD collection of Antarctic AVHRR imagery is in jeopardy due to budget cuts. Few pre-1978 NOAA data are available. Coverage of specific areas of Antarctica can be scheduled, giving 2–3 months notice, by contacting the Interactive Processing Branch of SDSD – LAC coverage for an image dataset of about 2,500km swath width by 3,000km along-track.

Argentina has an APT receiving station at Marambio Base (64°14'S, 56°43'W), receiving NOAA (and Russian METEOR) imagery.

xi) The University of Rhode Island has put together a comprehensive listing of available high resolution AVHRR data (from all sources). The inventory is on-line and menu-driven. For further information, contact George Milkowski, Graduate School of Oceanography, University of Rhode Island, Narragansett, RI, 02882, USA, telephone (401)–792–6939.

xii) Weekly Snow and Ice Charts digitised with the SIGRID output format are available from National Snow and Ice Data Center (NSIDC), WDC-A for Glaciology, Campus Box 449, University of Colorado, Boulder, CO 80309, USA, telephone (303)–492–5171, telex 257673 (WDCA UR), telemail MAIL/USA [NSIDC/OMNET], VAX mail via SPAN: KRYOS: NSIDC. E and W Arctic data for January 1973–December 1984, and Antarctic data for January 1972–December 1984, with annual updates. The ice charts derive from the US Naval Polar Oceanography Center (see above).

xiii) Sea Ice Charts derived from NOAA data are produced by other sources, including AES, Ottawa, Canada. These charts include information on ice extent, concentrations of each ice category, concentrations of floes within each category, an indication of iceberg presence, and surface features of the ice.

xiv) Snow cover maps for the N and S hemispheres, based on the VIS band of the AVHRR, are available from SDSD. The digital archive includes data for the N hemisphere from conventional observations dating back to 1966.

xv) Twice daily global SST coverage (8km resolution) when cloud-free, composite weekly analyses (on a $1 \times 1°$ grid) and monthly means ($2.5 \times 2.5°$ bins). Quality control data from buoys, etc included in the archive. Before March 1990, several linear combinations of the TIR channels were used to calculate SSTs from previously cloud-cleared data. NESDIS now apply the non-linear Cross-Product SST (CPSST) atmospheric correction algorithm, which better accounts for varying water vapour conditions and scene temperature. For further information, contact Dr Paul McClain at NESDIS, telephone (301)–763–8078. Monthly global SST products, $2.5 \times 2.5°$ resolution, available from National Space Science Data Center (1 January 1979 to present). For full address, see TIROS I-X.

xvi) AVHRR PA and OLR data (since 1974) available from SDSD.

xvii) AVHRR aerosol products (since June 1987) available from Product Code AER2, National Climatic Data Center, Federal Building, Asheville, NC 28801, USA.

ARGOS:

Service ARGOS, Centre National d'Études Spatiales, 18 Av. Edouard Belin, 31055 Toulouse Cedex, France, telephone 61274351, telex 531752 F, FAX 61 75 10 14. In the USA, contact: Service ARGOS Inc., Suite 10, 1801 McCormick Drive, Landover, MD 20785, USA, telephone 301–925–4441, telex 898146, FAX 301–925–8995. In Australia contact: CLS ARGOS Australia, Bureau of Meteorology, GPO Box 1289 K, Melbourne, Victoria 3001, Australia, telephone +61 3 669–4650, telex 30434, FAX +61 3 669–4168. In Japan, contact: Cubic I Ltd, 2–9–7–312 Nishi Gotanda, Shinagawa Ku, Tokyo 141, Japan, telephone 3 779 5506, FAX 3 779 5783. User guides, newsletters, data sheets, etc are available.

Data from the CAOBP are available from NSIDC. 9-track EBCDIC, odd parity, 6250bpi, 80 characters per record, 4800 characters per block. Antarctic Automatic

Weather Station data are available from the Department of Meteorology, University of Wisconsin, Madison.

SBUV/2:

Level Ib, Historical Instrument File and ozone products (from 14 March 1985 to present) are available from NCDC (see above for address). TOVS data (from 19 October 1978 to present) are also available from this source.

References and important further reading

Bessis, J. 1979. Operational data collection and platform location by satellite. *Proceedings of the 13th International Symposium on Remote Sensing of Environment*, 23–27 April 1979, Volume 1: 485–504. Environmental Research Institute of Michigan, Ann Arbor, Michigan.

Bindschadler, R. and Vornberger, P. 1990. AVHRR imagery reveals Antarctic ice dynamics. *EOS* 71, 23: 741.

Bromwich, D. 1989. Satellite analyses of Antarctic katabatic wind behaviour. *Bulletin of the American Meteorological Society* 70, 7: 738–49.

Burke, S. and Anderson, J. 1989. Oceanographic measurements from drifting buoys. *ARGOS Newsletter* 37: 13–15.

Chandra, S., McPeters, R., Hudson, R. and Planet, W. 1990. Ozone measurements from the NOAA-9 and the NIMBUS-7 satellites: implications of short and long term variabilities. *Geophysical Research Letters* 17, 10: 1,573–6.

Dey, B. 1981. Monitoring winter sea ice dynamics in the Canadian Arctic with NOAA TIR images. *Journal of Geophysical Research* 86, 4: 3,223–35.

Ebert, E. 1987. Classification and analysis of surface and clouds at high latitudes from AVHRR multispectral satellite data. Unpublished PhD thesis, University of Wisconsin, Madison.

Emery, W., Brown, J. and Novak, V. 1989. AVHRR image navigation: summary and review. *Photogrammetric Engineering and Remote Sensing* 8: 1,175–83.

French, J. 1988. Power budget improvements for miniature PTTs. *ARGOS Newsletter* 34: 13.

Fusco, L., Muirhead, K. and Tobiss, G. 1989. EARTHNET's coordination scheme for AVHRR data. *International Journal of Remote Sensing* 10, 4 and 5: 625–36.

Gesell, G. 1989. An algorithm for snow and ice detection using AVHRR data. An extension of the APOLLO software package. *International Journal of Remote Sensing* 10, 4 and 5: 897–905.

Kidwell, K. 1988. *NOAA Polar Orbiter Data (TIROS-N, NOAA-6, NOAA-7, NOAA-8, NOAA-9, and NOAA-10) Users' Guide.* Available from SDSD.

Matson, M., Ropelewski, C. and Varnadore, M. 1986. *An atlas of satellite-derived northern hemisphere snow cover frequency.* NOAA, Washington DC.

McClain, E.P. 1989. Global sea surface temperatures and cloud clearing for aerosol optical depth estimates. *International Journal of Remote Sensing* 10, 4 and 5: 763–9.

McClain, C., Chen, J., Darzi, M., Firestone, J. and Endres, D. 1989. The SEAPAK User's Guide. *NASA Technical Memo TM 100728.* NASA GSFC, Greenbelt, MD 20771. SEAPAK software is available from NASA's Computer Software Management Center (COSMIC), University of Georgia, 382 East Broad St., Athens, GA 30601, USA, telephone (404)–542–3265.

McClain, C., Fu, G., Darzi, M. and Firestone, J. 1990. PCSEAPAK User's Guide. NASA GSFC, Greenbelt, MD 20771. Available from COSMIC (see above).

Ng, C. and Stonesifer, G. 1989. Data catalog series for space science and applications flight missions. Volume 4B: descriptions of data sets from meteorological and terrestrial applications spacecraft and investigations. *NSSDC/WDC-A-R&S 89–10.* Available from NSSDC.

Ninnis, R., Emery, W. and Collins, M. 1986. Automated extraction of pack ice motion from advanced very high resolution radiometer imagery. *Journal of Geophysical Research* 91, C9: 10,725–34.

Oslik, N, (ed.). 1984. *Solar Backscattered Ultraviolet Radiometer/2 Users' Guide.* NOAA, Washington DC.

Parkinson, C., Comiso, J., Zwally, H., Cavalieri, D., Gloersen, P. and Campbell, W. 1987. Arctic sea ice, 1973–6: satellite passive-microwave observations. *NASA Special Publication SP-489*. NASA Scientific and Technical Information Branch, Washington DC.

Planet, W. 1988. Data extraction and calibration of TIROS-N/NOAA radiometers. *NOAA Technical Memo Ness 107*. NOAA SDSD.

Prada, K. and Baggeroer, A. 1988. An Arctic remote autonomous measurement platform. In *Instrumentation and Measurements in the Polar Regions. Proceedings of the Workshop*, January 1988: 373–83. Marine Technology Society, Berkeley, CA 94702.

Susskind, J., Reuter, D. and Chahine, M. 1985. Multispectral remote sensing of climatic parameters from HIRS-2/MSU data. In Deepak, A., Fleming, H. and Chahine, M. (eds), *Advances in remote sensing retrieval methods*: 205–20. A. Deepak Publishing.

Takabe, H. and Yamanouchi, T. 1989. NOAA data processing system. *Antarctic Record* 33, 1: 73–87.

Tomkiewicz, S. and Beaty, D. 1988. Wildlife satellite telemetry. *ARGOS Newsletter* 34: 7–11.

Untersteiner, N. and Thorndike, A. 1982. Arctic Ocean Data Buoy Program. *Polar Record* 21: 127–35.

Walton, C. 1987. The AVHRR/HIRS operational method for satellite-based sea surface temperature determination. *NOAA Technical Report NESDIS 28*. US Department of Commerce, Washington, DC.

Warren, D. 1989. AVHRR channel-3 noise and methods for its removal. *International Journal of Remote Sensing* 10, 4 and 5: 645–51.

WMO. 1989. Information on meteorological and other environmental satellites. *WMO-411*, World Meteorological Organization, Geneva.

Yamanouchi, T. and Seo, Y. 1984. Discrimination of sea ice edge in the Antarctic from NOAA MSU. *Polar Memoirs, Special Issue* 34: 207–17. National Institute of Polar Research, Tokyo.

17. Insat

Operational lifespan and orbital characteristics

Dates: 10 April 1982 to present
Semimajor axis: 41,693–42,153km.
Orbital height: 33,476–37,154km.
Inclination: 0.1–0.5°.
Period: 1,412–1,435 minutes (geostationary).
Longitude: 94°E.
Notes: Operated by India.

Sensor characteristics

VHRR
  i) VIS (0.55–0.75μm). GR 2.75km. 10-bit quantisation. Imagery every 3 hours.
 ii) TIR (10.5–12.5μm). GR 11km.
Also carries a Data Collection System (DCS).

Polar applications and limitations

See GOES, Meteosat and GMS.

Data format

CCTs.

Data availability

Director General of Meteorology, Indian Meteorological Department, Mausam Bhavan, Lodi Road, New Delhi 110003, India, telephone (11)–626021, telex 3166494 MDGM IN. For further information, see WMO (1989).

References and important further reading

WMO. 1989. Information on meteorological and other environmental satellites. *WMO-411*, World Meteorological Organization, Geneva.

18. Seasat

Operational lifespan and orbital characteristics

Dates: 27 June 1978–10 October 1978.
Semimajor axis: 7,166km.
Orbital height: 776–800km.
Inclination: 108.0°.
Period: 100.63 minutes. Exactly-repeating orbit. Repeat period 17 days for commissioning phase, then 3 days from 14 September onwards.
Notes: Near circular orbit, not sun-synchronous. SMMR and SASS repeat coverage 36 hours.

Sensor characteristics

Radar Altimeter (ALT):
   Active microwave: frequency 13.5GHz (K$_u$-band). RF bandwidth 320MHz. Antenna beamwidth 1.6°. Pulsewidth 3.2μs (uncompressed), 3.125ns (effective/compressed). PRF 1,020Hz. Altitude precision ±10cm rms. SWH of 1–20m accuracy ±10% or 0.5m, whichever is greater. Pulse-limited footprint diameter 2.4km (for calm seas or flat surfaces) to 12km (for rough seas), centred on nadir. Data rate 10 Kbps.
Visible and Infrared Radiometer (VIRR):
   i)   Band 1: VIS/NrIR (wavelength 0.47–0.94μm). GR 3km. Swath width aproximately 2,280km, centred on nadir (both bands).

  ii) Band 2: TIR (wavelength 10.5–12.5μm). GR 5km.
Scanning Multi-channel Microwave Radiometer (SMMR):
  i) Passive microwave radiometer. Channel 1: frequency 6.63GHz, wavelength 4.55cm. GR
      149 × 87km. Swath width 600km for all channels (40km to the left of nadir, 560km to
      the right).
      Looks northward in both hemispheres.
  ii) Channel 2: frequency 10.69GHz, wavelength 2.81cm. GR 89 × 58km.
  iii) Channel 3: frequency 18.00GHz, wavelength 1.67cm. GR 53 × 31km.
  iv) Channel 4: frequency 21.00GHz, wavelength 1.43cm. GR 42 × 27km.
  v) Channel 5: frequency 37.00GHz, wavelength 0.81cm. GR 27 × 16km.
  vi) HH and VV polarisations for all channels. SST accuracy ±2K. Wind speed accuracy
      ±2ms^{-1} or 10%, whichever is greater for speeds of <30ms^{-1} in non-rainy ocean areas.
      Identical to the SMMR launched on Nimbus-7, but the Seasat SMMR was aft-viewing
      with the swath centre 22° from nadir to the right of the satellite to permit overlap with
      the footprints of the SASS and SAR; Nimbus-7 SMMR was forward-viewing with the
      swath centred on the sub-satellite track. Spatial resolution for ice products is 25km, and
      125km for SSTs.
Seasat-A Radar Scatterometer System (SASS):
    Active microwave: frequency 14.595GHz (K$_u$-band). Bandwidth ±500kHz. Incidence
    angle 25–65°. Four dual polarisation fan beam antenna, giving an X-shaped illumination
    on the surface. 15 backscatter measurements every 1.89s. Antenna switching cycle every
    7.56s. Receiver noise 5.7dB. Transmitted pulselength 4.8ms^{-1}. Eight modes of operation
    (different combinations of transmmitting antennae and polarisations). Wind speed from
    7–50ms^{-1}, ±2ms^{-1} or 10%, whichever is greatest. 50km resolution cell (100km cross-
    track and along-track spacing between cell centres).
Synthetic Aperture Radar:
    Active microwave: frequency 1.275GHz (L-band), wavelength 23.5cm. RF bandwidth
    19MHz. Pulse width 33.8μs. Contrast ratio 9dB. Pulse repetition range 1463–1640pps.
    Transmit time bandwidth product 634. HH polarisation. Incidence angle 23°. Four look
    directions. Azimuth resolution 25m (four look). Range resolution 25m. Ground resol-
    ution dependent upon processing. Swath width 107km, centred 290km from nadir (Figure
    II.10). Data rate 110 Mbps.

Polar applications

ALT:
  i) *Sea ice*: potentially accurate delineation of ocean-sea ice and sea ice-ice sheet boundaries
      (the latter to an accuracy of ±1km). Ulander (1988) has shown that different sea ice
      types can be separated on the basis of their normal-incidence backscatter coefficients.
      Data from the Beaufort Sea showed that three ice types could be separated: new ice (with
      a mean backscatter coefficient of 33.0 ±2.8dB), and two types of MY ice, one consisting
      of large relatively underformed floes with clearly delineated ridges (16.6 ±1.3dB) and
      the other composed of small and deformed floes (24.1 ±3.2dB) (open water has a
      backscatter coefficient of 8–15dB for wind speeds between 15 and 2 ms^{-2}). Potential
      detection of swell penetration and attenuation in sea ice, and wave and wind field
      measurements near the ice edge. ALTs can measure both the baroclinic and barotropic
      geostrophic flow. Data collection is not limited by weather, darkness or ice surface

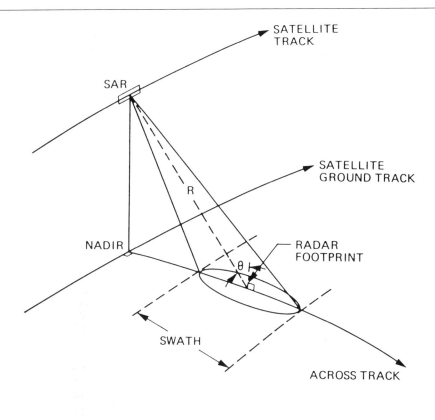

Figure II.10   Seasat SAR imaging geometry. From Fu and Holt (1982).

wetness. Large-scale variations of ocean circulation and the determination of ocean surface currents. See also Skylab, GEOS-3, GEOSAT, TOPEX/Poseidon and ERS-1.

ii) *Land ice*: accurate surface elevation measurements, with information on changes in ice volume, accumulation rates, the delineation of major ice flow features, including ice divides, ice domes, drainage basins, ice streams, and outlet glaciers (Figure II.11). Measurement consistency ± 25cm over the smoother portions of Antarctica. Overall data consistency estimated at ± 2.7m over the ice sheet covered (by comparing measurements at orbit crossing points). Showed the potential for long-term monitoring of ice sheet mass balance and dynamics. See also Skylab, GEOS-3, GEOSAT, TOPEX/Poseidon and ERS-1.

VIRR:

primarily to aid in feature identification for the microwave data, and to act as a means of linking the performance of microwave sensors to previous experience with optical sensors. Design inheritance from ITOS SR. See also NOAA AVHRR.

SMMR:

identical to the SMMR flow on Nimbus-7. It provided liquid water and vapour corrections for ALT and SASS. Channel 1: SST. Channel 2: ocean surface wind speed. There is a greater contrast in the TBs of ice and water with H polarisation. V TBs are always greater

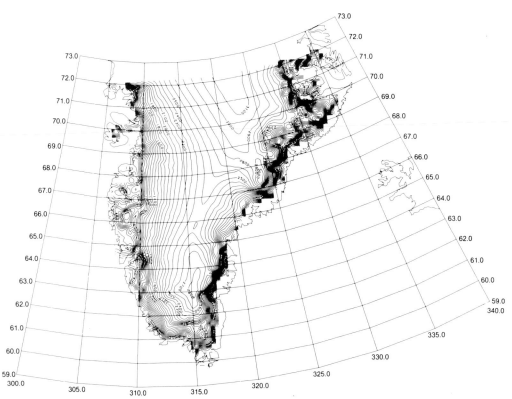

Figure II.11   Topographic contour map (100m intervals) of the S Greenland ice sheet constructed from Seasat ALT data by Zwally and others (1990a). Dashed lines signify less reliable data.

than those with H. H is more sensitive to changes in surface roughness. See also DMSP SSM/I.

SASS:

primarily designed to give surface wind speed and direction over oceans.

  i) *Sea ice*: potential large-scale sea ice classifications (based on backscatter differences), sea ice-ocean boundaries, and broad-scale ice surface roughness characteristics. Spatial resolution and swath width are similar to passive microwave radiometers. Onstott and Larson (1986) have measured active microwave properties of summer sea ice, at frequencies from 1–17GHz, angles 0–70° from vertical, and with like and cross-antenna polarisations. Their results indicate that snow thickness, melt-water, snowpack morphology, snow surface roughness, ice surface roughness and deformation characteristics are the fundamental properties that govern the summer backscatter response. A thick, wet snow dominates the backscatter response, reducing the latter and masking any features below. The backscatter coefficient increases when surface melt ponds refreeze. There is an indication of reduced anisotropy in SASS backscatter in cold ice conditions compared to wet ice conditions (Carsey 1985). Sensitive to ice roughness and dielectric properties, which are determined by the ice age and thermal history. As the instrument views at angles off-nadir, the percentage of the signal associated with roughness is greater than that of the ALT. High backscatter coefficient in old ice, moderate in new/young, and low in open water. The return signal from ice may be 10–15dB greater than that from calm water (Onstott and others 1979).

ii) *Land ice*: the penetration depth is far less than that of the L-band SAR frequency, and therefore SASS measurements contain information primarily on surface roughness and snow characteristics within a few metres of the surface. Regions of strong backscatter are assumed to be areas of surface melt.

See also ADEOS NSCAT and ERS-1 AMI-SAR Wind Scatt Mode.

SAR:

i) *Sea ice*: information on meso-scale behaviour. High resolution, all-weather, all-season, day and night capability. Pixels in open ocean can be located with an rms error of ±250m (±100m if land is in the image). Detailed information on ice concentration, floe size or shape distribution, pressure ridge extent, pressure zones, and lead or polynya delineation and orientation. Classification (MY versus FY). Very accurate delineation of the ice/water boundary, and fast ice edge. The data are particularly useful for detailed studies of complex MIZs, where it is possible to track ice-traced eddies, bands and other ice edge features. Tracking the relative and absolute position of features to derive ice floe motion vectors. Tracking of icebergs and ice islands.

L-band allows an easy distinction between open water and ice, and identification of surface roughness features/ridges, but provides little information on ice type. Ice production rates in polynyas, as grease and pancake ice have distinctive signatures (more research is necessary). Internal horizons due to deformation are discernible. Two dimensional surface wave spectra at the ice edge, with information on the reflection, refraction, swell direction, penetration into the pack and attenuation. Identification of current shear zones and internal waves.

In summer, the surface morphology (eg ridge number and size) dominates the backscatter; in winter, volume scattering is more important, and ice microscopic physical structure is apparent. Operational applications include ship routing, and monitoring the threat of ice and icebergs to the hydrocarbon and fishing industries.

ii) *Land ice*: accurate mapping and monitoring of coastal margin positions and glacier termini. Ice motion can be studied by tracking identical features on sequential imagery. Flow streamlines on outlet glaciers and ice streams, ice rises, ice shelves, crevasse fields, flowbands and melt zones. Rate of ice discharge, with data on the size, distribution and calving rate of icebergs. SAR can identify and delineate lakes and streams (even those covered by snow). Potential data on the extent of ablation and wet snow zones on ice sheets. The location of blue ice zones, which mark the presence of meteorites. See also SIR-C/X-SAR, JERS-1, ERS-1 and Radarsat.

Limitations

Coverage was limited to the N hemisphere summer and early autumn (90 days in all), and only up to a maximum latitude of 72°. Seasat failed before many of the scheduled *in situ* ground truthing and calibration observations could be made.

ALT:

the tracking circuitry, designed for use over open ocean, was not well suited to operation over rougher ice surfaces, and was only successful in areas of ≤1° slope. Suffered frequent loss of track even in flat ice sheet areas. Over sea ice, the intense, peaked and variable echo waveforms caused instrument anomalies, including saturation and height glitch. The beam-limited footprint size limits the scale of individual surface features that can be studied. Spatial coverage limited by the narrow swath. The resulting apparent ice surface is derived from the measured ranges representing a smoothed envelope biased slightly above the actual slope; this is dependent upon local slope, undulation amplitude and ALT characteristics, and can be corrected to some extent by a dense network of tracks. Accurate orbit ephemeris unavailable. Wind speed measurements impossible over ice. Atmospheric water vapour lengthens the electromagnetic propagation path, and horizontal water vapour variations cause variations in observed ALT height (can be corrected using simultaneous SMMR data). The ALT and SASS do not produce imagery. Simultaneous intercomparisons with other microwave sensors are restricted. See also Skylab, GEOS-3, GEOSAT, TOPEX/Poseidon and ERS-1.

VIRR:

no direct polar applications.

SMMR:

the Seasat project evaluation teams did not develop algorithms for the extraction of sea ice parameters from Seasat SMMR data, although several Nimbus-7 SMMR algorithms have been applied to the data. See Nimbus-7 SMMR and DMSP SSM/I.

SASS:

most of the retrieved wind vectors had four directional ambiguities due to the fixed frequency Doppler filtering design. Thick cloud or rain cause an underestimate of wind speed. No measurement of wind speed or direction is possible over ice. In defining the ice edge, the SASS data are susceptible to a loss of ice/water contrast when the ice surface is wet or the seas are under the influence of high winds ($\geqslant 16\,\text{ms}^{-1}$) – the latter give a backscatter coefficient similar to that of new/young ice. Wet ice has a low backscatter coefficient, dominated by ice morphology. Snow and melt water are not distributed uniformly, and the stage of melt may be variable. These non-uniformities related to ice type are not well understood, and produce unique microwave signature characteristics. Performance uncertain over regions with a mixture of ice and open water (operates best over consolidated ice cover or open water). SASS operation was not continuous. The mode setting in polar regions was random. Coarse spatial resolution. Backscatter variations over ice sheets are poorly understood. See also ADEOS NSCAT and ERS-1 AMI-SAR Wind Scatt Mode.

SAR:

insufficient onboard storage for SAR data; only operational in real-time when in range of specialised receiving stations. No SAR data were collected over Antarctica. Polar coverage is limited to the Beaufort and Bering Seas, along with one strip of coverage over Iceland and Denmark Strait and two strips over S Greenland. Coverage limited by the narrow swath width (often too small to contain the same ice features on a three-day repeat interval), duty cycle, very high data rate and the complexity and expense of processing. Seasat SAR did not function long enough to study and readjust the parameters affecting the geometric calibration. The iceberg detection objective was not met because of the absence of surface truth and the season of operation.

It is difficult to classify sea ice at L-band based on backscatter alone – shape and texture information must also be used. The contrast between FY and MY ice is not strong at L-band, as volume scattering by air bubbles is not a dominant process at this long wavelength (Shuchman and others 1988). The dynamic range of FY and MY ice equals the range of new ice at L-band. Small leads are not resolved. The returns from open water, young, grease and slush ice have overlapping dynamic ranges. Further ambiguities may be caused by upward migration of brine on to the ice surface to form highly saline frost flowers. This, together with the effect of depth hoar on the ice surface and salt wicking, may drastically affect radar backscatter. Small incidence angles make the definition of certain floe characteristics difficult. At L-band, slush on surface melt areas produces strong backscatter which could be misinterpreted as rubble fields, and which also have a bright signature similar to that of MY ice.

Under summer melt conditions, L-band SAR imagery of sea ice is relatively featureless, showing ridge-like features but little small-scale structure. Surface wetness prevents penetration of the signal. The masking effect and seasonal variations of snow cover (especially when wet) are not well understood. Strong winds or heavy seas can cause ambiguities in interpretation, particularly in the MIZ, large leads and polynyas; often difficult to distinguish open water areas from areas of smooth thin ice. Backscatter from open sea clutter often greater than that of adjacent sea ice.

A problem common to all SARs is that they generate images by coherent processing of scattered signals, making them susceptible to speckle noise. This noise is proportional to the reflected signal power, and is therefore multiplicative. Brighter areas of the imagery are much noisier than darker areas (speckle can be reduced by using multiple look data). Ionospheric and tropospheric propagation effects may be significant over polar regions at L-band. Surface ocean wave measurements have a 180° directional ambiguity.

SAR over land ice must be corrected for elevation distortion effects. Shadowing largely precludes its use as a means of monitoring valley glaciers. Surface melt effects and the nature of the surface as it relates to backscatter are not well understood. See also SIR-C/X-SAR, JERS-1, ERS-1 and Radarsat.

Data format

ALT:
CCTs only. ALT data recorded at the receiving station as Sensor Data Records (SDRs), which were then processed and averaged to produce Geophysical Data Records (GDRs). SDR data are intended for application to oceanic research in regions not covered with ice. GDRs are further divided into two types:
  i) Geophysical Files (GF): each CCT contains global data for approximately six days.
  ii) Sensor Files (SF): each CCT contains global data for about three days.
The principal data sets produced by Zwally and others (1990a, b and c) and retained by NSIDC and NSSDC are:
  i) Level 0: Condensed Sensor Data Records.
  ii) Level 1: Waveform Data Records (WDRs). Orbital format records, including waveform amplitudes by gate, AGC, ranges and latitude/longitude positions; the averaged radar return pulses contained in the SDRs.
  iii) Level 2: Ice Data Records (IDRs). Orbital format records, including ALT parameters, AGC, and latitude/longitude positions. AGC, applied corrections, corrected elevations, retracking beta parameters, and estimates of along-track and across-track slope corrections.
  iv) Level 3: Georeferenced data base, including all individual elevation measurements (together with time, latitude/longitude, and slope-correction estimates); accessible by geographic cells or 'bins'.
  v) Level 4: contour maps and gridded elevations with respect to Earth ellipsoid and sea level.

As data over the ice sheets are not available from the GDRs, Zwally and others (1990a) applied various corrections and adjustments to the SDR data to obtain ice sheet elevation measurements. Methods to derive sea ice information from SDR data are described by Laxon (1989).

SASS:

CCTs only. 9-track, 1600bpi. Note: ALT and SASS do not produce imagery.

SMMR:

9-track, 1600bpi CCTs. Both GDRs and SDRs.

SAR:

signal data, recorded on HDDTs at receiving stations, can be converted into imagery in two ways:

i) Optical: GR 40m, swath width 25km, for standard processing. The slower and more expensive digital processing give 25m resolution. Advantages (in comparison with digitally processed data): lower cost, high throughput, ideal for scanning data quality, intercomparing datasets and recognition of large-scale phenomenon. Once an area of interest has been identified from optically processed data, the data can be digitally processed and quantitatively analysed. Disadvantages: suitable only for qualitative analysis, not as flexible as digitally processed imagery (various minor distortions remain), degraded resolution and the difficulty of intensity calibrations and geometric distortions. The bulk of Seasat SAR data have been processed optically at JPL. Standard data products include 70mm black and white 1:500,000 scale paper prints (100km × 100km coverage), duplicate negatives, and positive transparencies. 9-track 1600bpi CCTs.

ii) Digital/rectified digital processing: has only recently become practical. GR 25m, 100km swath width. Advantages: preservation of full dynamic range, correction of minor distortions, and suitable for quantitative analysis. Disadvantages: low through-put rate and correspondingly higher cost. Standard data products include 1:500,000 scale paper prints, duplicate negatives, positive transparencies, and 9-track, 1600bpi CCTs.

Data availability and archiving data

ALT:

data are archived at JPL/NODS, Mail Stop 300–323, Jet Propulsion Laboratory, California Institute of Technology, Pasadena, CA 91109, USA, telephone (819)–354–6980 (Elizabeth Smith), telex 675429 (attention NODS), and FAX (819)–393–6720 (attention NODS). The NODS data system also contains, in machine readable form, the extensive sets of surface observations used to calibrate the Seasat instruments, handbooks, and a bibliography of descriptions of the satellite, instruments, the calibrations and the application of the satellite data to oceanography. GDRs and SDRs are also available for global ocean surface height and wind speed. Subroutines to read data are included with orders from JPL/NODS.

Data are also available from Satellite Data Services Division (SDSD), NOAA/NESDIS, Princeton Executive Center, 5627 Allentown Road, Camp Springs, MD 20746, USA, telephone (301)–763–8399, telex 248376 OBSWUR, FAX 7638443. SDSD has the capability to extract specific geographic regions and/or time periods from CCTs. Global coverage is available for: 7 July 1978–17 July 1978; 24 July 1978–28 August 1978; 1

September 1978; 6 September 1978–7 September 1978; 10 September 1978; 13 September 1978; 15 September 1978–10 October 1978.

Georeferenced ALT data collected over Greenland and Antarctica are available both from NSSDC (see TIROS I-X for address) and NSIDC, WDC-A for Glaciology, University of Colorado, Boulder, CO 80309, USA, telephone (303)–492–5171, telex 257673 (WDCA UR), telemail MAIL/USA [NSIDC/COMNET], VAX mail via SPAN: KYROS::NSIDC. These data are a derived product from retracked and atmospherically-corrected data, and are available as discrete elevations or on an interpolated 20km grid (Zwally 1990a). Zwally and others (1990a, b and c) document technical details and provide guidance to users of these data products.

Under an agreement between NASA and ESA, the Royal Aircraft Establishment (RAE) acquired a subset of the Seaset ALT SDR covering the polar regions. This is divided into Land-ice, Greenland, Sea-ice (North) and Sea-ice (South) categories, and has been extracted by the ERS-1 Algorithm Development Facility team for one 17-day repeat cycle from the Seasat mission. For further details, contact Mr N. Hammond, National Remote Sensing Centre, RAE, Farnborough, Hampshire GU14 6TD, UK, telephone 0252 24461, telex 858442.

VIRR:
no polar imagery archived. SST products are archived by JPL/NODS (see above for address).

SMMR:
data are archived at JPL/NODS (see above for address). Global coverage for 7 July 1978–10 October 1978. CCTs contain data products of surface wind speed, SSTs, integrated air column liquid water and vapour, rain rate, and individual channel TBs. An entire data set of sea ice concentration and extent, produced on daily 100×100km grids by Dr F. Carsey, is also available. User's handbooks are available from JPL/NODS.

SASS:
data and data products (surface wind and wind stress vectors) are archived at JPL/NODS (see above for address). CCTs containing latitude, longitude and time-located geophysical data of surface wind velocity and fully corrected backscatter coefficients. SASS GDRs are contained in three datasets:
  i) complete Geophysical and Sensor files on the same tape, each tape containing about six hours of continuous data for 7 July 1978–10 October 1978;
  ii) basic GDRs. Each CCT contains about 48 hours of continuous global data for 7 July 1978–10 October 1978; and
  iii) basic and supplementary GDRs. Each CCT contains about 24 hours of global data for 7 July 1978–10 October 1978. Note: Seasat carried two tape recorders for non-SAR data, with sufficient capacity to store data from two orbits at 25 Kbps for subsequent playback to a US receiving station. Data from SASS, ALT and SMMR therefore exist for the S hemisphere.

SAR:
data collected by the following receiving stations: Rosman (North Carolina), Goldstone (California), Fairbanks (Alaska), Shoe Cove (Canada), and Oakhanger (England). A total of about 2,500 minutes of SAR data were recorded and stored on CCTs (representing 500 passes). See Fu and Holt (1982). Data are archived at the following centres:
  i) SAR Data Catalog System (contact Amy Pang), NODS, MS 300–235, Jet Propulsion Laboratory, 4800 Oak Grove Drive, Pasadena, CA 91109, USA, telephone (818)–354–3386, SPAN KAHUNA::AAP. Requests from non-NASA scientists must

be addressed to Dr John Curlander, MS 320–235, JPL, telephone (818)–354–8262. Compressed and stored in an online system, thereby offering easy access.

ii)   SDSD, NOAA/NESDIS (see above for address): US data products and final GDRs. Optically processed imagery have been routinely processed by JPL. Fewer data have been processed digitally. SDSD, through an arrangement with JPL, can make available, on request, areas not processed by the Seasat Project. Data products are available for 7 July 1978–10 October 1978: about 478 passes of optically processed imagery as 70mm film products; about 250 passes of digital film products and CCTs.

iii) Some photos and 1600bpi CCTs of data from the Canadian Arctic are archived at Canada Centre for Remote Sensing, 2464 Sheffield Road, Ottawa, Ontario K1A OY7, Canada, telephone (613)–952–0500, telex 0533777 CA. At Shoe Cove, 6 passes were recorded optically, and 29 were recorded using both optical and digital techniques.

iv)   Oakhanger (UK): recorded only two passes optically, with polar coverage limited to Iceland and the Denmark Strait. Photographic products are available from ESRIN/Earthnet Programme Office, Via Galileo Galilei, 00044, Frascati, Italy, telephone 06 94011, telex 610637 ESRIN I.

v)   Under an international agreement with NASA, ESA have obtained the full set of Sensor Data Records from Seasat SASS and SMMR, and an increased amount of SAR. A set of 15 SAR images has been processed at RAE. These are located in areas for which simultaneous surface measurements are available. Each image has been chosen to contain a particular feature, such as internal and swell waves. Five CCTs containing geophysical parameters extracted from the Seasat GDR have also been acquired. For further details, contact Mr M. Hammond, National Remote Sensing Centre, RAE, Farnborough, Hampshire GU14 6TD, UK, telephone (0252) 24461, telex 858442.

The *Seasat Synthetic Aperture Radar Data User's Manual* (JPL Publication 82–90) is available from JPL/NODS.

ALT, SAR, SASS, SMMR and VIRR data and data products are also available from the National Climatic Data Center (NCDC), Federal Building, Asheville, NC 28801–2696, telephone (704)–259–0682, telex 6502643731, FAX 704–259–0876.

References and important further reading

Bernstein, R. 1982. Sea surface temperature mapping with the Seasat microwave radiometer. *Journal of Geophysical Research* 87, C10: 7,865.

Bindschadler, R., Zwally, H., Major, J. and Brenner, A. 1989. Surface topography of the Greenland ice sheet from satellite radar altimetry. *NASA SP–503.* NASA Scientific and Technical Information Division, Washington DC.

Born, G., Held, D., Lame, D., Lipes, R., Montgomery, D., Rygh, P. and Scott, J. 1982. SEASAT data utilization project report. *JPL Report D-36.* Contains an extensive SEASAT bibliography.

Carsey, F. 1985. Summer Arctic sea ice character from satellite microwave data. *Journal of Geophysical Research* 90, C3: 5,015–34.

Fedor, L. and Brown, G. 1982. Waveheight and wind speed measurements from the SEASAT radar altimeter. *Journal of Geophysical Research* 87: 3,254–60.

Ford, J. 1984. Mapping of glacial landforms from SEASAT radar images. *Quarternary Research* 22: 314–27.

Fu, L.-L. and Holt, B. 1982. SEASAT views oceans and sea ice with Synthetic Aperture Radar. *JPL 81–120.* Jet Propulsion Laboratory, Pasadena, California.

Journal of Geophysical Research. 1983. SEASAT Results Special Issue. 88, C3.

Laxon, S, 1989. *Satellite radar altimetry over sea ice.* Unpublished PhD Thesis, University College, London, UK.

Martin, S. 1980. Use of SEASAT (SMMR) imagery to study the Bering Sea ice. *Interim Report A80–21, NTIS, PB81–106643.* Department of Oceanography, Washington State University, Seattle.

McClain, E. 1980. Visible and Infrared Radiometer on SEASAT-1. *IEEE Journal of Oceanic Engineering* OE-5: 164–8.

Onstott, R. and Larson, R. 1986. Microwave properties of sea ice in the marginal ice zone. *Proceedings of IGARSS '86* 1: 353–6.

Onstott, R., Moore, R. and Weeks, W. 1979. Surface-based scatterometer results of Arctic sea ice. *IEEE Transactions of Geoscience and Electronics* GE-17: 78–85.

Peterherych, S., Davies, A. and Muttit, G. 1986. Application of the SEASAT scatterometer to observations of wind speed and direction and direction and arctic ice/water boundaries. In *Proceedings of IGARSS '86* 1: 585–8.

Remy, F., Mazzega, P., Houry, S., Brossier, C. and Minster, J. 1989. Mapping of the topography of continental ice by inversion of satellite altimeter data. *Journal of Glaciology* 35, 119: 98–107.

Shuchman, R., Onstott, R., Sutherland, L. and Wackerman, C. 1988. Intercomparison of synthetic- and real-aperture radar observations of Arctic sea ice during Winter MIZEX '87. *Proceedings of IGARSS '88*, Edinburgh, 13–16 September 1988. *ESA SP-284 (IEEE 88CH2497–6)*, 3: 1,419–22.

Swift, C., Cavalieri, D., Gloersen, P., Zwally, H., Mognard, N., Campbell, W., Fedor, L. and Peteherych, P. 1985. Observations of the polar regions from satellites using active and passive microwave techniques. *Advances in Geophysics* 27: 335–92.

Ulander, L. 1987. SEASAT radar altimeter and synthetic aperture radar: analysis of an overlapping sea ice data set. *Research Report 159* Department of Radio and Space Science, Chalmers University of Technology, Gothenburg, Sweden.

Zwally, H., Brenner, A., Major, J., Martin, T. and Bindschadler, R. 1990a. Satellite radar altimetry over ice. Volume 1: processing and corrections of SEASAT data over Greenland. *NASA Reference Publication 1233, Volume 1*. NASA Office of Management, Scientific and Technical Information Division (OMSTID), Washington DC, USA.

Zwally, H., Brenner, A., Major, J., Martin, T. and Bindschadler, R. 1990b. Satellite radar altimetry over ice. Volume 2: Users' guide for Greenland elevation data from SEASAT. *NASA Reference Publication 1233, Volume 2*. NASA OMSTID, Washington DC, USA.

Zwally, H., Brenner, A., Major, J., Martin, T. and Bindschadler, R. 1990c. Satellite radar altimetry over ice. Volume 4: Users' guide for Antarctic elevation data from SEASAT. *NASA Reference Publication 1233, Volume 4*. NASA OMSTID, Washington DC, USA.

Zwally, H., Brenner, A., Major, J., Martin, T. and Bindschadler, R. In preparation. Satellite radar altimetry over ice. Volume 3: processing and corrections of SEASAT data over Antarctica. *NASA Reference Publication 1233, Volume 3*. NASA OMSTID, Washington DC, USA.

19. Nimbus 7

Operational lifespan and orbital characteristics

Dates: 24 October 1978 to the present (depends on sensor – nominal operation of some sensors at present, ie SAM-II, ERB and SBUV/TOMS).
Semimajor axis: 7,327km.
Orbital height: 943–953km.
Inclination: 99.3°.
Period: 104.08 minutes. Exactly repeating orbit, repeat period 6 days.
Notes: Sun-synchronous. Equatorial crossings noon (ascending node), midnight (descending node) LST.

Sensor characteristics

THIR:
  i) Band 1: TIR (wavelength 6.5–7.0µm). Dynamic range 0–270K. GR 22.6km. Swath width 3,000km.
 ii) Band 2: TIR (wavelength 10.5–12.5µm). Dynamic range 0–330K. GR 7.7km. Swath width 3,000km.
Coastal Zone Colour Scanner (CZCS):
  i) Band 1: VIS, blue (wavelength 0.433–0.453µm). GR 0.825km (all bands). Swath width 1,600km (all bands). Data rate 800Kbps (all bands combined). Scan tilt ±20° (for sun glint avoidance).
 ii) Band 2: VIS, green (wavelength 0.51–0.53µm).
iii) Band 3: VIS, yellow (wavelength 0.54–0.56µm).
 iv) Band 4: VIS, red (wavelength 0.66–0.68µm).
  v) Band 5: NrIR (wavelength 0.70–0.80µm).
 vi) Band 6: TIR (wavelength 10.5–12.5µm).
Scanning Multichannel Microwave Radiometer (SMMR):
  All channels passive microwave. All channels: incidence angle 50°; H and V polarisations; RF bandwidth 220MHz; dynamic range 10–330K; swath width 783km; scan period 4.1s. Data rate 2Kbps (all channels combined).
  i) Channel 1: frequency 6.63GHz, wavelength 4.52cm. Geometric footprint dimension (GFD) 136 × 89km. Antenna beamwidth (AB) 4.2°. Specified temperature resolution (STR) 0.9K.
 ii) Channel 2: frequency 10.69GHz, wavelength 2.81cm. GFD 87 × 58km. AB 2.6°. STR 0.9K.
iii) Channel 3: frequency 18GHz, wavelength 1.67cm. AB 1.6°. GFD 54 × 35km. STR 1.2K.
 iv) Channel 4: frequency 21GHz, wavelength 1.43cm. AB 1.4°. GFD 44 × 29km. STR 1.5K.
  v) Channel 5: frequency 37GHz, wavelength 0.81cm. AB 0.8°. GFD 28 × 18km. STR 1.5K.

vi) SST measured to ±2K. Wind speed measured to ±2.5ms^{-1} or 10%, whichever is greater for winds of <30ms^{-1} for non-rainy ocean areas. Identical to the SMMR launched on Seasat-A, except that the latter was aft-viewing; the Nimbus-7 SMMR was forward-viewing, with the swath centred on the sub-satellite track. Note: channel 4 turned off in March 1985.

SCAMS:

as for Nimbus-6.

Earth Radiation Budget (ERB) sensor:

22 bands in range of 0.2–50μm, limb to limb. Spatial resolution 150km (narrow FOV mode) to 1,500km (wide FOV mode).

Solar Backscatter Ultraviolet and Total Ozone Mapping Spectrometer (SBUV/TOMS):

SBUV 12 bands in range of 160–400nm; TOMS six bands in range of 312.5–380nm. Swath widths 2,700km for TOMS, 200km for SBUV. TOMS GR 50km at nadir.

Limb Infrared Monitoring of the Stratosphere (LIMS):

six bands in TIR range of 6.2–15.0μm, vertical resolution 2km from upper troposphere to lower mesosphere (10–65km).

Stratospheric and Mesospheric Sounder (SAMS):

12 bands in range of 4.1–100μm, vertical resolution 10km to a height of 90km.

Stratospheric Aerosol Measurement-II Experiment (SAM-II):

waveband 0.98–1.02μm, vertical resolution 1km, height range 5–40km.

Polar applications

THIR:

as for Nimbus-5. Research is continuing to obtain estimates of ice/snow surface physical temperature from TIR data (Figure II.12). THIR data have been used to provide surface temperature measurements to derive emissivity values for input into sea ice algorithms using SMMR data (Comiso 1983). See also NOAA AVHRR.

CZCS:

primarily designed to measure concentrations of chloropyll-*a* and phytoplankton in the upper 10m of the water column. VIS bands provide data on solar energy and water colour as affected by the absorption and scattering of radiation due to sediment, chlorophyll-*a*, and waste materials. The TIR band (band 6) provided concurrent SSTs. The measurement of phytoplankton blooms (primary productivity) close to the ice edge. The data reveal patterns related to residual flows, eddies, frontal mixing and instabilities close to ice edge, eg those related to the Antarctic circumpolar current. Large-scale ice extent and large polynyas at a finer ground resolution than AVHRR during daylight hours. Narrow band widths enhance detection of very subtle reflectance differences. CZCS data appear to be more useful over sea ice than was previously believed, revealing a fairly high degree of morphological detail. See also JERS-1 OCTS and SeaWiFS.

SCAMS:

as for Nimbus-6.

SMMR:

designed primarily to measure SST (156km resolution), wind speed (but not direction) over open ocean (from 10.7GHz and 37GHz data, resolution 97.5km), and atmospheric liquid water and water vapour. All-weather, day and night operation. Atmospheric parameters can be retrieved over open ocean because the high reflectivity of water

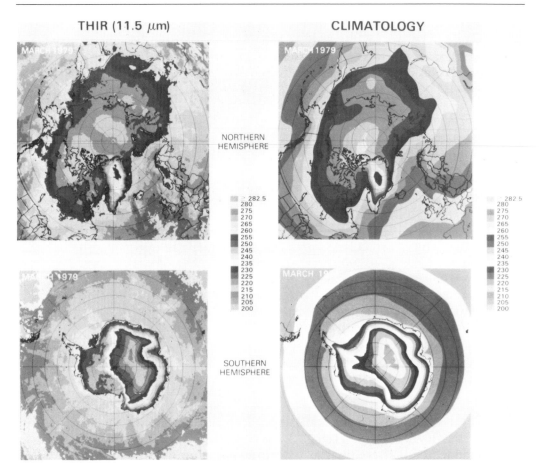

Figure II.12   Average surface temperatures over both polar regions, March 1979, derived from Nimbus-7 THIR data and climatic records. Use of daily TIR data to study surface effects is more problematical due to cloud contamination. Courtesy of Joey Comiso, NASA GSFC.

provides a relatively cold background. This retrieval is far more difficult over ice due to the low reflectivity of the latter at microwave frequencies.

i) *Sea ice*: sea ice concentration (open water extent within the pack), classification (multi-year [MY] fraction) and extent. Information on annual and interannual variability in the overall extent of sea ice, possibly related to climatic change (see Figure 4.19). The TB contrast between FY and MY ice increases with decreasing wavelength (under freezing conditions); that between ice and open water decreases with decreasing wavelength. Ice features need not be resolved to be quantified.

Data from the 18 GHz and 37 GHz channels have been used to solve for both total concentration of sea ice and MY fraction using a number of specialised algorithms (Cavalieri and others 1984; Comiso 1983; Rothrock and others 1988; Rubinstein and others 1985; Svendsen and others 1983; Swift and others 1985 and Walters and others 1987). Estimated overall accuracies are of the order of 10% (increases to 20% under melt conditions, heavy rain clouds or extensive areas of newly formed ice). The movement of the MY ice edge can be monitored under certain conditions to derive information about sea ice dynamics and convergence/divergence (Comiso 1990; Zwally and Walsh 1990). Large polynya growth and decay. Determination of inertial and tidal oscillations in ice cover.

ii) *Land ice and snow*: potential in distinguishing extent, SWE and the onset of melt for seasonal snow cover. Several algorithms are available to evaluate and retrieve seasonal snow cover and depth parameters for specific regions and seasonal conditions. Can detect regions of summer melt on the Greenland and Antarctic ice sheets, and possible deduction of accumulation rates. For land ice, the choice of V polarisation minimises the effect of surface reflection and maximises the ability to observe radiation from the volume layer of the snow (Figure II.13). H channels more sensitive to surface geometric and dielectric roughness – may be useful in comparing surface properties of different glacial regimes. Detects fluctuations in large-scale surface temperature field of ice sheet. Changes in TB field are complicated and occur at most time scales (Jezek and others 1990). There is a phase lag between surface temperature and weighted average physical temperature over depth, particularly at 18GHz (of order of 30 days). 37H may be best for monitoring changes in near-surface physical temperature. Over Antarctica, difference between TBs at different frequencies but like polarisations is maximum in summer and minimum in winter. Difference in TBs between like polarisations is greater for H than V channels. Particularly useful when combined with meteorological data.

iii) *Sea and land ice*:
   a) Channel 1: good delineation of ice edge, physical ice temperature and SST. Channels 1 and 2 are essentially unaffected by dry snow cover and atmospheric effects short of heavy rain.
   b) Channel 2: good delineation of ice edge, wind field observations over open ocean, with little atmospheric contamination. The contrast of TB between open ocean and ice is strongest at 6.63 and 10.69GHz and at H polarisations. Relatively insensitive to spring/summer melt conditions; wet snow layers are substantially more transparent to channel 1 and 2 than higher frequencies.
   c) Channel 3: best delineation of ice edge, ice concentration and MY ice fraction (in combination with 37GHz). Provides the most reliable description of MIZ conditions during spring, due to less sensitivity to surface effects than channel 5 and higher resolution than channels 1 and 2. Provides a more clear ocean signature at V polarisation than 37GHz V (ie a higher contrast between open water and ice). Most accurate separation of ice from open water (along with the lower frequency channels).
   d) Channel 4: detects atmospheric water vapour. For atmospheric corrections.
   e) Channel 5: ice concentration and MY fraction (latter in Arctic only), in combination with 18GHz data. Sensitive to wet ice; information on the onset/progression of melt. Use of the two 37GHz channels is especially well suited to the determination of total sea ice concentration values (particularly in central Arctic), as they show the least errors and offer the best spatial resolution. In MIZs and areas of predominantly FY ice cover (eg the Antarctic), more accurate retrievals of ice concentration can be gained by use of the 37GHz V in combination with the 18GHz V data (Comiso and Sullivan 1986). Volume scattering (in Arctic MY ice) is more pronounced at 37GHz than 18GHz; larger contrast between ice types at higher frequencies. Below a certain threshold thickness, the radiation observed at 37GHz comes largely from the ice, whereas that observed at 18GHz contains a

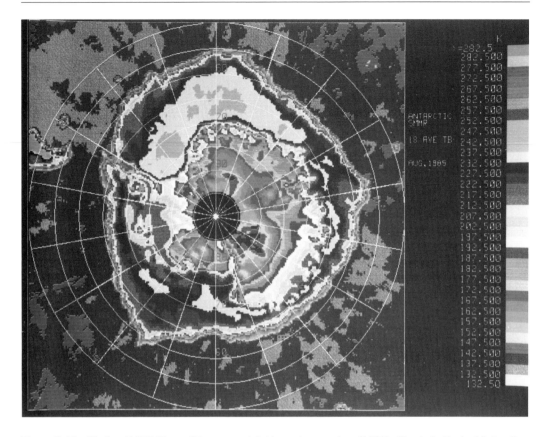

Figure II.13   Nimbus-7 SMMR monthly average brightness temperature (18GHz, V polarisation) of Antarctica, August 1985. Courtesy of Jay Zwally, NASA GSFC.

contribution from the underlying seawater. Thus, the emissivity of 'new ice' is similar to that of FY ice at 37GHz but lower at 18GHz. Due largely to snow cover variations, Arctic MY ice shows larger polarisation at 37GHz than FY ice in late spring (and at the same time of day), thereby offering a potential means of sea ice classification under melt conditions.

See also Nimbus-5 and -6 ESMR and DMSP SSM/I.

ERB:

short- and longwave upwelling radiances and fluxes. Snow/ice angular reflectance, and snow/ice albedo. Cloud BRDFs, which determine their hemispherical albedos (Taylor and Stowe, 1984). Broadband scanning radiometers can be used to measure typical BRDF values for large homogeneous surfaces (ie sea ice and Antarctic ice sheet).

SBUV/TOMS, LIMS, SAMS and SAM-II:

monitoring the vertical and horizontal variability of atmospheric constituents and albedo (SBUV). Ozone holes over both polar regions were discovered in SBUV/TOMS data (Figure II.14). See also Figure II.15. Polar stratospheric cloud data base produced by SAM-II has been crucial to understanding the Antarctic ozone hole dynamics. For a more detailed description, see Madrid (1979).

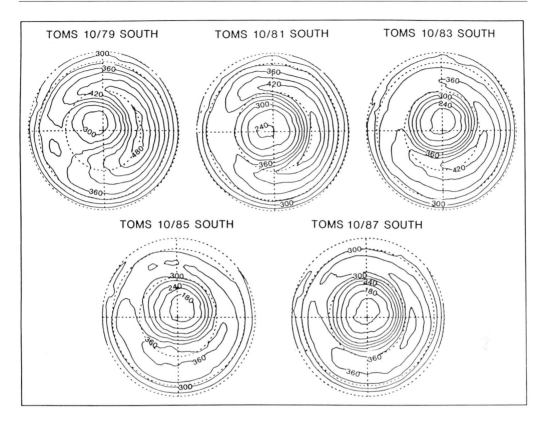

TOMS 10/79 SOUTH   TOMS 10/81 SOUTH   TOMS 10/83 SOUTH

TOMS 10/85 SOUTH   TOMS 10/87 SOUTH

Figure II.14 Progressive thinning of Antarctic ozone minimum documented on October monthly averages, 1979–87, by the Nimbus-7 TOMS. From NASA (1988).

Limitations

THIR:
   as for Nimbus-5. See also NOAA AVHRR.

CZCS:
   bands one–five limited by cloud and darkness, band six by cloud. Bands one–four tended to saturate over highly reflective surfaces, ie snow and ice – data often unusable up to 100km from point of saturation, always to to the right of a bright target. Data from leads or along ice edge/land must be interpreted/processed with caution. Sun glint problems (alleviated to some extent by the tilt capability). Elimination of atmospheric contamination critical. Band seven data noisy. The imagery is distorted, and requires geometric correction (eg by using SEAPAK software [McClain and others 1989; 1990]). Validation of CZCS data has been limited by a lack of knowledge of the role of subpixel scale, cloudiness, pigment/sediment variability, and various instrument (electronic) and calibration problems. Limited to the study of oceanic features with spatial scales of ⩾5km. Studies of the spatial and temporal variability in primary productivity have been limited to large-scale events, eg spring blooms. Two fundamental problems:
   i) at high concentrations of suspended sediments or pigments, atmospheric correction is difficult because significant amounts of radiance can emerge from the ocean in the red band – improved algorithms have alleviated this problem recently; and

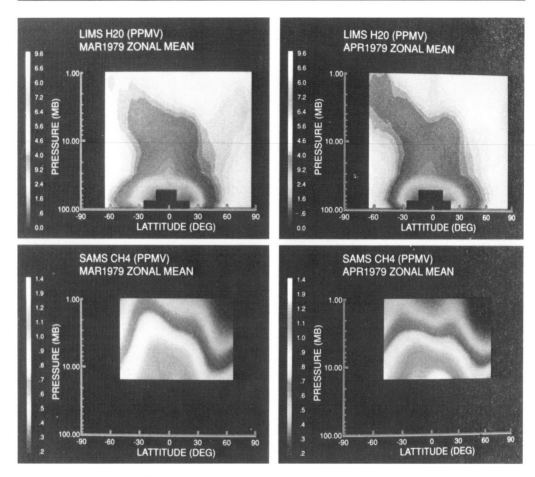

Figure II.15   Mean distribution of upper atmosphere water vapour and methane, recorded in March–April 1979 by the Nimbus-7 LIMS and SAMS. From NASA (1988).

   ii)   at high pigment concentrations, the radiance backscattered out of the water in the blue is very small, requiring use of bands two and three, which are rather insensitive to phytoplankton pigments, to estimate the concentration.

During the last two years of operation, the system's scan and momentum compensation motors failed to achieve proper synchronisation, and this resulted in an automatic shutdown of the sensor. See also JERS-1 and ADEOS OCTS and SeaWiFS.

SCAMS:
   as for Nimbus-6.

SMMR:
   no data collection poleward of 84° (in contrast to the almost complete global coverage of ESMR). Poorer temporal resolution than both Nimbus-5 and Nimbus-6 ESMRs – SMMR operated on alternate days. 'Three-day average' (actually two days with an intervening no data day) images are needed to eliminate data gaps in daily data at lower latitudes. Under certain circumstances, such as during and immediately after the passage of a storm, the more frequent coverage offered by the current DMSP SSM/I (ie twice daily) is desirable.

Very coarse ground resolution – individual features are not resolved. Difficult to validate data with *in situ* observations. Difficult to distinguish new and young ice types (thickness <5cms) from open water. Since slightly thicker (>5cms) ice types have emissivities intermediate between those of open water and thick FY ice, areas of thin ice translate as areas of lower concentration older (thicker) ice. Continuous (ie 100%) covers of pancake ice may thus be misinterpreted as lower concentrations of older ice. Most of the algorithms are based on the assumption that the surface within IFOV consists of only three components, namely open water, FY ice and MY ice. Evidence that algorithms can extract intermediate ice types remains inconclusive.

SSTs and wind speeds cannot be retrieved from ice-covered areas, open ocean within 300km of a land mass, or regions to the south of 55°S latitude. SSTs cannot be retrieved if severe rainstorm is in the IFOV. Wind-induced ocean surface roughness and atmospheric water and vapour can affect ice concentration retrievals in MIZs at low latitudes (ie at maximum extent) and in regions of low ice concentration; spurious high values can result outside the ice edge, ie in areas known to be ice-free. Wind speed affects the emissivity of the surface of open water by the generation of waves and foam; the complexity of this relationship makes it difficult to develop precise models of TB-dependence on wind speed. Can counteract this to some extent by applying 'weather' filters, which can also enhance the ice edge detail. Dispersed ice edges (banding, etc) are not easily resolved.

Geolocation problems. Pixels containing coastlines can show erroneous values – antenna pattern and sidelobe effects; these vary depending on the direction of approach of the satellite relative to the feature of interest, and are particularly severe at the ice edge and close to the coastline. Gloersen and others (1980) have reported other problems, including the effect of direct solar heating on the feed horn leading to abnormal values for cold reference signals at some latitudes in the southern hemisphere, and thus a few degrees error in TB. Sensor drift is discussed by Comiso and Zwally (1989).

Large spatial and temporal variations in observed sea ice emissivities, even under relatively stable winter (freezing) conditions – diurnal freeze-thaw cycles cause increased snow cover granularity and layering and surface crust formation, particularly in MIZs, thereby suppressing the effective emissivity and masking the signature of the underlying ice (Garrity 1988). Greater variance in H than V radiances in MIZs, caused by presence of thin layers of solid ice in dry snow cover (Mätzler and others 1984).

Melting during the austral spring and summer causes serious ambiguities in the extraction of ice concentration and MY fraction; moisture in the freeboard and snowcover greatly increases the loss tangent of the target. The high absorption coefficient of a lossy, wet snow cover drastically decreases the penetration depth, and becomes a significant scatterer, masking out the signature of the underlying ice. The emissivity of MY ice then approaches that of FY ice. FY sea ice with a highly granular snow cover has a similar emissivity to MY ice with a wet snow cover. Melt ponds drastically reduce the emissivity; when they refreeze, meltponds assume an emissivity similar to that of FY ice, thereby hampering the retrieval of MY ice fraction. Melt at the snow/ice interface can give areas of spuriously high MY fraction within areas of predominantly FY ice. Even outside the melt season, Antarctic second year/MY sea ice is difficult to distiguish from FY ice by its microwave signature (at SMRR frequencies) due to its relatively high salinity.

Flooding by seawater at the snow-ice interface may cause spurious values by drastically increasing the salinity (Comiso and others 1989). Moreover, highly saline frost flowers may persist to form an anomolously bright passive microwave target. Wind blown snow may accumulate salt. Few emissivity data exist. Radiative transfer modelling is difficult due to the extreme complexity of the medium. Little work so far on the analysis of SMMR data collected over the great ice sheets.

i) Channel 1 (6.63GHz) and channel 2 (10.69GHz): relatively small contrast between emissivities of FY and MY ice at frequencies <10GHz. The potential usefulness of channels 1 and 2 offset by the fact that they offer the poorest resolution. Use of channel 1 complicated by antenna pattern problems. The retrieval of surface physical temperature hindered by a poor knowledge of ice/snow emissivity variability.

ii) Channel 3 (18GHz): summer melt ambiguities if these data are used alone. For further details regarding this frequency, see Nimbus-5 and -6 ESMR.

iii) Channel 4 (21GHz): sensitive to atmospheric water vapour. Channels 4 and 5 contain little ice information during the summer melt season, except for a wet snow signal. These data drift with time.

iv) Channel 5 (37GHz): more sensitive to ocean effects than channel 3. Radiation has the least penetration depth at 37GHz, and is most sensitive to volume scattering and melt conditions (compared to other channels). Frequencies $\geq$37GHz sensitive to snow cover metamorphosis effects (described above), particulary in mainly FY ice regions. Often difficult to resolve snow cover and other surface effects from openings of leads and polynyas using 37GHz data alone. Polarisation schemes (combining 37GHz H and V data) tend to underestimate concentrations in MIZs and areas of mainly FY ice, due to the sensitivity of 37GHz to snow and surface effects. Although the 37GHz channels offer the highest spatial resolution, their higher sensitivity to surface and atmospheric effects, together with the poorer contrast that they offer between ice and ocean, render ice edge analysis and delineation using these channels alone subject to considerable error.

See also Nimbus-5 and -6 ESMR and DMSP SSM/I.

ERB:

limited by cloud, darkness and poor spatial and spectral resolution.

SBUV/TOMS, LIMS, SAMS and SAM-II:

as for ERB. Also, these sensors were not designed to study surface features.

Data format

THIR:

i) calibrated Earth-located TBs (CBDTs) for both bands. The 9-track, 1600bpi CCTs also contain Earth location information and engineering/housekeeping data. 12 hours of data per CCT. For further details regarding cloud products, see Oakes and others (1989).

ii) 241mm black and white positive or negative transparencies, containing individual swaths for each day in grey scale fashion. Day and night data are displayed on separate films for both bands.

CZCS:

i) Scientists in the Oceans and Ice Branch of NASA GSFC and the University of Miami/Rosenstiel School of Marine and Atmospheric Science have processed the global CZCS dataset (previously archived by NOAA) to level 2–3, including various types of composite imagery. The entire level 1 data set has been copied to digital optical disc. Data products:

a) Level 1: standard NASA CRTT format data, containing calibrated radiances for all six bands, and Earth location information for a single CZCS scene (maximum two minutes of data) at full resolution in swath projection.

b) Level 1a: subsampled level 1 data (every fourth pixel, every fourth line) in DSP (University of Miami image display) format for bands 1–5 (GR 4km).

c) Level 2: derived geophysical parameters for a single CZCS scene in DSP format, 4km resolution and swath projection. The six derived products are diffuse attenuation coefficient, normalised water-leaving radiance (nlw520), phytoplankton pigment concentration, normalised water-leaving radiance (nlw440), normalised water-leaving radiance (nlw550) and aerosol radiance at 670 (la670).

d) Level 3: composited Earth-gridded (binned) data of the above listed level 2 parameters, including compositing statistics binned to a fixed, linear latitude–longitude array of dimension 1024 × 2048 (18.5km resolution at the equator). These files are stored in compressed form at daily, weekly (5-day) and monthly intervals. Fields consist of those for level 2 described above, the standard deviations of the above parameters, and numbers of valid level 2 pixels and days for a total of 14 fields.

Distribution media are 9-track CCT, 8mm cassette or optical disc. Image file/hard-copy representations of these products can be reproduced, but are not done so routinely, nor are they proposed as archive products.

The computer hardware (and software) requirements necessary to access the CZCS browse facility (see below) are:

a) Browse programme: provides the search, display and order routines for the data set. It can be used with the video disc to preview the images if desired.

b) Panasonic 8-inch video disc player: model TQ2024 (player), or TQ2023 or TQ2026 (player/recorders).

c) Colour monitor: any monitor capable of accepting video input.

d) Computer: VAX or Microvax, IBM or IBM-compatible PC, Apple Macintosh, or Sun Workstation. PC systems need at least 10mb hard disc space. They need a serial port to connect to a Panasonic player, and should ideally be connected to electronic mail to NASA.

e) An RS232 cable: used for computer control of the video disc player. NASA will provide the pin configurations for this cable.

ii) EARTHNET: standard photographic products and raw or processed data on CCT.

iii) Dundee: standard photofacsimile, full pass with or without latitude/longitude and land outline grids, or sectorised enlargements. Linearised to correct Earth curvature distortion, but not rotation deskewed. Any band. Browse files. CCT: 1600bpi, unprocessed.

SCAMS:

as for Nimbus-6.

SMMR:

i) Photographic products:

a) MATRIX-LO (Mapped Land-Ocean Data Film): 105mm colour film products, maps of six-day and monthly averages of land/ocean geophysical parameters, Mercator projection. Each map corresponds to one six-day or monthly average on the MAP-LO CCT.

b) MATRIX-SS (Mapped Sea Ice and Snow and Ice on Land Data Film): 105mm colour film products, maps of six-day and monthly averages of sea ice/ice sheet geophysical parameters, polar stereographic projection. Each map corresponds to one six-day or monthly average on the MAP-SS CCT.

c) MATRIX-30 (Mapped 37GHz Channel Data Film): 105mm colour film products, maps of six-day averages of 37GHz TBs, polar steographic projection. Each map corresponds to one six-day or monthly average on the MAP-30 CCT.

ii) Digital data on CCT (1600bpi, 9-track) include:

a) Antenna Temperature Tape (TAT): calibrated antenna temperatures and Earth locations for each IFOV for each polarisation, plus ephemeris, attitude and SMMR housekeeping information. The most basic form of the SMMR data available, ie it contains uncalibrated TBs. As such, TAT data are of limited use to most investigators.

b) CELL-ALL CCTs: H and V polarisation calibrated TBs and seasonal geographical filters. The preprocessing scheme inputs raw, Earth-located data from the TATs, applies a radiometric calibration, corrects for polarisation mixing, removes sidelobe effects and remaps into evenly-spaced, equal area cells across the swath of the SMMR scan. Data are averaged into four cell systems (156 × 158km, 97.5 × 98.5km, 60 × 60.6km and 30 × 30.3km) – correspond roughly to the resolutions at 6.6, 10.7, 18 and 37GHz respectively. For each cell, the appropriate antenna temperatures, incidence angle, and latitude and longitude are provided, together with the day/night status and the intrument engineering values. One data record on each CCT contains the information for a 30-scan period (122.88s), covering an area of 780 × 780km, blocks in orbital format. The purpose of the remapping is to assure that, when performing geophysical parameter retrievals, the data from each channel are equally weighted. One 1600bpi CCT contains three days of data. The data on CELL-ALL CCTs are used to derive geophysical parameters such as sea ice concentration and SST (see PARM tapes below).

c) Temperature Calibrated Tapes (TCT): calibrated TBs for each channel in its original pixel resolution. The TCT radiances are different from the corresponding CELL radiances because a different calibration procedure was used.

d) Half-degree TCT Map Tapes: six-day averages (ascending and descending orbits separated) of TCT TBs mapped onto a 55km resolution grid. One 1600bpi CCT contains one month of data.

e) Quarter-degree TCT Map Tapes: six-day averages (ascending and descending orbits separated) of TCT TBs of the 37GHz channels, mapped onto a 27.5km resolution grid. One 1600bpi CCT contains three months of data.

f) PARM: extracted geophysical parameters for each IFOV, retrieved from CELL-ALL data. Include PARM-LO, PARM-SS. MY fraction data have a 60km resolution.

   i) PARM-LO: land/ocean geophysical parameters, computed from the CELL-ALL CCTs. Earth-located cells with spatial resolutions of 156km, 97.5km and 60km. One 1600bpi CCT contains six days of data.

   ii) PARM-SS: sea ice/ice sheet geophysical parameters as computed from the CELL-ALL CCTs. Earth-located cells with spatial resolutions of 156km, 97.5km and 60km. One 1600bpi CCT contains six days of data.

   iii) PARM-30: contain values of sea ice concentration computed from the CELL-ALL CCTs. 30km cells in the polar regions. Six days of data on 1 or 2 1600bpi CCTs.

Note: on the PARM CCTs, ice concentrations available for both hemispheres, 60km and 30km resolution; MY fraction computed for N hemisphere only; and ice temperature is computed at a 156km resolution. The sea ice retrievals are made using the NASA 'Team' algorithm developed by Cavalieri and others (1984).

g) MAP: the same parameters as PARM but in digital format of map projections, Mercator/polar stereographic, three- and six-day and monthly (30-day) formats. Generated from PARM and CELL-ALL CCTs.

i) MAP-LO: contain six-day and monthly averages of geophysical parameters (SST, sea surface winds, total water vapour, etc), Mercator projection. The mapped form of PARM-LO CCTs. One 1600bpi CCT contains one month of data.

ii) MAP-SS: contain six-day and monthly averages of sea ice/ice sheet geophysical parameters, polar stereographic grids. The mapped form of PARM-SS CCTs. 1 1600bpi CCT contains one month of data.

iii) MAP-30: contain six-day averages of 37GHz TBs (30km resolution), polar stereographic grids. The mapped form of PARM-SS CCTs. One 1600bpi CCT contains one month of data.

h) PARMAP CCTs. Beginning with the sixth year of data, MAP products have been replaced by PARMAP, whereby each CCT contains individual geophysical parameters (sea ice, SST, sea surface wind speed, and total water vapour) mapped on a $1/2° \times 1/2°$ spatial resolution longitude–latitude grid. One 6250bpi CCT contains four months of data.

Users' Guides are available from NSSDC, NASA/GSFC. Further information regarding media and general format etc is provided in Oakes and others (1989).

SEA ICE PRODUCTS FROM NSIDC

The SMMR data set available from NSIDC (see below) includes MY fraction and sea ice concentration, retrieved from dual-polarised 18 and 37GHz data, regridded into the SSM/I data format (304 × 448 cells) (see DMSP SSM/I). These products were generated using the algorithm developed by Cavalieri and others (1984) and upgraded by the inclusion of a weather filter (Gloersen and Cavalieri 1986), the NASA 'Team' algorithm. Data are also available in ESMR grid format (293 × 293 pixels, with each map cell representing a surface area varying from about $32km^2$ near the poles to about $28km^2$ near 50° latitude). The fixed-length records comprising the N hemisphere SMMR grids are 6080 bytes (2 bytes × 304 cells × 10 parameters per cell). The regridding procedure involves rebinning of the slightly coarser 30km SMMR data to the 25km SMM/I grids. This introduces a slight distortion, but allows direct comparison with follow-on SSM/I data. Data from north of 30°N and south of 50°S are available (orbital and mapped format, CCTs and some hard copy products). TB data (all channels) are also available for any researcher who wishes to apply an alternative algorithm.

Sea ice products have also been generated in the same formats (ESMR and SSM/I grids) from the CELL-ALL data at the Oceans and Ice Branch, NASA GSFC by Dr Joey Comiso, using an alternative cluster analysis approach (Comiso and Sullivan 1986). Contact Dr Joey Comiso, Code 971, NASA Goddard Space Flight Center, Greenbelt, MD 20771, USA, telephone (301)–286–9135. This algorithm is also available from NSIDC.

NSIDC is distributing SMMR data on CD-ROM. In addition to the TB data from all channels, the CD-ROMs contain a landmask, coastal outline map, latitude and longitude pairs, sensor and grid documentation, and references. Also supplied, on two floppy diskettes, is software to display and manipulate the grids on a PC-based system (and to extract daily ice concentrations/MY fractions), and a User's Guide. Sensor drift corrections will be supplied on diskette when available. Future publication of the remainder of the data on CD-ROM will depend on user response. Minimum hardware requirements to access, manipulate and display the gridded data on CD-ROM are:

i) A '286 processor, such as a PC/AT;
ii) 640K memory (more if you plan to use RAM discs);
iii) DOS 3.3;

iv) 20Mb hard disc drive (needs to be larger if you plan to extract and store a lot of data on the workstation);

v) CD Player and File Manager supporting ISO 9660 standard (investigate the Sony and Hitachi brands);

vi) EGA monitor and adapter card (VGA is necessary if higher resolution and more colours are required); and

vii) Display software: IMDISP. Supports CGA, EGA and VGA graphics adapters.

Requirements for CD-ROM Readers:

i) CD Reader must comply with the Phillips/Sony CD-ROM hardware specifications;

ii) CD Reader must be capable of reading ISO 9660 and High Sierra formats – this is usually accomplished via PC file management software.

Note: for further details, contact Vince Troisi at NSIDC.

ERB, SBUV/TOMS, LIMS, SAMS and SAM-II:

CCTs (9-track, 1600bpi or 6250bpi). Also 16mm negative transparencies.

Data availability and archiving dates

CZCS:

November 1978–23 June 1986. The central archive and distribution centre: Code 636 (Dr Gene Feldman), NASA Goddard Space Flight Center, Greenbelt, MD 20771, USA, telephone (301)–286–9428, OMNET: G.FELDMAN, SPAN: MANONO::GENE. Offers a browse facility (see above). See Feldman and others (1989) for further details. SEAPAK software, developed by NASA to read and display/manipulate the data (McClain and others 1989; 1990), is available from NASA's Computer Software Management Center (COSMIC), University of Georgia, 382 East Broad St., Athens, GA 30601, USA, telephone (404)–542–3265.

In addition, a number of academic and research institutions in the USA have been established by NASA to serve as regional browse, distribution and analysis centres for Level 1a, 2 and 3 CZCS data: NASA GSFC Laboratory for Oceans (Code 971) (contact Chuck McClain), Bigelow Laboratory for Ocean Sciences (Janet Campbell), University of Miami (Otis Brown), University of South Florida (Frank Muller-Karger), University of California at Santa Barbara (Ray Smith), NASA Jet Propulsion Laboratory (Don Collins), Oregon State University (Mark Abbott), University of Washington (David English), and the University of Rhode Island (Peter Cornillon).

Other national archives include:

i) Lannion receiving station in France covered the Greenland Sea up to 81°N. These data (April 1979–June 1986) are archived at and available from ESA-EARTHNET Programme Users Services, Via Galileo Galilei, 00044 Frascati, Rome, Italy, telephone 396–9426285, telex 610637 ESRIN I, or from EARTHNET National Points of Contact. The UK National Point of Contact: Mr M. Hammond, National Remote Sensing Centre, RAE, Farnborough, Hampshire GU14 6TD, telephone (0252)–24461, telex 858442.

ii) Department of Electrical Engineering (Dr P. Baylis), University of Dundee, Dundee DD1 4HN, Scotland, telephone 0382–23181, telex 76293 ULDUND G. Coverage limited to E Greenland south of the Fram Strait (3 August 1979–June 1986).

Note: CZCS was only switched on as required by NASA approved experiments (with observation periods of 2–10 minutes up to a maximum of two hours per day), and coverage of the polar regions is therefore sporadic.

The following data are available from NSSDC (for full address etc, see TIROS I–X):

THIR:
16 November 1978–9 May 1985.

SCAMS:
November 1978–August 1985.

LIMS:
daily maps, November 1978–29 May 1979.

SAMS:
November 1978–13 June 1983.

SAM-II:
1 November·1978–1 May 1989.

SBUV/TOMS:
31 October 1978 to present (daylight only for TOMS). Data beyond 30 March 1988 processed to level 1 only (due to instrument malfunction).

ERB:
November 1978 to present.

SMMR:
  i) Oribital data (scan-on mode):
      a) PARM October 1978–October 1985;
      b) CELL-ALL, TAT and TCT 19 November 1978–20 August 1987.
 ii) Gridded data:
      a) MAP to October 1983;
      b) TCT in various formats until 29 August 1987.
iii) MATRIX photographic products: for first five years of data.

The SMMR was turned off on 21 August 1987, and turned back on 25 August 1987 (looking forward without scanning). Operated continuously every day, instead of the 2-day on/off cycle, until 5 July 1988, when it was turned off permanently. The steady long-term drift behaviour of the 21GHz radiometer changed in May 1983; it was turned off in March 1985.

TB data and sea ice products (see data format) are available from: Cryospheric Data Management System (CDMS), NSIDC, University of Colorado, Boulder, CO 80309, USA, telephone (303)–492–1834, telex 257673 (WDCA UR), telemail MAIL/USA [NSIDC/OM-NET], VAX mail via SPAN: KYROS::NSIDC.

Mid-high latitude SMMR snow cover data are available from Pilot Land Data System, Code 634, NASA Goddard Space Flight Center, Greenbelt, MD 20771, USA, telephone (301)–286–9761.

ERB:
data are available from: SDSD, NOAA/NESDIS, Princeton Executive Center, 5627 Allentown Road, Camp Springs, MD 20746, USA, telephone (301)–763–8399, telex 248360BSWUR, FAX 7638443. Summaries of satellite data archived at SDSD are printed in *Environmental Satellite Imagery* and in the *Oceanic Monthly Summary*. More general information is contained in the *Satellite Data Users' Bulletin*, which is published at irregular intervals.

References and important further reading

American Geophysical Union. 1984. NIMBUS-7 Special Issue: Scientific results. *Journal of Geophysical Research (Oceans)* 89, C3.

Cavalieri, D., Gloersen, P. and Campbell, W. 1984. Determination of sea ice parameters with the NIMBUS-7 SMMR. *Journal of Geophysical Research* 89, D4: 5,355–69.

Chang, A., Foster, J. and Hall, D. 1987. NIMBUS-7 SMMR derived global snow cover parameters. *Annals of Glaciology* 9: 39–44.

Comiso, J. 1983. Sea ice effective microwave emissivities from satellite passive microwave and infrared observations. *Journal of Geophysical Research* 88: 7,686–704.

Comiso, J. 1990. Arctic multi-year ice classification and summer ice cover using passive microwave satellite data. *Journal of Geophysical Research* 95, C8: 13,411–22.

Comiso, J. and Sullivan, C. 1986. Satellite microwave and *in situ* observations of the Weddell Sea ice cover and its marginal ice zone. *Journal of Geophysical Research* 91, C8: 9,663–81.

Comiso, J. and Zwally, H. 1989. Polar microwave brightness temperatures from NIMBUS-7 SMMR: time series of daily and monthly maps from 1978–1987. *NASA Reference Publication 1223*. NASA Office of Management, Scientific and Technical Information Division.

Comiso, J., Maynard, N., Smith, W. and Sullivan, C. 1990. Satellite ocean color studies of Antarctic ice edges in summer/autumn. *Journal of Geophysical Research* 95: 9,481–96.

Feldman, G., Kuring, N., Ng, C., Esaias, W., McClain, E., Elrod, J., Maynard, N., Endres, D., Evans, R., Brown, J., Walsh, S., Carle, M. and Podesta, M. 1989. Ocean color: availability of a global data set. *EOS* 70, 23: 634–5 and 640–1.

Garrity, C. 1988. Shipborne passive microwave sea ice experiment in the East Greenland Sea: May–July 1987. *Proceedings of IGARSS '88*, Edinburgh, 13–16 September 1988. *ESA SP-284 (IEEE 88CH2497–6)*, Volume 3: 1691–2.

Gloersen, P., Campbell, W., Cavalieri, D., Comiso, J., Parkinson, C. and Zwally, H. (in press). Arctic and Antarctic sea ice, 1978–1987: satellite passive microwave observation and analysis. *NASA Special Publication*.

Gloersen, P. and Cavalieri, D. 1986. Reduction of weather effects in the calculation of sea ice concentration from microwave radiances. *Journal of Geophysical Research* 91, C3: 3,913–9.

Gordon, H., Brown, J. and Evans, R. 1988. Exact Rayleigh Scattering calculations for use with the NIMBUS 7 Coastal Zone Color Scanner. *Applied Optics* 27: 862–71.

Jezek, K., Hogan, A. and Cavalieri, D. 1990. Antarctic ice sheet brightness temperature variations. In Ackley, S. and Weeks, W. (eds), Sea ice properties and processes. *CRREL Monograph 90–1*: 217–23.

Kyle, H.L. 1990. The Nimbus-7 Earth Radiation Budget (ERB) data set and its uses. *SPIE* 1299: 27–39.

Krueger, A., Penn, L., Larko, D., Doiron, S. and Guimares, P. 1989. The 1988 Antarctic ozone monitoring NIMBUS-7 TOMS data atlas. *NASA Scientific Report RP–1225*. NASA, Washington DC.

Madrid, C. (ed.). 1978. *NIMBUS-7 Users' Guide*. NASA GSFC, Greenbelt, MD 20771.

Mätzler, C., Ramseier, R. and Svendsen, E. 1984. Polarization effects in sea-ice signatures. *IEEE Journal of Oceanic Engineering* OE9, 5: 333–8.

McClain, C., Chen, J., Darzi, M., Firestone, J. and Endres, D. 1989. The SEAPAK User's Guide. *NASA Technical Memo TM 100728*. NASA GSFC, Greenbelt, MD 20771. Available from COSMIC (see above).

McClain, C., Fu, G., Darzi, M. and Firestone, J. 1990. PCSEAPAK User's Guide. NASA GSFC, Greenbelt, MD 20771. Available from COSMIC (see above).

Millman, A. and Wilheit, T. 1985. Sea surface temperatures from the Scanning Multichannel Microwave Radiometer on NIMBUS-7. *Journal of Geophysical Research* 90, C6: 11,631–41.

NASA. 1988. Earth Science Applications Division. The program and plans for FY 1988–1989–1990. NASA, Washington DC.

Oakes, A., Han, D., Kyle, H., Feldman, G., Fleig, A., Hurley, E. and Kaufman, B. 1989. NIMBUS-7 data product summary. *NASA Report RP 1215*, NASA GSFC, Greenbelt, MD 20771.

Parkinson, C. and Cavalieri, D. 1989. Arctic sea ice 1973–1987: seasonal, regional and interannual variability. *Journal of Geophysical Research* 94: 14, 499–523.

Rothrock, D., Thomas, D. and Thorndike, A. 1988. Principal component analysis of satellite passive microwave data over sea ice. *Journal of Geophysical Research* 93, C3: 2,321–32.

Rubinstein, I., Bunn, F. and Ramseier, R. 1985. NIMBUS-7 microwave radiometry of ocean surface winds and sea ice. In *Proceedings of the 19th International Symposium on Remote Sensing of the Environment* 2: 961–70. Ann Arbor, Michigan, 21–5 October 1985.

Smith, G., Rutan, D. and Bess, T. 1990. Atlas of albedo and absorbed solar radiation derived from NIMBUS-7 Earth Radiation Budget data set – November 1978 to October 1985. *NASA Reference Publication 1231*. NASA, Washington DC.

Svendsen, E., Kloster, K., Farrelly, B., Johannessen, O., Johannessen, J., Campbell, W., Gloersen, P., Cavalieri, D. and Mätzler, C. 1983. Norwegian remote sensing experiment: evaluation of the NIMBUS-7 Scanning Multichannel Microwave Radiometer for sea ice research. *Journal of Geophysical Research* 88, C5: 2,781–91.

Swift, C., Fedor, L. and Ramseier, R. 1985. An algorithm to measure sea ice concentration with microwave radiometers. *Journal of Geophysical Research* 90, C1: 1,087–99.

Taylor, V.R. and Stowe, L.L. 1984. Reflectance characteristics of uniform Earth and cloud surfaces derived from Nimbus-7 ERB. *Journal of Geophysical Research* 93, C5: 5,093–5,099.

Walters, J., Ruf, C. and Swift, C. 1987. A microwave radiometer weather-correcting sea ice algorithm. *Journal of Geophysical Research* 92, C6: 6521–34.

Wilheit, T., Greaves, J., Gatlin, J., Han, D., Krupp, B., Milman, A. and Chang, E. 1984. Retrieval of ocean surface parameters from the Scanning Multichannel Microwave Radiometer (SMMR) on the NIMBUS-7 satellite. *IEEE Transactions on Geoscience and Remote Sensing* GE–22: 133–42.

Zwally, H., Comiso, J. and Gordon, A. 1985. Antarctic offshore leads and polynyas and oceanographic effects. In Jacobs, S. (ed.), *Oceanology of the Antarctic continental shelf*, *Antarctic Research Series* 43: 203–26. American Geophysical Union, Washington DC.

Zwally, H. and Walsh, J. 1990. Multiyear sea ice in the Arctic: model- and satellite-derived. *Journal of Geophysical Research* 95, C7: 11,613–28.

20. Landsat 4 and Landsat 5

Operational lifespan and orbital characteristics

Dates: Landsat 4: 16 July 1982 to present. Landsat 5: 1 March 1984 to present.

Semimajor axis: 7,070km.

Orbital height: 683–700km.

Inclination: 98.3°.

Period: 98.63–98.64 minutes (exactly repeating orbit). Repeat period 16 days for each satellite.

Notes: Circular, sun-synchronous orbit. Landsat 4 crosses the equator at about 09:30 local time.

Sensor characteristics

MSS:

Spectral bands as for Landsats 1 and 2 (Note: bands 4–7 are renamed bands 1–4). GR 82m. Swath width 185km, with a 5.4% forward lap and a 7.3% sidelap at the equator, increasing towards the poles. Data rate 15.06Mbps. 6-bit quantisation.

Thematic Mapper (TM):

i) Band 1: VIS (wavelength 0.45–0.52μm). Band 2: VIS (0.52–0.60μm). Band 3: VIS (0.63–0.69μm). Band 4: NrIR (0.76–0.90μm). Band 5: IR (1.55–1.75μm). Band 6: TIR (10.40–12.50μm). Band 7: IR (2.08–2.35μm).

ii) Ground resolution: all bands 30m, except band 6 (120m). Swath width 185km (all bands), overlaps as for MSS. Data rate 84.9Mbps. 8-bit quantisation.

Polar applications

MSS:

as for Landsats 1, 2 and 3. See Figure II.16.

TM:

i) bands 1–4 similar to MSS bands 1–4, but are narrower, radiometrically more sensitive, and better calibrated, with more separation between bands. TM is less prone to saturation in the 0.53–0.69μm region. Large ice sheet features such as flow lines, ice streams and crevasse patterns show up best in these bands due to photometric effects on slopes; low sun angles enhance many subtle topographic features that may not be visible on the ground. Ice sheet velocity from sequential images. Good contrast between snow and land ice. Band 4 can distinguish new snow from wet/melted snow. These bands are most easily interpreted, as they are similar to the familiar MSS bands. Enhanced data from band 4 have even been used to detect penguin rookeries on the Antarctic coast as a result of the unique spectral characteristics of their guano deposits. Sea ice applications similar to MSS, but finer resolutions, more detail and higher latitude coverage.

ii) Band 5: with band 7, offers a reasonable discrimination between cloud and snow – snow appears darker than cloud at this waveband.

iii) Band 6: winter/night-time detection and identification of features. Fracture patterns and stages of sea ice development. Delineation of leads and polynyas. Surface features and topographic detail. Mapping of ice margins, thin ice areas and SSTs. Boundaries between new and MY ice are more reliably identified than those between open water and thin ice. Information on the ground radiation temperatures on ice sheets. Detects onset of melt by albedo change; may serve to detect snow liquid water content. Can discriminate between features having similar reflectivities in the other bands, eg blue ice.

iv) Band 7: with band 5, offers better discrimination between clouds and snow than either the MSS or the lower TM bands (snow appears darker than cloud at these wavelengths). Bands 5 and 7 also reveal ice sheet surface snow properties, including grain-size variations (these features are not detected by VIS wavelengths).

v) Can produce maps at a scale of 1:100,000 using TM data (as opposed to 1:250,000 for MSS data), equivalent to a 1/2° × 1° sub scene. The TM offers improved detection of the location of ice margins and the annual equilibrium line separating ice areas from regions of residual snow accumulation. Accurate mapping of the Antarctic coastline. Detection of calving of large tabular icebergs, and tracking their drift. TM data are useful in the monitoring of snow and glacial features in regions of dramatic topography, eg valley

ALASKA   LANDSAT-4   BAND ■■■4
P75   R9   87-09-17   4188921063
LA 72.15N LO 147.09W   SUN EL 19 AZ 170   1:1000000

Figure II.16   Landsat MSS data are useful in providing detailed 'snapshot' data in areas of low concentration sea ice. This image (band 4) of an area to the north of Prudhoe Bay, Alaska (17 September 1987) was obtained from the University of Alaska. Courtesy of Konrad Steffen, University of Colorado.

glaciers. Different bands can be digitally combined to enhance subtle detail. Digital Landsat data can be readily regridded for co-registration with coarser resolution data, eg from Nimbus-7 SMMR and DMSP SSM/I. Landsat data have been useful in validating sea ice products from the latter.

See also SPOT HRV, MOS MESSR, NOAA AVHRR and IRS LISS.

Limitations

MSS:
    as for Landsats 1, 2 and 3.

TM:

bands 1–5 and 7 limited by cloud, fog, darkness, and saturation (at certain times of the year, ie November to January in the Antarctic. Band 1 is the worst affected). Band 6 limited by cloud and fog. Limited coverage of polar regions, and Antarctica in particular. FY and MY sea ice cannot be readily distinguished. Ambiguities arise when the physical temperature of the ice is similar to that of the surrounding water, and when there is meltwater on the ice surface; under these circumstances, wet ice may appear as open water on TIR imagery. Image geolocation is difficult in the relatively featureless interiors of ice sheets. Problems arise with the identification of thickly debris-covered ice. Expense, high data rate and thus large volume of data per scene. Coverage possible up to a maximum latitude of 82.5°, permitting coverage of Ice Streams C, D and E (but not A and B).

Note: X-band downlink failure and loss of half the solar array on Landsat 4 have led to restricted data availability (Landsat 4 is able to transmit 40 MSS scenes per day). TM coverage was terminated in February 1983 for Landsat 4, although MSS acquisition continues. This led to the premature launch of Landsat 5 with a modified electrical system and an X-band transmitter. Landsat 5 has lost one of its redundant X-band downlinks and one K_u-band transpoder, but is able to transmit about 150 TM scenes per day to international stations and a further 100 to US receiving stations, as well as MSS data. Relay system problems have also occurred, although both TDRS-East and -West are now in orbit (Note: the first TDRS satellite to be launched, in 1983, soon encountered difficulties. TDRS-2 was destroyed in the Challenger Shuttle disaster. TDRS-4 was launched on the Shuttle *Discovery* in February 1989).

Data format

MSS:

i)   9-track 1600bpi CCTs, raw (after June 1981) or radiometrically and geometrically corrected (from January 1979 to present). Note: 1600 and 6250bpi 9-track CCTs, Tape Set format, are available from EOSAT. Several enhancement and correction procedures can be applied by EOSAT upon request, including geocoding. Geocoded products are unenhanced digital image tapes processed to a simple, usually affine, relationship to a predetermined base map. Users can select the map projection, rotation and Earth ellipsoid; allows GIS operators to easily use the data with less preprocessing. Data acquired before 1 January 1979 are available only in band-interleaved-by-pixel-pairs (BIP-2) format. BSQ and BIL formats also available.

ii) Standard Photographic Data Products: as for Landsats 1 and 2 plus microfiche. Note: colour film products are not available in the film negative format.

iii) Since 1 April 1988, MSS data collected in Japan have been available from RESTEC on floppy disc – size 5.25 × 8 inch, BSQ and BIL format, 512 pixels × 400 lines × 4 bands.

TM:

i)   9-track 1600/6250bpi CCTs, Tape Set format (raw or system corrected data): full scene with all seven spectral bands; full scene, one spectral band; or quarter scene with all seven spectral bands. Both false and natural colour film composites. Full scene digital data in BSQ format also available. Geocoded products and moveable digital scenes and sub-scenes are available.

ii) Standard Photographic Data Products: as for the MSS of Landsats 1 and 2, at scales of 1:1,000,000, 1:500,000 and 1:250,000. Colour composites combine three spectral bands of the user's choice. Note: colour film products are not available in the film negative format.

iii) TM data are now available on 5.25 inch floppy discs in MS/PC-DOS format and are usable with all PC-XT and PC-AT systems, or compatible computer systems capable of handling colour graphics and supporting image processing software – 512 × 512 pixels (213km^2). With respect to ground location, these data are accurate to within 6–8 pixels. Since 1 April 1988, TM data collected in Japan have been available from RESTEC on floppy disc – size 5.25 × 8 inch, BSQ and BIL format, 512 pixels × 400 lines × 7 bands.

Since January 1985, EOSAT has provided microfiche images to subscribers which illustrate MSS (band 2) and TM (bands 3 or 6; band 6 is for night-time imagery). There are about 1,200 scenes in a set of 20 microfiche cards. Microfilm (16mm) is available for MSS (only) bands 2 or 5 covering 1972–84 inclusive. Note: images on each microfiche or microfilm are not geographically arranged. The Landsat Reference MicroCatalogue (a non-image microfiche reference system) is available on request. For further details on data product formats, see the EOSAT publication *Landsat Products and Services*.

Data availability and archiving dates

MSS and TM data are available from EOSAT (see Landsats 1 and 2 for address). Data listings, coverage maps (World Reference System) and order forms are also available. Computer listings of images contain information on corner coordinates, dates and an estimate of cloud cover. Note: cloud cover estimates are subjective and often inaccurate in the presence of snow and ice; also, no information is given on their location. Cloud cover assessments are made digitally on TM scenes, but manually on MSS scenes. *Landsat Data Users Guides* and *Landsat Data Users Notes* are also available. The user can acquire specific data with a specific time scheduling.

Data are also available from EOSAT international sales representatives. The UK representative is: ERSUN, British Aerospace (Space Systems) Ltd, FPC 311, PO Box 5, Filton, Bristol B5R 7QW, England, telephone (0272)–366832/366416, telex 449452, FAX 0272 366819/363812. Details of other representativs are available from EOSAT.

Data are also available from European, Canadian, Japanese and Alaskan receiving stations: details (addresses, etc) as for Landsats 1, 2 and 3. The single X-band antenna at Prince Albert must be shared for both Landsat and SPOT-1 reception. Selection of one satellite due to conflicting acquisition times for the two satellites results in a loss of data if careful planning has not taken place. Japan has collected TM data since 1 April 1988.

The Geophysical Institute of the University of Alaska in Fairbanks holds a regional archive of Landsat data, offering coverage of the Bering, Chuckchi and Beaufort Seas (W Arctic). Although no digital data are available for parts of the Bering Sea, photographic transparencies are available, displayed in a Space Oblique Mercator projection. For address, see Landsats 1 and 2.

The US Geological Survey (USGS) has, in a cooperative project with the Norsk Polarinstitutt, been compiling detailed maps of an area in Dronning Maud Land, Antarctica using digitally enhanced TM imagery. Under the auspices of the USGS, the Scientific Committee on Antarctic Research (SCAR) have put together a collection of over 5,000 scenes of Antarctica, acquired from 1982–90 and covering 60% of the coastal regions. USGS have compiled a set of 22 microfiche and a listing of 219 Landsat 4 and 5 scenes from 1984–8. The complete photographic collection can be inspected at the US SCAR Library at

USGS. For further information, contact the USGS Headquarters at 12201 Sunrise Valley Drive, National Center, Reston, VA 22092, USA.

Note: all data products produced by the EROS Data Center before the EOSAT contract (27 September 1985) remain in the public domain. All data products produced by the EDC after this date fall under the strict copyright restrictions of the agreement, regardless of when the datasets were originally recorded by the satellites.

References and important further reading

Ahlnäs, K. and Royer, T. 1989. Application of satellite visible band data to high latitude oceans. *Remote Sensing of Environment* 28: 85–93.

Dozier, J. 1989. Spectral signature of alpine snow cover from the LANDSAT Thematic Mapper. *Remote Sensing of Environment* 28: 9–22.

EOSAT. 1985. *User's guide for LANDSAT Thematic Mapper computer-compatible tapes.* EOSAT, Lanham, Maryland.

Hall, D., Ormsby, J., Bindschadler, R. and Siddalingaiah, H. 1987. Characterization of the snow-ice reflectance zones on glaciers using LANDSAT Thematic Mapper data. *Annals of Glaciology* 9: 104–8.

Lucchitta, B. and Ferguson, H. 1986. Antarctica: measuring glacier velocity from satellite images. *Science* 234, 4780: 1105–8.

Lukowski, T. 1987. *CCRS Satellite Acquisition Request Processing (SARP) Specification.* Energy, Mines and Resources Canada, Ottawa, Canada.

Orheim, O. and Lucchitta, B. 1988. Numerical analysis of LANDSAT Thematic Mapper images of Antarctica: surface temperatures and physical properties. *Annals of Glaciology* 11: 109–20.

Stephenson, S. and Zwally, H. 1989. Ice-shelf topography and structure determined using satellite-radar altimetry and LANDSAT imagery. *Annals of Glaciology* 12: 162–9.

Swithinbank, C. and Lucchitta, B. 1986. Multispectral digital image mapping of Antarctic ice features. *Annals of Glaciology* 8: 159–63.

US Geological Survey, 1984. *LANDSAT 4 Data Users' Handbook.* Available from the Distribution Branch, Text Productions Section, US Geological Survey, 604 South Pickett Street, Alexandria, VA 22304, USA.

21. Shuttle Imaging Radars (SIR-B and SIR-C/X-SAR)

SIR-B

Operational lifespan and orbital characteristics

Dates: 5 October 1984–13 October 1984.
Semimajor axis: 6,730km.
Orbital height: 345–359km.
Inclination: $57.0°$
Period: 91.55 minutes.
Note: Space Shuttle Mission 41-G, Challenger F6.

Sensor characteristics

Synthetic Aperture Radar (SAR):
    Active microwave: frequency 1.282GHz (L-band), wavelength 23.4cm. Radar pulse bandwidth 12MHz. HH polarisation. Incidence angles variable in the range of 15–65°. Azimuth GR 20m (4 look). Range resolution 17–58m. Swath width variable in the range of 20–50km. Data rate 46 Mbps.

SIR-C/X-SAR

Operational lifespan and orbital characteristics

Dates: 3 launches are scheduled in different seasons in 1993, 1994 and 1996.
Orbital height: Average 225km.
Inclination: $57.0°$
Period: 91.55 minutes. 1-day repeat cycle.
Notes: A joint US-German venture.

Sensor characteristics

SAR:
    Frequency: 1.25GHz (L-band, wavelength 24cm) and 5.3GHz (C-band, wavelength 5.7cm). Altitude 225km. Radar pulse bandwidth 12MHz for both frequencies. HH, VV, VH and HV polarisations. Incidence angle range 15–55°; swath width 15–65km for calibrated data, 40–90km in mapping mode. Azimuth resolution 40m (4 look); range resolution 10–60m. Under an agreement between NASA and the Science Ministry of the Federal Republic of Germany, it will also carry an X-band SAR operating at 9.6GHz (VV polarisation only), bandwidths 20 and 10MHz.

Polar applications (for all SIRs)

SIR-B:

icebergs and sea ice were imaged on one pass only in the S hemisphere, near the South Sandwich Islands, permitting limited observations of the MIZ of the Weddell–Scotia Seas (Carsey and others 1986). The ice cover showed band-like features at the outermost margin similar to those observed in the Bering Sea, but on a larger scale. Further south, more fully developed floes were observed. Comparisons of ice concentrations derived from this imagery and Nimbus-7 SMMR are favourable, except at the outermost margin where changes in ocean surface roughness resulting from air-ice band interactions tended to blur the ice-water separation on the SAR imagery (Martin and others 1987).

SIR-C/X-SAR:

multi-spectral and multi-polarisation capabilities. Sea ice feature verification, ice drift and deformation, spatial and temporal variations in ice characteristics, the nature and occurrence of ice margins and bands, and improved characterisation of sea ice types in Labrador Sea, Gulf of St Lawrence, S Hudson Bay, S Bering Sea, Sea of Okhotsk, N Weddell Sea and Gulf of Bothnia (all at maximum extent only). See also Seasat, JERS-1, ERS-1 and Radarsat.

Limitations

SIR-B:

offered very limited polar coverage, due to problems with the Shuttle-to-TDRS communications antennae system. Poor ice/water contrast, with even a contrast reversal between images acquired at steep and shallow incidence angles (Carsey and others 1986). Image interpretation problems also occurred due to the poor signal strength received, which in turn resulted from electronic failure and weak backscatter from the area of sea ice imaged. Icebergs were not resolvable in the L-band images when embedded in sea ice (only in the open ocean). Few supporting data are available, the only 'ground truth' data from the Antarctic being hand-held photography taken by the astronauts.

SIR-C/X-SAR:

at near-nadir look angles, the swath width is limited by antenna illumination. At large look angles, it is limited by the data rate that can be accommodated by the digital data relay link. See also Seasat, JERS-1, ERS-1 and Radarsat. Coverage limited to maximum latitude of only 57°.

Data format

SIR-B – both digitally and optically processed data:
 i) SAR imagery on film, 10 × 8 inch black and white negatives;
 ii) image data and annotation on CCT, 9-track 1600/6250bpi. Images are about 40 × 90km. Signal data also available.

SIR-C/X-SAR:

digital correlation to full resolution will result in images with 25m (20MHz bandwidth) and 40m GR (10MHz bandwidth). 8 and 4 bits per sample. Standard products will be single-look complex, compressed polarimetric, and four-look detected data. CCT and film products, with data also distributed on CD-ROM.

Data availability and archiving dates

Data from Shuttle missions available from NSSDC (see TIROS I-X for address, etc). SIR-B data are archived as film for 9 October 1984 and 1 CCT for 8 October 1984. Contact Amy Pang, SAR Data Catalog System, NASA Ocean Data System, MS 300–235, Jet Propulsion Laboratory, 4800 Oak Grove Drive, Pasadena, CA 91109, USA, telephone (818)–354–3386, SPAN KAHUNA::AAP. Requests from non-NASA scientists must be addressed to Dr John Curlander, MS 320–235, JPL, telephone (818)–354–8262. Prints, slides and transparencies of Shuttle Earth-looking photography can be obtained from Technology Applications Center, University of New Mexico, Albuquerque, NM 87131, USA, telephone (505)–277–3662.

Nominally, 50 hours of SIR-C data will be recorded onboard on each of the four channels for each flight, in addition to 50 hours of X-SAR data. Data will be digitised and formatted onboard for direct downlink via TDRSS, or buffered through two onboard high density digital recorders (storage capacity 50 Mbps each) for storage. SIR-C will record the amplitude of H, V and both cross-polarised radar echoes, as well as the electrical phase angle between the HH and VV echoes. These amplitude and phase data are expected to greatly increase the information content of the data, thereby leading to expanded geoscientific applications. The data tapes will be shipped from NASA GSFC to JPL for processing.

References and important further reading

Carsey, F., Holt, B., Martin, S., McNutt, L., Rothrock, D., Squire, V. and Weeks, W. 1986. Weddell-Scotia Sea marginal ice zone observations from space, October 1984. *Journal of Geophysical Research* 91, C3: 3920–4.

Cimino, J., Holt, B. and Richardson, A. 1988. The Shuttle Imaging Radar B (SIR-B) Experiment. *NASA JPL Publication* 88–2. NASA JPL, Pasadena, California.

Curlander, J. 1986. The SIR-C ground data system: digital processor, data products, information flow. In *The Second Spaceborne Imaging Radar Symposium*, JPL, Pasadena, California: 149–56.

Martin, S., Holt, B., Cavalieri, D. and Squire, V. 1987. Shuttle Imaging Radar-B (SIR-B) Weddell Sea ice observations: a comparison of SIR-B and SMMR ice concentrations. *Journal of Geophysical Research* 92, C7: 7,173–9.

NASA. 1986. The SIR-C Science Plan. *NASA JPL Publication* 86–29. NASA JPL, Pasadena, California.

Roth, A., Craubner, H. and Bayer, T. 1989. Prototype SAR geocoding algorithms for ERS-1 and SIR-C/X-SAR images. *Proceedings of IGARSS '89*, Vancouver, Canada, 10–14 July 1989: 604–7.

Velten, E. 1989. Technical status of X-SAR. *Proceedings of IGARSS '89*, Vancouver BC, 10–14 July 1989: 1,719.

22. GEOSAT

Operational lifespan and orbital characteristics

Dates: 13 March 1985–January 1990.
Semimajor axis: 7,167km.
Orbital height: 760–817km.
Inclination: 108.1°
Period: 100.67 minutes.
Notes: A dedicated US Navy military oceanographic satellite, the name being a contraction of GEOdetic SATellite. Repeat cycle three days for the first 18 months, the Geodetic Mission (GM). On 8 November 1986, the satellite moved into an exact repeat mission (ERM) orbit with a repeat period of 17.05 days, and with a ground track designed to correspond closely to the tracks followed by the Seasat ALT. ERM lasted until 18 November 1987. A programme of follow-on satellites is planned by US Navy for launch in 1994 (possible); will carry almost identical ALTs, and passive microwave radiometers for atmospheric correction.

Sensor characteristics

Altimeter (ALT):

Active microwave: frequency 13.5GHz (K_u-band). Linear FM pulse. RF bandwidth 320MHz. Antenna beamwidth 2.1°. Footprint 6.8km (on smooth surface). Pulsewidth (uncompressed) 102µs, 3.125ns (compressed). Interpulse period 980µs. PRF 1,020Hz. Altitude precision <5cm rms for 1s average measurements over open ocean. Over ice, altitude precision is approximately 1.5m overall (about 30cm in regions of smooth topography). Accuracy over open ocean of the order of 3.5cm at 2m SWH. SWH of 1–20m, ±10% or 0.5m, whichever is greater. Wind speed measurement (over open ocean) to an accuracy ±1.8ms^{-1} over the range of 1–18ms^{-1}. Nadir looking.

Polar applications

GEOSAT covered the same latitude band as Seasat (maximum 72°).
  i) *Land ice*: mapping of ice sheet elevation, to detect possible thinning or thickening of ice sheets. The dense high latitude convergence of orbits has permitted the testing and development of analysis techniques for optimal use of data from satellites providing less dense coverage.
 ii) *Sea ice*: Ulander (1988) has separated different sea ice types based on the normal-incidence backscatter coefficient using GEOSAT data from the brackish Bothnian Sea: fast ice, with a mean backscatter coefficient of 37.8 ± 2.8dB, and thin ice (32.5 ± 2.1dB) could be separated. These two ice types are also separated from the open ocean backscatter coefficient by ±10dB for wind speeds of 2ms^{-1} or more.
    The US Naval Oceanographic and Atmospheric Research Laboratory (NOARL) provided an operational product, the 'GEOSAT ice index', to sea ice mapping centres in

the USA (Lybanon and others 1990; Laxon 1989). This technique is based upon that developed by Dwyer and Godin (1980) for application with GEOS-3 data. The ice index provides a simple ocean/ice discrimination, and is thus designed primarily for the delineation of ice-ocean boundaries (ie the ice edge) rather than features of the interior pack. Values are <1 over open ocean, but >1 over sea ice. In the E Greenland Sea, ice index values show a sharp rise at the ice edge, and drop off further into the pack (Hawkins and Lybanon 1989). Fetterer and others (1988) observe more peaked and high power values in the MIZ than in the interior pack. In the Kara Sea, however, a much lower degree of variation is observed in the ice index as it transects different ice types. Laxon (1989) has mapped Antarctic sea ice extent (Figure II.17) and measured the freeboard of giant tabular icebergs; it may be possible to measure ablation rates of the latter with time series data.

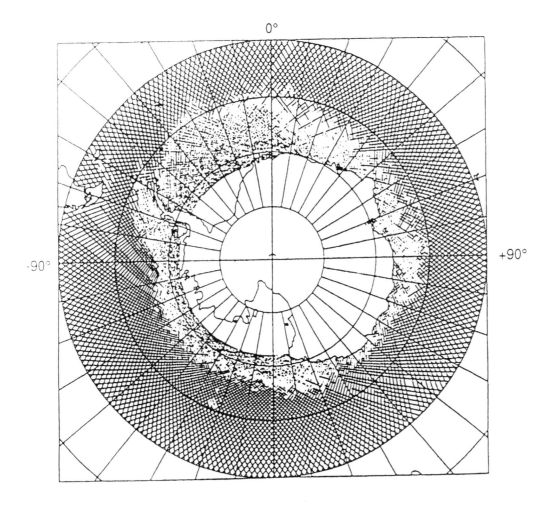

Figure II.17 GEOSAT Ocean Mode GDR Southern Ocean coverage for ERM cycle 1 (November 1986). The presence of sea ice is detectable as it results in data gaps where the GDR processing algorithm has classified echoes as 'Land/ice'. From Laxon (1989).

Simultaneous surface experiments were conducted to compare GEOSAT data with other satellite, *in situ* and aircraft data. Details of these, and data availability, have been compiled by Dr Jeff Hawkins of NOARL, who has convened a GEOSAT Sea Ice Working Group. A mini proceedings is available from him at the Remote Sensing Branch, Code 321, at NOARL, Stennis Space Center, MS 39529, USA, telephone (601)–688–5270, OMNET J.Hawkins.

iii) *Open ocean*: data extensively used in detection of ocean fronts and eddies, surface geostrophic currents, SWHs and surface wind speed. Marine gravity and geoid measurements.

See also Skylab, GEOS-3, Seasat, TOPEX/Poseidon and ERS-1.

Limitations

Polar coverage is limited to a maximum latitude of 72° (S Greenland Ice Sheet and N Antarctic Ice Sheet). Onboard tracking algorithms used by GEOSAT were optimised for operation over open ocean, and loss of track tended to occur over rougher, steep ice surfaces. Loss of track over sea ice more frequent with GEOSAT than Seasat. Spatial coverage is limited by the narrow swath width. GEOSAT was not equipped with a radiometer for atmospheric correction; atmospheric effects are compensated for by integrating 22GHz data from DMSP SSM/I, although these were not available for the first two years of the mission. Some data remain classified. The first 18 months of the mission devoted to completing that element of US Department of Defense activity originally planned for SEASAT. Comparison between GEOSAT and Seasat ALT data is hindered by problems with ensuring that the orbits are compatible. This stresses the need for accurate ephemerides.

One major problem in the use of a single sea ice index, outlined by Laxon (1989), is that changes caused by distinct physical properties cannot be distinguished in the data. Changes in waveform shape are caused largely by changes in small (mm) scale surface roughness; changes in waveform power, on the other hand, depend not only on small scale surface roughness but also on dielectric properties. See also Skylab, GEOS-3, Seasat, TOPEX/Poseidon and ERS-1.

Data format

CCTs only. 3 primary data products were provided by the APL:

i) Sensor Data Record (SDR): a classified CCT containing measured altitude, SWH, wind speed, and AGC and corrections for satellite and instrument errors. Data sampled at 1s intervals.

ii) US Naval Ocean Research and Development Activity (NORDA) Data Record: a classified data record with contents similar to the SDR.

iii) Waveform Data Record (WDR): an unclassified CCT containing the raw waveform samples from the ALT. Once generated, the CCTs were given to the Naval Research Laboratory for ice studies. These are different from the standard Seasat data, but may be associated with SDR data by applying a frame counter (Laxon 1989).

iv) NORDA transmitted a graphic sea ice product daily to the US Naval Polar Oceanography Center, where it was used routinely to compile daily ice maps. Digital and hardcopy format.

During processing, data from ocean areas and land/ice areas are now segregated and stored on separate magnetic CCTs in a standard Geophysical Data Records (GDR) format (Cheney and others 1987). Each GDR archive CCT contains two 17-day sequences.

For the format of ice sheet data products produced by Zwally and others (1990), see Seasat. GM and ERM data are being processed for both the Antarctic and Greenland ice sheets. Users guides will be published by NASA, and the data will be available on CD-ROM. For further details, contact Dr Jay Zwally, Code 971, NASA GSFC, Greenbelt, MD 20771, USA.

Data availability and archiving dates

Control Centre: Applied Physics Laboratory, Johns Hopkins University. Data stored onboard for 12 hours were dumped here for preprocessing, then transmitted to the Naval Surface Weapons Center for geodetic processing, and to NOARL for oceanographic processing. Then transmitted to FNOC. These data products are tailored for operational use and are not widely distributed. The geoid data are classified. All data collected after 8 November 1986 have been made available by NOAA, beginning in early 1987, completed unclassified GDRs being produced by the National Geodetic Survey and distributed to the NOAA National Oceanographic Data Center, Washington, DC 20235, telephone (202)–673–5549. Users should address all enquiries to: Dr R. Cheney, Mail Code E/OC21. Non-US citizens must request data via the scientific officer of their country's embassy in the USA.

Potential users should now contact the centre appropriate to their primary interest:

i) Snow and Ice Applications data are archived at the Cryospheric Data Management System (CDMS), NSIDC, University of Colorado, Boulder, CO 80309, USA, telephone 303–492–1834 or 303–492–5171, telex 257673 (WDCA UR), telemail MAIL/USA [NSIDC/OMNET], VAX mail via SPAN: KRYOS::NSIDC. These georeferenced ALT data collected over Greenland and Antarctica are derived from retracked and atmospherically corrected data, and are available as discrete elevations or on an interpolated 20km grid (Zwally and others 1990). These data are also available from NSSDC (see TIROS I-X for address).

ii) Oceanographic Applications: National Ocean Data Center (see address above).

References and important further reading

Results from GEOSAT are presented in special editions of i) *Journal of Geophysical Research (Oceans)*, 1990, Volume 95, C3: 2,833–3180; and ii) *Special Issue, Johns Hopkins Applied Physics Laboratory Technical Digest* 1987, 8.

Cheney, R., Douglas, B., Agreen, R., Miller, L., Porter, D. and Doyle, N. 1987. GEOSAT Altimeter Data Record User Handbook. *NOAA Technical Memorandum NOS NGS-46*. Rockville, Maryland.

Dwyer, R. and Godin, R. 1980. Determining sea ice boundaries and ice roughness using GEOS-3 altimeter data. *NASA Contract Report CR 156862*.

Fetterer, F., Johnson, D., Hawkins, J. and Laxon, S. 1988. Investigations of sea ice using coincident GEOSAT altimetry and synthetic aperture radar during MIZEX 87. *Proceedings of IGARSS '88*, Edinburgh, 13–16 September 1988. *ESA SP-284 (IEEE 88CH2497–6)*, Volume 2: 1,125–6.

Hawkins, J. and Lybanon, M. 1989. GEOSAT altimeter sea ice mapping. IEEE Journal of Oceanic Engineering 14, 2: 139–48.

Koblinsky, C., Klacka, S. and Williamson, R. 1989. Interannual global sea level change between 1987 and 1978 from GEOSAT and SEASAT altimetry. *EOS 70*, 43: 1,052.

Laxon, S. 1989. *Satellite radar altimetry over sea ice*. Unpublished PhD Thesis, University College, London, UK.

Lybanon, M., Crout, R., Johnson, C. and Pistek, P. 1990. Operational altimeter-derived oceanographic information: the NORDA GEOSAT Ocean Applications Program. *Journal of Atmospheric and Oceanic Technology* 7, 3: 357–76.

Lybanon, M. and Hawkins, J. 1988. Sea ice detection by GEOSAT. In *Proceedings of the Workshop on Instrumentation and Measurements in the Polar Regions*: 103–13. Marine Technology Society, San Francisco Bay Region Section, 2411 Valley Street, Berkeley, California 94702, USA.

Partington, K. 1988. *Studies of the ice sheets by satellite altimetry*. Unpublished PhD thesis, University College, London, UK.

Ulander, L. 1988. Observations of ice types in satellite altimeter data. *Proceedings of IGARSS '88*, Edinburgh, 13–16 September 1988. *ESA SP-284 (IEEE 88CH2497–6)* Volume 2: 655–8.

Zlotnicki, V., Hayashi, A. and Fu, L. 1989. The JPL-Oceans-8902 version of the GEOSAT altimeter data. *NASA JPL Technical Report*. Available from JPL/NODS.

Zwally, H., Major, J., Brenner, A. and Bindschadler, R. 1987. Ice measurements by GEOSAT Radar Altimetry. *Johns Hopkins Applied Physics Laboratory Technical Digest* 8, 2: 251–4.

Zwally, H., Brenner, A., Major, J., Martin, T. and Bindschadler, R. 1990. Satellite radar altimetry over ice. Volume 1: processing and corrections of SEASAT data over Greenland. *NASA Reference Publication 1233, Volume 1*. NASA Office of Management, Scientific and Technical Information Division, Washington DC, USA.

23. Système Probatoire pour l'Observation de la Terre (SPOT) series

SPOT-1

Operational lifespan and orbital parameters

Dates: 22 February 1986. 3-year life expectancy. Phased out in June 1990.
Semimajor axis: 7,204km.
Orbital height: 824–828km.
Inclination: 98.7°
Period: 101.48 minutes. Exactly repeating orbit with a repeat period of 26 days (this can be varied by adjusting the tilt of the sensor – see next section).
Notes: An operational and commercial satellite. Sun-synchronous. The off-nadir viewing means that imagery can be acquired of the same area on successive days, and that the average imaging interval can be as little as 2.5 days for priority areas. Equatorial crossing time 10:30 LST.

Sensor characteristics

2 identical High Resolution Visual (HRV) sensors:

 i) Panchromatic mode: wavelength 0.51–0.73μm. Sampling interval 10m. Swath width 60km at vertical inclination. Pointing capability 27°E/W off-nadir in 45 steps of 0.6°, allowing the observation of any area of interest within a 950km wide strip (475km on either side of the subsatellite track). This enables stereoscopic images to be collected from several orbits with different angles. Data rate 25Mbps (all channels combined). 6-bit quantisation. 6,000 detectors in a linear CCD array.

 ii) Multispectral modes: wavelengths 0.50–0.59μm (VIS), 0.61–0.68μm (VIS) and 0.79–0.89μm (NrIR). Sampling interval 20m. 8-bit quantisation. 3,000 detectors in a linear CCD array.

 iii) 2 onboard recorders (each 23 minutes). One recorder failed in September 1986, and one of the HRV sensors failed in late 1986.

SPOT-2

Operational lifespan and orbital characteristics

Launched 22 January 1990 into an 850km sun-synchronous orbit. Orbital parameters identical to SPOT-1, except 180° apart.

Sensor characteristics

 i)   Two identical HRVs.
 ii) A DORIS (Determination d'Orbité et Radiopositionnement Integrés par Satellite) precise location experiment.

SPOT-3

Operational lifespan and orbital parameters

Possible launch mid-1992. Orbital parameters as for SPOT-1 and -2. Two-year design life.

Sensor characteristics

As for SPOT-2 (ie two identical HRVs), plus a Polar Ozone and Aerosol Measurement (POAM) instrument supplied by the USAF Space Test Program.

SPOT-4 and -5

Operational lifespan and orbital parameters

Planned for launch in 1995 and 1999 respectively. Orbital parameters as for SPOT-1.

Sensor characteristics

Paired High Resolution Visible Infra-Red (HRVIR) sensors with spectral ranges: Band 1 0.5–0.59μm, Band 2 0.61–0.68μm, Band 3 0.79–0.89μm, Band 4 1.58–1.75μm, yielding

20m GR in the spectral mode. In the monospectral mode (ie Band 2), the GR is 10m. Viewing characteristics as for the HRVs of SPOTS 1–3. Will also include a new imaging sensor (1km resolution, 2,000km swath).

Polar applications

HRV and HRVIR:

Similar applications to Landsat MSS and TM. More flexible coverage.

i) *Land ice*: accurate mapping of ice sheet margins, glacier termini and ice shelves (Figure II.18). Monitoring of iceberg calving, size distribution and drift rates. Detailed delineation of flow lines, streams, crevasses and measurement of flow velocity from sequential imagery. Snow cover mapping. Low sun angles highlight topographic features not visible on the ground. Stereo-photogrammetry. Not reliant on having receiving stations in line of sight for data collection. Can cover the region 1.5° of latitude further poleward than Landsat TM, enabling coverage of ice stream B in Antartica.

ii) *Sea ice*: accurate ice/water boundaries. Sea ice concentration. Floe size and shape distributions. Lead/polynya dynamics. Short-term floe dynamics and velocity. The distinction of thin versus thicker ice, based upon textural and tonal variations. Fast ice distribution. Large pressure ridge delineation. Oceanic features adjacent to ice edge. Validation of poorer resolution sensor data. Very fine detail under cloud-free, day-light conditions. Particularly useful in mesoscale study of complex MIZs.

Limitations

HRV:

limited by cloud/fog/darkness. Narrow swath width. Not useful for macro-scale studies. Cloud is often difficult to distinguish from snow and ice. Measurements are limited to VIS and NrIR. Relatively poor distinction between thicker sea ice forms, and poor response to thinner ice forms. The repeat cycle is too long to study many important dynamic geophysical processes, eg sea ice floe tracking. Fine resolution stereo-photogrammetry from SPOT is difficult over the interior regions of ice sheets, due to the lack of ground control points and the smooth topography. Image geolocation is difficult in the relatively featureless interiors of ice sheets. Problems arise with the identification of thickly debris-covered ice. Expense, high data rate and thus large volume of data per scene. Coverage is limited to a maximum latitude of 84°. See also Landsat MSS and TM.

Data format

HRV: data products include panchromatic data (1 band, 10m resolution) or multi-spectral data (3 bands, 20m resolution). The basic photographic medium is 241 × 241mm film corresponding to a full scene, scale of 1:400,000, radiometrically and geometrically corrected. Black and white or colour composite. Level 2 products are radiometrically and precision corrected to map the image into a cartographic projection. The basic format is 350 × 350mm (1:400,000). Available, by full scene or sub-scene, as a photographic film/print. Scales range from 1:400,000 to 1:500,000, level 1b and 2 products. CCTs: 1600 or 6250bpi, BIL or BSQ, ASCII or EBCDIC character coding. Quick look prints as 1:400,000 (240mm).

New products include quarter scenes, registered Panchromatic Linear Array and Multi-spectral Linear Array (MLA) imagery, a scene anywhere along a track, a double scene

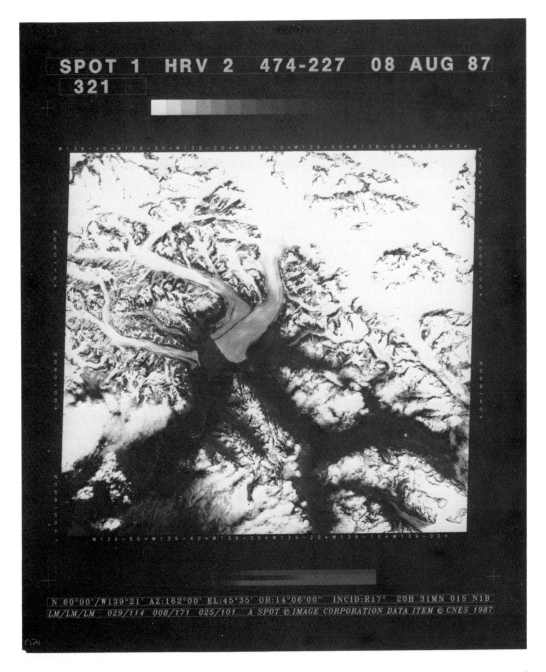

Figure II.18 SPOT-1 HRV image of the Hubbard Glacier, Alaska, 8 August 1987. In the past, the advance of the glacier has temporarily cut off the flow of water between Disenchantment Bay and Russell Fjord, threatening the local fishing industry and native land use. Copyright 1990 CNES. Provided by SPOT Image Corporation.

(60 × 120km) and a quadruple scene (117 × 120km). Digital Terrain Models (DTM) are available. The SPOT Grid Reference System designed to allow for the geographic location of images – used to allocate each SPOT scene a pair of scene designators corresponding to precise coordinates. In the two polar regions above latitude 71.7°, the distance between adjacent nodes is constant at 26km.

Note: SPOT panchromatic and multi-spectral scenes can be preprocessed to any of five levels:
   i) Level 1: basic radiometric and geometric corrections (does not involve ground control points or satellite attitude restitution data).
   ii) Level 1a: essentially raw data with no geometric correction.
   iii) Level 1b: includes full radiometric correction and geometric correction of system-related effects (rotation and curvature of the Earth, viewing angle and desmearing). Absolute location accuracy is 1,500m (rms) for vertical viewing, and internal distortion is $<10^{-2}$.
   iv) Level 2: equivalent to 1b plus bidirectional geometric correction using ground control points. Map projections available are conformal Lambert, transverse Mercator, polar stereographic, and polyconic. Location accuracy is 50m (rms) for vertical viewing.
   v) Level S: rectified using control points for registration with a SPOT reference scene. Registration accuracy of 0.5 pixels when 2 scenes are recorded at the same viewing angle. Level S products are intended primarily for multi-date use.
See the *SPOT User's Guide* (1987) for further details.

Data availability and archiving dates

   i) The two master receiving stations are at Aussaguel-Issus in France and Kiruna in Sweden. Main data sources: SPOT IMAGE, 16 Bis, Avenue Edouard Belin BP, 31055 Toulouse, Cedex, France, telephone 33 61273131, telex 531081; or SPOT IMAGE Corporation, 1987 Preston White Drive, Reston, VA 22091–4326, USA, telephone (703)–620–2200. The user can acquire specific data with a specific time scheduling. Single frames or stereo pairs are available. The *SPOT Users Handbook* is available upon request. The SPOT Image Corporation publish a quarterly newsletter called *SPOTLIGHT*.
   ii) UK distributor of world data: Nigel Press Associates, Edenbridge, Kent, UK (NRSC, RAE supply UK data only).
   iii) SPOT data have also been received at the Prince Albert Station in Canada (covering the North West Passage) and Gatineau (covering Hudson Bay, Labrador and S Greenland) since spring 1986. Canadian data are available from: Order Desk, Prince Albert Satellite Station, PO Box 1150, Prince Albert, Saskatchewan S6V 5S7, Canada, telephone (306)–764–3602, FAX 306–763–1773, telex 074 29123; or Data Acquisition Division, Canada Centre for Remote Sensing, 2464 Sheffield Road, Ottawa K1A OY7, Canada, telephone (613)–952–2717, telex 0533777 CA. A catalogue of SPOT imagery is regularly published by the CCRS. SPOT general information and applications-oriented advice are available from: The User Assistance and Marketing Unit, Canada Centre for Remote Sensing, 1547 Merivale Road, Nepean, Ontario K1A OY7, Canada, telephone 613 952 0500, FAX 613 952 7353, telex 053 3777. Note: the single X-band antenna at Prince Albert must be shared for both SPOT-1 and Landsat data reception. Careful planning is necessary prior to required acquisition as selection of one satellite due to conflicting acquisition times for the two satellites results in a loss of data.

iv) SPOT data have been received at Hatoyama station in Japan, which offers coverage of almost the entire Sea of Okhotsk, since October 1988. Data can be obtained from Remote Sensing Technology Center, Uni Roppongi, Building 7–15–17, Roppongi, Minato Ku, Tokyo 106, Japan, telephone 03 403 1761, telex 2426780 RESTEC J, FAX (81)–3–403–1766.

References and important further reading

CNES/SPOT Image. 1989. *SPOT User's Handbook, Volumes 1 and 2*. Spot Image/Spot Image Corporation.

Courtois, M., and Weill, G. 1985. The SPOT satellite system. In Schnapf, A. (ed.) *Monitoring Earth's ocean, land, and atmosphere from space – sensors, systems and applications*. American Institute of Aeronautics and Astronautics, New York: 493–523.

NRSC. 1987. *SPOT User's Guide*. National Remote Sensing Centre Space Department, Royal Aircraft Establishment, Farnborough, Hampshire, UK.

Lukowski, T. 1988. Satellite data acquisition planning at the Canada Centre for Remote Sensing. *Proceedings of IGARSS '88*, Edinburgh, 13–16 September 1988. *ESA SP-284 (IEEE 88CH2497–6), volume 3: 1,471–4.*

24. Marine Observation Satellites (MOS)

MOS-1 and MOS-1b

Operational lifespan and orbital characteristics

Dates: 19 February 1987 to present. MOS-1b launched on 7 February 1990.
Orbital height: 908.7km.
Inclination: 99.1°
Period: 103.2 minutes. Exactly repeating orbit, repeat period 17 days.
Notes: Experimental oceanographic satellites. Sun-synchronous orbit. Descending node time between 10:00 and 11:00LST for MOS-1; MOS-1b in a 180° phase orbit relative to that of MOS-1. MOS-2 and MOS-3 to follow on. Orbit designed to compatible with Landsat-4 and -5 orbits.

Sensor characteristics

Multispectral Electronic Self-Scanning Radiometer (MESSR):
i) Band 1: VIS (wavelength 0.51–0.59μm). GR 50m (for all bands). Swath width 100km × 2 (for all bands). Pushbroom scanning method; four sets of linear CCD arrays corresponding to 4 spectral bands. Radiometric resolution S/N 15–39dB. Scanning period 7.6μs. Data rate 8Mbps (all bands combined). 6-bit quantisation levels.
ii) Band 2: VIS (wavelength 0.61–0.69μm).

iii) Band 3: NrIR (0.72–0.80μm).
iv) Band 4: NrIR (0.80–1.10μm).
Visible and Thermal Infrared Radiometer (VTIR):
 i) Band 1: VIS (wavelength 0.5–0.7μm). GR 0.9km. Swath width 1,500km (for all bands). Mechanical rotating mirror scanning method. Scanning period 1/1.7s. Radiometric resolution S/N 55dB. Data rate 0.8Mbps (all bands combined). 8-bit quantisation levels.
 ii) Band 2: TIR (wavelength 6.0–7.0μm). GR 2.7km (bands 2–4). Radiometric resolution 0.5K (Bands 2–4).
iii) Band 3: TIR (10.5–11.5μm).
iv) Band 4: TIR (11.5–12.5μm).
Microwave Scanning Radiometer (MSR):
 i) Channel 1: passive microwave, frequency 23.8 ± 0.2GHz, H polarisation. Beam width 1.89° ± 0.19°. GR 32km. Swath width 317km. 2 different integration times 10 and 47μs (for both channels). Mechanical conical scan. Radiometric resolution 1.5K (both channels). Dynamic range 30–330K. Scanning period 3.2s. Data rate 2 Kbps (both channels combined). 10-bit quantisation levels.
 ii) Channel 2: passive microwave, frequency 31.4 ± 0.25GHz, V polarisation. Beam width 1.31° ± 0.13°. GR 23km. Swath width 317km.
Data Collection System (DCS) Transponder:
  A forerunner of a Japanese Tracking and Data Relay Satellite System (TDRSS), used to collect and relay information from surface Data Collection Platforms (DCPs), and locates DCPs based on the Doppler frequency of received signals (similar to ARGOS on the NOAA satellites). DCS functional specifications:
 i) DCP – Uplink. Frequency 401.5MHz ± 40kHz. Transponder PCM-PSK. Bit rate 400bps. Effective radiated power 35dBm. Transmission method 1s burst radiation per 60s.
 ii) Transponder. Bandwidth 90kHz ± 10kHz. Random access phase remodulation method. Local oscillator frequency stability: short 1×10^{-9}/s; medium 1×10^{-8}/10 minutes; long 1×10^{-6}/year.
iii) Downlink frequency 1702.4848MHz. Remodulation method PM.

MOS-2

Operational lifespan and orbital characteristics

Dates: possible launch in 1992. Other parameters to be determined.

Sensor characteristics

Possible payload:
MSR:
  as for MOS-1 and -b.
SAR:
  Frequency 1.2GHz (L-band). 4 look, 25m GR. Design based upon JERS-1 SAR. Radar Scatterometer (SCATT): Frequency 13.99GHz (K_u-band). GR 25–50km. Swath width 200–700km. Wind speed measurement in range of 4–25ms^{-1}, direction accuracy ±20°.
Ocean Colour and Temperature Scanner (OCTS):

5 VIS and 2 TIR bands – wavelengths 0.43–0.45, 0.50–0.53, 0.54–0.58, 0.65–0.67, 0.75–0.79, 10.5–11.5 and 11.5–12.5μm. GR 0.7km. Swath width 1,500km.

SEASAT-class Radar Altimeter (ALT):

Frequency 13.8GHz (K_u-band). Linear polarisation. Pulse width 3.125nsec. Pulse compression ratio 1,024.

Polar applications

MESSR:

ocean current and cloud cover measurements. Medium–high resolution, intermediate between Landsat MSS and SPOT HRV/Landsat TM. Snowpack mapping. Ocean colour measurements are also possible, but not in any great detail, and this is not a primary ocean colour sensor. Floe size distribution and sea ice extent (particularly band 4). For further discussion of applications of sensors operating at similar wavelengths, see Landsat MSS and RBV, and SPOT HRV.

VTIR:

ocean current and SST measurements. Bands 2–4 can penetrate polar darkness. Large-scale eddy structure and circulation pattern detection. Sea ice extent. Theoretically possible to derive ice surface temperatures from the data (in absence of cloud). Snowpack mapping. Day, night and winter cloud cover mapping. Similar to the NOAA AVHRR, DMSP OLS, HCMM HCMR and the non-microwave bands of ERS-1 ATSR-M.

MSR:

all-weather/all-season large-scale SSTs, snow cover and sea ice extent (channel 2). Channel 1: atmospheric liquid water and water vapour content. Channel 2: sea ice concentration and MY fraction. Ice concentrations are retrieved using a linear algorithm similar to the Nimbus-5 ESMR technique. Information on the onset of spring melt. Potential in the study of broad-scale properties of land/ice snow. Wind speed (but not direction) over open ocean. For further comments regarding the use of similar frequencies, see Nimbus-5 and -6 ESMR, Seasat and Nimbus-7 SMMR, and DMSP SSM/I. Very similar to the microwave channels of ERS-1 ATSR-M.

DCS:

cost-effective method of regularly collecting data from surface instrument packages, buoys, etc.

SAR:

see JERS-1 and SEASAT.

SCATT:

see ERS-1 AMI Wind Scatterometer Mode, ADEOS NSCAT and Seasat SASS.

ALT:

see Seasat, GEOS-3, GEOSAT, TOPEX/Poseidon and ERS-1.

OCTS:

global ocean colour and SST data. Near-surface phytoplankton concentration/primary productivity. Monitor rapidly evolving oceanic phenomena with a high spectral resolution. May be useful in studying early stages of sea ice growth. See NIMBUS-7 CZCS and SeaWiFS.

Limitations

Polar coverage limited to a maximum latitute of 81–82°, and by the availability of power and suitably equipped receiving stations (Figure II.19). No onboard tape recorders for data storage. The repeat cycle of 16 days is too long for studies monitoring a number of dynamic geophysical processes. Even this repeat cycle is seldom achieved in an operational context as clouds often obscure the surface (for VIS/TIR sensors).

MESSR:

limited by cloud/darkness, and to measurements in the VIS and NrIR. Swath width too narrow to study large-scale phenomena effectively. For many oceanographic applications, bands 1 and 2 are too wide. The lack of a blue band (the chlorophyll absorption band) is a limitation for studies of water constituents, and makes it impossible to evaluate quantitatively the presence of pigments. There are certain technological constraints; for example, although the pushbroom scanner allows imaging within narrower spectral bands and at an improved spectral resolution, it does not allow the acquisition of data from many contiguous bands. See Landsat MSS and RBV, and SPOT HRV.

VTIR:

Band 1 limited by cloud/darkness. Bands 2, 3 and 4 limited by cloud. Moderate–narrow swath. Some technical problems, including pixel misalignment between the different detectors. For further comments regarding similar wavelengths, see NOAA VHRR and AVHRR, HCMM HCMR and DMSP OLS.

MSR:

coarse resolution and relatively narrow swath. Little surface information under summer melt conditions. Sensitive to atmospheric water vapour. Channel 2: sensitive to ocean effects, and may give spurious ice concentration results near the ice edge. Difficult to distinguish thin FY ice in summer/spring. Pixels containing coastlines may show erroneous values. For further discussions of the limitations of similar frequencies, see Nimbus-5 and -6 ESMR, Seasat and Nimbus-7 SMMR, and DMSP SSM/I.

SAR:

see JERS-1 and Seasat.

SCATT:

see Seasat SASS, ERS-1 AMI Wind Scatterometer Mode and ADEOS NSCAT.

ALT:

see Skylab, Seasat, GEOS-3, GEOSAT, ERS-1 and TOPEX/Poseidon.

OCTS:

Limited by cloud/darkness, sun glint effects, and possible intermittent sensor saturation over snow and ice masses. See Nimbus-7 CZCS and SeaWiFS.

Data format

The World Reference System similar to that of Landsat is adopted.
Data products are available at 5 processing levels:
 i) Level 0, no corrections applied;
 ii) Level 1, radiometric corrections only;
 iii) Level 2, radiometric and geometric corrections;
 iv) Level 3, precision correction (MESSR only);
 v) Level 4, precision registration (MESSR only).

MESSR:

available on Universal Transversal Mercator (UTM) or Space Oblique Mercator (SOM) map projections. Nearest neighbour (NN), bilinear (BL) and cubic convolution (CC) resampling methods.

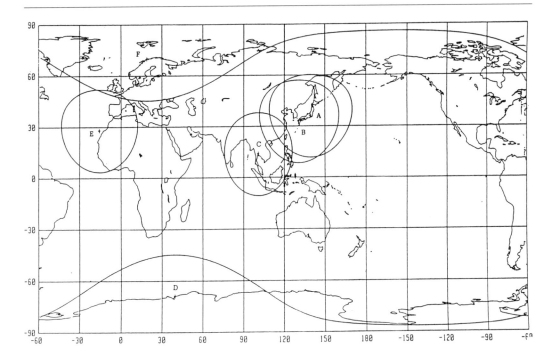

Figure II.19   MOS receiving station mask coverage. A: Hatoyama; B: Kumamoto; C: Bangkok; D: Syowa; E: Maspolamos; and F: Tromsø.

i) CCT products: unit of logical volume 90 × 100km, 4 bands. BSQ/BIL format. 6250/1600bpi, 9-track.

ii) Photographic products: unit of recording 90 × 100km. 70 or 240mm black and white or colour.

iii) From April 1 1988, MESSR data have been available from RESTEC on floppy disc – size 5.25 × 8 inch, BSQ/BIL format, 512 pixels × 400 lines × 4 bands.

VTIR:

available on Polar Stereographic (PS), Lambert Conformal Conic (LCC) or Mercator map projections. NN, BL and CC resampling methods.

i) CCT products: unit of logical volume 5,000 × 1,500km, 4 bands. BSQ/BIL format. 6250/1600bpi, 9-track.

ii) Photographic products: unit of recording 5,000 × 1,500km. 240mm black and white.

iii) From April 1 1988, VTIR data have been available from RESTEC on floppy disc – size 5.25 × 8 inch, BSQ and BIL format, 1 path × 4 bands.

MSR:

available on PS, LCC or Mercator map projections. NN, BL and CC resampling methods.

i) CCT products: unit of logical volume 5,000 × 317km, 2 frequencies × 2 integration time. BSQ/BIL format. 1600bpi, 9-tracks.

ii) Photographic products: unit of recording 5000 – 317km. 70mm black and white.

iii) From April 1 1988, MSR data have been available from RESTEC on floppy disc – size 8 × 8 inch, BSQ format, 1 path × 4 bands.

DCS:

products are CCT generated by 1.7GHz receiver and line printer output of the DCP data processing program, which contains DCP ID number, DCP data and position and velocity of the DCP. ESA/EARTHNET offer a number of basic products from MOS-1 sensors on 1600/6250bpi CCTs. MESSR data either in raw or in system-corrected format (similar to Landsat MSS and TM). VTIR data products are similar to existing EARTHNET AVHRR products. The raw data are complemented with radiometric and geometric correction parameters, although no actual correction is applied. One VTIR scene covers an area of 1,500 × 1,574km, or 4 minutes of data. The MSR data, available on full pass only, are in raw data format. Data from more than one sensor can be provided on 1 CCT. MOS-2 SAR, SCATT, ALT and OCTS data formats to be determined.

Data availability and archiving dates

Data receiving facilities:
   i) The Earth Observation Centre at Hatoyama, covering the Sea of Okhotsk and E Soviet Union. Data are also received by Kumamoto Station (since May 1987).
  ii) The Japanese Antarctic station at Syowa (at 69°S, 39.35°W), which has received data since February 1989.
 iii) Kiruna in N Sweden.
  iv) Trømso in N Norway (MSR data) – also an archive.
   v) Gatineau, Quebec (Data Acquisition Division, Canada Centre for Remote Sensing, 2464 Sheffield Road, Ottawa, Ontario K1A 0Y7, Canada, telephone [613]–952–0500, telex 053 3777). CCRS, under an agreement with NASDA, tracked 169 orbits of MOS-1, 1 May-31 December 1988. All three MOS-1 sensors were recorded, beginning on 1 June 1988. The present arrangement will be extended. It is proposed that the Prince Albert Satellite Station be upgraded to receive MOS-1 data. This would allow more complete coverage of the North American Arctic, and could serve as an alternative to Landsat, the future of which is uncertain. These data are not currently available to the wider scientific community.

Data are processed by: Data Processing Department, Meteorological Satellite Center, 3–235 Nakaiyoto Kiyose, Tokyo 180, Japan and the Earth Observation Centre, NASDA, 1401 Ohashi, Hatoyama-machi, Hiki-gun, Saitama-ken 350–03, Japan. Data products are distributed through the Data Service Department, Remote Sensing Technology Center (RESTEC), Uni-Roppongi Building, 7–15–17, Roppongi, Minati-Ku, Tokyo 106, Japan, telephone 03 403 1761, FAX (81)–3–403–1766, telex 2426780 RESTEC J.

ESA/EARTHNET access to data was granted by a memorandum of understanding signed between NASDA and ESA in 1986, and data have been received by ESA stations since July 1987 – MESSR and VTIR coverage of a substantial part of the European and Soviet high Arctic. EARTHNET User Services, ESRIN, Via Galileo Galilei, 00044 Frascati, Italy, telephone (39)–6–9426285, telex 610637 ESRIN I, or from EARTHNET National Points of Contact (NPC). The UK NPC is Mr M. Hammond, NRSC, RAE, Hampshire GU14 6TD, telephone (0252)–24461, telex 858442. No processing is carried out.

References and important further reading

Cho, K., Takeda, K., Maeda, K. and Wakabayashi, H. 1989. A study of sea ice monitoring using MOS-1 MSR. *Proceedings of IGARSS '89*, Vancouver BC, 10–14 July 1989: 991–4.

ESA. 1987. Investigation of European remote sensing user needs in accessing MOS-1 payload. *ESA Final Report under Contract 6630/86/HGE-1*. ESA Programme Office, ESRIN, Via Galileo Galilei, 00044 Frascati, Italy.

Koizumi, S. (ed.). 1987. *MOS-1 Data User's Handbook*. Earth Observation Centre, NASDA, Tokyo.

Lukowski, T. 1988. Satellite data acquisition planning at the Canada Centre for Remote Sensing. *Proceedings of IGARSS '88*, Edinburgh, Scotland, 13–16 September 1988. *ESA SP-284 (IEEE 88CH2497-6)*, Volume 3: 1471–4.

Okamoto, K. and Awaka, J. 1987. Statistic distribution of microwave brightness temperature of sea ice in the MOS-1 airborne verification experiment. In Matsuda, T. (ed.), *Proceedings of the NIPR Symposium on Polar Meteorology and Glaciology* 1: 154. National Institute of Polar Research, Japan.

Shibata, A. and others. 1989. Application of MOS-1 MSR data in meteorology and sea ice research. *Proceedings of the MOS-1 International Symposium: 79–89*. 2 March 1989, Tokyo.

Tsuchiya, K., Arai, K. and Igarashi, T. 1987. Marine Observation Satellite. *Remote Sensing Reviews* 3: 59–101.

25. Indian Remote Sensing Satellite-1A (IRS-1A)

Operational lifespan and orbital characteristics

Dates: 17 March 1988 to present.
Orbital height: 904km approx (reference orbit).
Inclination: 99.01°.
Period: 103.2 minutes. Exactly repeating orbit with repeat period 22 days.
Notes: An operational, resource evaluation satellite, launched by the Indian Space Research Organisation (ISRO). Sun-synchronous orbit. Equatorial crossing time 10:25 local. Design lifetime 3 years. The first of a 3-satellite series.

Sensor characteristics

Two solid state pushbroom CCD imaging systems comprising 3 Linear Imaging Self-Scanning (LISS) sensors:

i) LISS-I: 4 bands (all VIS): 0.45–0.52μm; 0.52–0.59μm; 0.62–0.68μm; and 0.77–0.86μm. GR 72.5m, and swath width 148km, for all bands. Each LISS-I scene corresponds to 4 LISS-II scenes. Data rate 5.2Mbps.

ii) LISS-IIA and B (identical sensors): identical bands to those of LISS-I. GR 36.25m, swath width 74km × 2 (A and B combined), for all bands. Data rate 2 × 10.4Mbps.

iii) The following specifications hold for all three sensors: 2,048 CCD elements per band; quantisation 128 grey levels; band-to-band registration ±0.25 pixels; NEP better than 1%.

iv) More spectral bands with a higher spatial resolution (15–20m) are being considered for IRS-1B. It is likely that IR bands will be included in sensors on future IRS-2 series satellites.

Data can be recorded on board for subsequent dumping at Shadnagar. Future satellites, launched after 1993, may carry improved sensors with TIR and microwave capabilities.

Polar applications

The monitoring of snow and glacier ice extent in the Himalayas. Polar coverage theoretically possible up to 81°. The LISS sensors are designed to be compatible with and complementary to the Landsat MSS/TM and the SPOT HRV in terms of spectral and spatial resolutions.

Limitations

Data not widely collected from polar regions.

Data format

Digital and photographic products. 7-bit data (128 grey levels). 70mm quick-look film products to level 4 Special Products (digital data processed through user-specified digital enhancements).

Data availability

Data products are disseminated through the Space Application Centre, Ahmedabad, India. Data reception and processing and distribution of standard and special products are carried out at the National Remote Sensing Agency (NRSA), Balanagar, Hyderabad 500037, Andhra Pradesh, India, telephone 262572 (ext 67), telex 0155–522. The ground station at Shadnagar comprises a Data Acquisition System, a real-time quick-look system and communications links. The IRS Data Products System (DPS) provides for the operational generation of data products and their dissemination to users. Catalogues of the data are generated according to the IRS-1A image reference system. The DPS has the following functions:

  i) payload data reception, quick-look display and filming (orbit, cloud cover, attitude, etc), ancillary tape generation and routine performance monitoring;
 ii) precision and special products generation and data quality evaluation;
iii) browse and standard product generation, archival and information management; and
iv) routine evaluation of data quality.

The Indian Government's Department of Space has published a *Data User's Handbook* for the Indian Space Programme (1986). Certain data are available from the US NSSDC (see TIROS I–X for full address).

References and important further reading

Joshi, K., Rathore, N. and Bhatt, V. 1987. Indian Remote Sensing Satellite (IRS-1): a step ahead, Indian Space Programme. *Proceedings Pecora XI Symposium*: 243–52.

26. European Remote Sensing Satellite-1 (ERS-1)

Operational lifespan and orbital characteristics

Dates: Launch May 1991. Life expectancy 3 years.

Orbital height: 780km approx (reference orbit).

Inclination: 98.52° (reference orbit).

Period: 100.5 minutes (reference orbit, repeat period 3 days). See below.

Notes: All orbits sun-synchronous and exactly repeating. The following orbits are planned in an attempt to satisfy the requirements of all users:

  i) Phase 0 – orbit acquisition: initial switch on and testing of the satellite, lasting about 2 weeks.

 ii) Phase 1 – commissioning phase: a 3-day repeat cycle, nominal duration 3 months;

iii) Phase 2 – first multi-disciplinary phase: repeat cycle 35 days, mean local solar time at ascending node 10:30 a.m.

 iv) Phase 3 – first ice phase: from January to March 1992, 3-day repeat cycle, with a longitudinal phase optimised for the specific requirements of planned Arctic sea ice experiments.

  v) Phase 4 – long multi-disciplinary phase: from April 1992 to December 1993, with same characteristics as Phase 2.

 vi) Phase 5 – second ice phase: from January to March 1994, with identical orbital characteristics to Phase 3.

vii) Phase 6 – long geodetic phase: 176-day repeat cycle, from April 1994 to the end of the mission.

ESA has produced an orbit simulation programme for use on a VAX or an IBM-compatible PC; this can be used to plan precise sensor coverage of areas of interest.

ERS-2 (also called the European Environment Monitoring Satellite or EEMS) has been approved for a 1994 launch, and will include a new Global Ozone Monitoring Experiment (GOME); may include a pushbroom, CCD ocean colour sensor. Plans exist for the launch of ERS-3.

Sensor characteristics

Along-Track Scanning Radiometer and Microwave Sounder (ATSR-M), composed of 2 parts:

  i) an IR Radiometer:

   a) Band 1: NrIR (wavelength 1.6μm). GR 1km (for all bands). Swath width 500km (for all bands). Data rate 110Kbps (all bands combined).

   b) Band 2: TIR (wavelength 3.7μm).

   c) Band 3: TIR (wavelength 11.0μm).

d) Band 4: (wavelength 12.0μm).
Bands 1–3 scan the surface at 0° and 55°. SST absolute accuracy 0.5K over 50 ×
50km in 80% cloud cover. Radiometric resolution 0.1K.
ii) Microwave Sounder:
   a) Channel 1: PMW frequency 23.8GHz. GR 22km (both channels), nadir-pointing.
   b) Channel 2: PMW frequency 36.5GHz.

Radar Altimeter (ALT):
frequency 13.8GHz (K$_u$-band). Pulse-limited footprint diameter 16–20km (dependent on
surface). Data rate about 10Kbps. Bandwidth 330MHz in ocean mode, 82.5MHz in ice
mode. Antenna beamwidth 1.3°. Pulse width 5000ns (uncompressed), 1000ns (compressed). Backscatter coefficient measurement ±0.7dB rms. Altitude precision <10cm
(goal <5cm). SWH accuracy ±0.5m or 10% (whichever is greater) in the range of 1–20m.
Wind speed (in range of 4–24ms^{-1}) accuracy ±2ms^{-1} or 10%, whichever is greater.
Tracking window (ocean mode) 64 gates × 3ns each; ice mode 64 gates × 12ns each.
Onboard processing of return echoes to give time delay, waveform slope and power level
every 0.05s; these data then averaged to yield 1s mean values. Pulse repetition frequency
1020Hz.

Active Microwave Instrument – Synthetic Aperture Radar (AMI-SAR) Imaging Mode:
frequency 5.3GHz ± 0.05MHz (C-band), wavelength 5.7cm, VV polarisation. GR 30m (4
looks). Swath 80–100km. Data rate 105Mbps. Incidence angle 23° mid-swath. Swath
located 250km to the right of orbital track. RF bandwidth 13.5MHz. FM chirp bandwidth 15.5MHz. PRF (nominal) 1700Hz. Pulse length 37.1μs, 64ns (uncompressed).
Radiometric resolution 2.5dB. Duty cycle about 10%, with a minimum switch-on time of
2 minutes. No onboard storage, therefore only operational within receiving station masks.

AMI-SAR Wave Mode:
Wave Scatterometer: frequency 5.3GHz (C-band), wavelength 5.7cm, VV polarisation.
FM chirp bandwidth 15.5MHz. PRF 1700Hz. Pulse length 12.3μs. Radiometric resolution 2.5dB. Incidence angle 23°, side-looking. GR 30m. Sample width 5 × 5km, every
200/300km along-track and programmable anywhere within the SAR swath. Data rate
800Kbps. Wave direction accuracy ±15° (in range of 0°–180°, with a 180° ambiguity).
Wavelength (in range of 100–1,000m) measured to an accuracy of ±20% in direction and
±25% in wavelength. Wave Mode data recorded on-board, global coverage.

AMI Wind Scatterometer (SCATT) Mode:
Frequency 5.3GHz (C-band), wavelength 5.7cm, VV polarisation. 3 beams, incidence
angle range 25–59° fore/aft, 18–47° mid beam 45°. PRF 205Hz (fore and aft), 230Hz
(mid). Pulse length 70μs (mid) and 130μs (fore and aft). Radiometric resolution 0.35dB.
GR 50km. Sample spacing (pixel size) 25km. Swath width (one sided) 500km,
200–700km to the side of orbital track, telemetered (400km within specification). Data
rate 1Kbps. Wind speed accuracy (in the range of 4–24ms^{-1}) 2ms^{-1} or 10%, whichever is
greater. Wind direction accuracy ±20% (within the range 0–360°). Data from an entire
orbit can be stored on board.

The payload includes a Precise Range and Range-Rate Experiment (PRARE), a dual frequency (2.2 and 8.5GHz) transponder system, and a Laser Retro-Reflector (LRR) (Figure
II.20). In conjunction with ground stations, these experiments allow accurate orbit ephemeris determination, and as such enhance the ALT data.

The sensors can operate in the following combinations:
 i) AMI in SAR Imaging Mode, alone or with the ALT;
 ii) AMI in Wave and Scatterometer interleaved mode, with the ALT;

iii) ATSR-M – can function continuously (as can the ALT).

Polar applications

ATSR-M:

an IR radiometer system designed for SST and cloud top temperature measurement, combined with 2 nadir-looking passive microwave channels to provide water vapour corrections for SST and ALT data (range correction of latter to within 4cm if sea state known). Cloud statistics a by-product. Good potential for ice and oceanographic studies. Ice and snow 'surface' radiances and temperatures from TIR data during clear conditions/darkness to better than 0.5K, spatial resolution 1km. Near-complete polar coverage. Mapping of thin sea ice, leads and polynyas (for heat flux computation). Boundaries between new and MY ice are more reliably identified in TIR than those between open water and thin ice. Band 1 allows good discrimination between snow and cloud, as snow appears dark and cloud bright at this waveband. See also NOAA AVHRR and Landsat TM.

The 36.5GHz microwave channel is similar to the Nimbus-5 and -6 ESMR, channel 5 of the Nimbus-7 SMMR and channel 3 of the DMSP SSM/I, and should be well suited to (although not optimised for) land and sea ice diagnostics. As such, it is useful for measuring ocean/ice boundaries and possibly for distinguishing FY and MY ice, albeit at nadir. Supplementary data to SSM/I. The footprint size (22km) is comparable to that of ALT, and ATSR-M may complement ALT sea ice products by providing directly comparable data on ice type, except when the ALT receives near specular reflections.

ALT:

SWH and wind speed (but not direction) at nadir. Atmospheric corrections from the microwave sounder of ATSR-M. Has an 'ice mode' of operation to help compensate for greater mean slope/rougher topography over ice, and to maintain tracker lock; incorporates a ×4 increase in range window width, with individual bins corresponding to 1.5m of range (0.4m in standard oceans mode). All-season/all-weather, day and night capability.

i) *Sea ice*:

likely continual operation over oceans. Potential applications:

a) detection of ice edge position at each crossing;
b) ice surface roughness data (drag coefficients);
c) ice concentration and compactness in MIZs;
d) detection and quantification of lead/polynya distribution;
e) wind and wave field information outside the ice edge, and swell penetration, attenuation and propagation into the ice;
f) estimates of sea ice thickness distributions (averaged over large areas), derived from measurements of the ice freeboard. Geoid measurement in sea ice cover;
g) recent research into the separation of sea ice types based upon the normal-incidence backscatter coefficient is showing great promise (Ulander 1988).
    Note: many of these are still research topics.

ii) Potential applications over *land ice*:

a) changes in ice sheet surface topography/elevation, ie ablation or accumulation;
b) accurate mapping of ice sheet/ice shelf margins;
c) ice sheet stability/mass balance/dynamics;
d) delineation of ice streams, drainage basins and outlet glaciers;

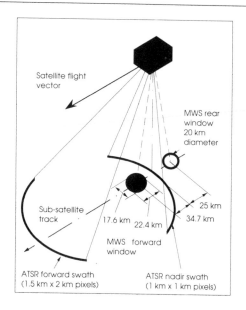

Figure II.20 (i)   ATSR/M viewing geometry. Courtesy of Chris Rapley, MSSL and UK Earth Observation Data Centre.

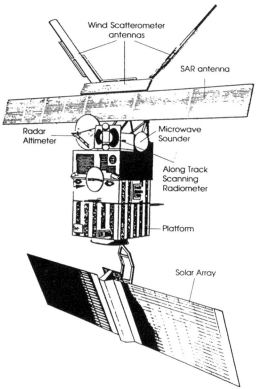

Figure II.20 (ii)   ERS-1 satellite and instrument configuration. The multi-mission platform is based on the SPOT satellite design.

e) identification of equilibrium line (boundary between accumulation and ablation zones) through the associated change in return signal;

f) the delineation of dry and wet snow regions, and bare ice zones on large ice masses from the ALT wave-form characteristics;

g) detection of large tubular icebergs, and measurement of their freeboard and therefore thickness.

See also Skylab, GEOS-3, Seasat, GEOSAT and TOPEX/Poseidon.

AMI-SAR Image Mode:

all-season/all-weather, day and night capability.

i) *Sea ice*: routine mesoscale ice velocity, motion and deformation rates in areas of special interest. Polynya/lead dynamics. Information on the correlation between deformation, wave-ice interaction and acoustic noise generation. Data on relationships between the ice strain and wind/ocean current fields. Complementary information for wider swath, poorer resolution sensors (eg AVHRR). Separation of snow-free from snow-covered areas, and wet and dry snow covers. Ice type discrimination, based on tonal and textural differences. Pressure ridge detection and delineation. Sea ice surface roughness characteristics, ie region with different drag coefficients. Full digital analysis required to obtain gravity wave spectral information and backscatter statistics for ice type discrimination and automated ice concentration algorithms. Penetration depth at C-band less than that at longer wavelengths, and backscatter determined both by volume scattering and surface roughness. Airborne C-band data suggest that this frequency should discriminate ice types (FY and MY ice) better than L-band (Shuchman and others 1988), but not as well as X-band (Figure 4.31). Ice edge detail, and better estimates of sea ice concentration and floe size distribution in areas of low concentration. Detection of thin/new ice types. Detailed data for heat, momentum and salt flux equations.

ii) *Land ice/snow*: detailed, accurate mapping and monitoring of coastal margins/ice shelves, glacier termini, crevasse fields, ice motion and discharge rates. Tonal and textural variations allow the detection of lakes, streams, residual snow patches and topographic (orientation) effects, snow lines and ablation zones, and the position of the equilibrium line. Detection of icebergs and measurement of their velocities. Data in support of the ALT mission over ice sheets. Terrestrial snow cover extent and state.

iii) *Oceanography*: detection and evolution of ice edge jets, eddy features and internal waves. Wave spectra at the ice edge (reflection and refraction) and within the pack (attenuation). Current boundaries, frontal boundaries, internal waves, cold water regions and bathymetric regions. Atmospheric conditions and features, indirectly detected by characteristics variations in the wind field associated with wind-generated short gravity waves. Coverage between latitudes 85°N and 78°S.

See also Seasat, SIR-B, SIR-C/X-SAR, JERS-1 and Radarsat.

AMI-SAR Wave Mode:

i) wave and swell refraction/reflection near ice edges. If 'true' directional wave spectra are available for the open ocean adjacent to the sea ice boundary, it may be possible to calculate the direction and extent to which wave energy is propogated through the sea ice cover using these data alone. The occurrence and energy distribution of internal waves.

ii) The statistical nature of wave fields (wavelengths and directions), eg in MIZs, in relation to the effect of storms on ice edge deformation and floe size distribution.

AMI-Wind SCATT Mode:

primarily to give surface wind vectors over open ocean.

i) gridded winds for use in large-scale ocean-atmosphere models.

ii) Winds near ice edge, suitable for driving regional and mesoscale MIZ models.
iii) Potential use in conjunction with microwave radiometers for snowcover mapping, particularly during melt periods. The higher sensitivity and the larger incidence angle of the wind SCATT as compared to the SAR in imaging mode are advantages for snow detection and mapping. This is still a research topic.

See also Skylab, Seasat SASS and ADEOS NSCAT.

Limitations

No coverage poleward of about latitude 85° (in the Arctic).
ATSR-M:

limited by 80% cloud cover. At TIR wavelengths, floes may appear as open water during melt periods and when the snow/ice temperature is close to that of the surrounding water or extensive meltponding, and it is difficult to distinguish thick FY from MY ice. Differences of 1–3K between surface and air temperatures common in polar regions, and temperature inversions of up to 20K can be encountered. Further research necessary to investigate the relationship between surface and air temperature; both are needed for the modelling of radiance effects, and there appears to be no prospect of measuring boundary layer air temperatures from space. Uncertainties in surface temperature retrievals due to emissivity variations and cloud. Microwave channels limited by poor spatial resolution and narrow swath. Channel 1 affected by atmospheric water vapour (which it is designed to measure). The use of the 36.5GHz channel alone to derive sea ice concentration and type is prone to ambiguities due to emissivity variability. See also NOAA AVHRR, Landsat TM, Nimbus-5 and -6 ESMR, Nimbus-7 SMMR and DMSP SSM/I.

ALT:

the beam-limited footprint size limits the scale of surface features that can be extracted from the data. Being a nadir-viewing instrument, any simultaneous intercomparisons with other microwave instruments, apart from ATSR-M, are restricted. Difficult to extract ice concentration values from the data due to the non-Gaussian nature of the sea ice surface and spatial inhomogeneity. The delineation of sea ice edge boundaries difficult in diffuse MIZs; this is still a research topic. Ambiguities remain in understanding the mechanisms influencing the scattering from ice mass surfaces, and how these affect mean ALT wave-form characteristics. Changes in penetration depth with season (ie freeze-up versus ablation), and their effect on the accurate monitoring of ice sheet elevation, are largely unknown (although modelling studies are taking place). See also Skylab, GEOS-3, Seasat, GEOSAT and TOPEX/Poseidon.

AMI-SAR Image Mode:

coverage only within receiving station masks. Duty cycle limited to about ten minutes per orbit (maximum). Antarctic coverage limited to the masks of a German receiving station at the Chilean base Bernardo O'Higgins, and the Japanese base Syowa (although Australia is looking into the possibility of building a dedicated receiving station at Casey, Davis or Hobart – even if approved, these will not be ready for the launch of ERS-1). Due to power restrictions, SAR reception at O'Higgins may be restricted during winter. The German SAR data may be recorded on a system that cannot easily be read by other systems (eg the Alaska SAR Facility or ASF). Complexity and expense of data processing.

C-band SAR is reasonable for ice type discrimination, but old pressure ridges often do not show up well. At the VV polarisation, the contrast between FY and MY sea ice is not as great as at HH (see Radarsat) (M. Drinkwater personal communication). Ambiguities in interpretation arise due to melt and freeze-thaw effects, depth hoar and saline frost flowers. Brash ice, by presenting a rough target relative to the wavelength of the impinging radiation, gives a bright return similar to that of MY ice (although it can be distinguished by shape). Steep topography can cause image distortion and shadowing (elevation and slant corrections must be applied). The 23° inclination may be too steep for optimal sea ice classification (Shuchman and others 1988) and the detection of pressure ridges (Askne and Johansson 1988); at this angle and under freezing conditions, the return signal will tend to be dominated by surface rather than volume scattering.

SAR is a process-oriented sensor, and does not easily offer input into large-scale basin-wide meterological and oceanographic problems. Narrow swath may be too small to contain the same features on a 3-day repeat interval. Issues remain regarding the separation of new/thin sea ice, smooth water and wind-roughened water microwave signatures. Ice drift rates in MIZs tend to be too great to permit high density feature tracking using SAR imagery; floe rotations of >15° between passes cause problems.

External calibration is essential. Few direct calibrated SAR (or SCATT) data have been collected over polar ice and snow. More research is necessary before tonal variations on SAR imagery can be interpreted glaciologically. Cannot operate simultaneously with SCATT (only alone or with ALT); duty cycle tradeoffs in key areas must be addressed to maximise the research potential.

See also Seasat SAR, SIR-B and SIR-C/X-SAR, JERS-1 and Radarsat.

AMI-SAR Wave Scatt Mode:

operates in a sampling mode only. Cannot operate at the same time as the SAR in imaging mode. No measurement over ice (although some data may be collected in MIZs). The interpretation of SAR imagery is complicated by the fact that modulation of the backscattered return of short ripples by longer waves depends on local sea state and wind speed. The imaging mechanism can become strongly nonlinear for wind/sea spectra at weak to moderate windspeeds. For these reasons, the usefulness of SAR wave data is strongly dependent on the incorporation of algorithms in a general data assimilation model. Little experience to date in the assimilation of wave data in wave models. SAR image spectra can suffer from ambiguities in wave propagation and amplitude calibration. The objective of this mode is to measure ocean wave spectra, but specifications have been expressed in terms of image spectra. It is clear that peaks in 2-D spectra correspond to ocean wave and swell patterns. But the modulation transfer function between ocean wave and image spectra is not well developed, so ocean characteristics cannot make a sound foundation for the instrument specifications. Some doubt as to whether wave trains in certain directions or under certain wind conditions will appear at all on image spectra.

AMI-Wind SCATT Mode:

cannot operate at same time as SAR in imaging mode. No measurement over ice (although some data may be collected in MIZs). Dependence of C-band normalised radar cross section on wind speed and direction not well understood. Sea state and atmospheric boundary layer stability may have a strong influence on the backscatter cross section. Atmospheric water attenuates microwave radiation and may cause biases in derived winds – but smaller effect at C- than K_u-band, eg Seasat SASS and ADEOS NSCAT, although this advantage may be obviated by a reduced sensitivity of backscatter variation to wind speed. At lower latitudes, rainfall is correlated with windspeed, and can cause an under-sampling of high winds. Further ambiguity from the effect of rain roughening the sea surface. A correction for atmospheric attenuation may be possible using ATSR-M or

DMSP SSM/I data, but these are unreliable in areas of high precipitation rates or where rainfall is patchy.

Data format

An important mission goal is to provide operational Fast Delivery (FD) Products, with a <3 hour turnaround time after reception. Note: FD products are not quick-look (QL), but are geophysical products based upon proper instrument calibration.

Level 0: raw data. Level 1: engineering corrections applied. Level 1.5: engineering corrections, demulti-plexed, etc to give basic geophysical parameters. Level 2: precision-processed to obtain specific geophysical products.

ESA products:

i)   FD products:
  a) AMI-SAR Image Mode: full swath, 100 × 100km image, resolution <40m, 3 looks, turnround time 2–3 hours, location accuracy ±2km, no absolute calibration, 20 interpixel distance in ground range, 16m in azimuth direction. 16-bit and 8-bit/pixel data formats. Throughput of 3 high quality images in <3hours. A QL, regional survey product.
  b) AMI-SAR Wave Mode: 2-D power spectra of ocean waves images, both magnitude and direction, polar coordinates, 5 × 5km images. 12 8-bit amplitude levels in logarithmic form corresponding to spatial wavelengths between 100 and 1,000m. 12 angular sectors of 15° between 0° and 180°, with an angular resolution of 30°. Up to 150 spectra per orbit.
  c) AMI-Wind SCATT Mode: wind vectors, 19 × 19 point grid over a 500 × 500km area. Wind speed and direction for all points except those corresponding to resolution cells for which percentage of land is above a given threshold. Up to 70 datasets processed between 2 given passes.
  d) ALT: wind speed over open ocean, SWH, and altitude over ocean. Up to 80 datasets for 500km sections of ground track processed per orbit. Two FD sea ice products:
    i) Quick Dissemination (QD) Sea Ice Margin, a typed list of transitions between sea ice and open ocean, for use when transmitting large volumes of data is a restriction (eg communicating to ships in ice-infested waters); and
    ii) QD Sea Ice Indices product, produced from individual waveform profiles (at a rate of 20Hz). Problems caused by smoothing and instrumental effects eliminated by analysing individual waveforms. 6 categories of waveform shape and power; designed for post-processing. No FD products from ALT land ice data.
  e) ATSR-M: no ESA FD products, but data transcribed onto CCT with ancillary data for archiving and off-line processing. The UK may produce FD ATSR-M products from data received at West Freugh, Scotland.
ii) Off-line products: generated and produced on request in ESA's Processing and Archiving Facilities (PAFs). These include:
  a) AMI-SAR Image Mode: high resolution precision products, all calibrations and geometric corrections applied. Precision digital geocoded products, 80 × 80km, GR <30m, 6–6 looks, 2-week turnaround time, location accuracy ±50m. A regional analysis product. A number of user-requested image transforms available. Sea ice products include identification of leads, accurate geolocation and

geometric correction, and segmentation/classification of the image (into open water, new ice, FY and MY ice, etc.). Note: the most important level-1 product, in terms of generating derived products, is the 'Basic Multi-Look Image', 8-look 100 × 100km coverage, 30m GR (in azimuth and range) for the majority of applications. A complementary product, the 'Basic Full Resolution Image', offers maximum spatial resolution at the expense of speckle (1-look fully processed) for use in feature extraction and cartography. A third level 1 image product, the 'Regenerated Fast Delivery Bulk Image', offers 40m spatial resolution, coarser pixels and the processing of 3 of 8 looks for speckle reduction; matches the specifications of the ESA FD product; together, they are useful for browse purposes.

b) AMI-SAR Wave Mode: information on dominant wavelength, direction and energy. The main features of these products are:
   i) Basic Vignettes (level 1), from the raw data and geolocated using predicted orbit parameters.
   ii) Precision Vignettes (level 1.5), radiometrically corrected using calibration data and trends, geolocated using measured rather than predicted orbit parameters. GR 30m, 8-looks. Atmospheric corrections may be applied.
   iii) Image Spectrum (level 2), a 2-D FFT of the Precision Vignettes presented as wave number power spectra on a cartesian grid, 5 × 5km precision.
   iv) Directional Wave Spectra (level 2), an estimate of the wave spectrum. A developmental product.

c) AMI-Wind SCATT Mode: the off-line products are essentially the same as the FD products, but with directional ambiguities removed.

d) ALT Ocean Mode: all data are level 2. Special processing available upon request (waveform processing). Low- and high-resolution wave and wind speed (but not direction) products.

e) ALT Ice Mode: level 1.5, special processing on request (re-tracking). Atmospheric corrections using ATSR-M data. Level 2 products include:
   i) sea ice products: ice properties (eg surface roughness), precision ice margin location, swell penetration and ice surface elevation; and
   ii) land ice products: a 500km profile of ice sheet precision elevation and backscatter coefficient, Greenland and Antarctica, corrected for atmospheric and geometric effects, either ocean or ice mode. A number of potential ice products are still research topics.

f) ATSR-M: SST field, cloud top temperature, precision SST (500 × 500km, 1km resolution, accuracy 0.5°C, full atmospheric corrections applied) images, cloud classification, water vapour distributions plus corrected IR imagery, and micro-wave broad-scale sea ice boundary/sea ice type classification;

g) Note: algorithms to be used are less well defined than primary product algorithms. Research into new products is outside ESA's responsibility, but investigations, eg into sea ice concentrations and roughness using ALT data, are being conducted.

Under an agreement between ESA and NASA, the ASF provides the following ERS-1 data products (CCT is Computer Compatible Tape, DOD is 5.25 inch Digital Optical Disc, and FILM is 10 × 8 inch film):

i) Standard products:
   a) Computer compatible signal data. 12s segment. CCT, DOD.

    b)   Complex image data. 8m pixel spacing, 30 × 50km area, 10m resolution. CCT, DOD.

    c)   Full resolution images. 12.5m pixel spacing, 30 × 50km area. CCT, DOD, FILM.

    d)   Low resolution images. 100m pixel spacing, 240m resolution, 80 × 100km area. CCT, DOD, FILM.

ii) Geocoded products:

    a)   Geocoded full resolution. 12.5m pixel spacing, 30m resolution, 80 × 100km area. CCT, DOD, FILM.

    b)   Geocoded low resolution. 100m pixel spacing, 240m resolution, 80 × 100km area. CCT, DOD, FILM.

iii) Geophysical products:

    a)   Ice motion vectors. Sea ice displacement vectors, 5km grid, 100 × 100km (nominal). CCT, DOD. Grid aligned to the DMSP SSM/I grid.

    b)   Sea ice type classification. Ice type image, 100m pixels, 100 × 100km (nominal). CCT, DOD.

    c)   Sea ice type fraction. Fraction of ice classes, 5km grid, 100 × 100km (nominal). CCT, DOD.

    d)   Wave product. Wave direction and wavelength, 6 × 6km subsections, from full resolution imagery. CCT, DOD.

The Alaska SAR Processor is designed to process over 200 100 × 100km (SEASAT-like) frames per day from the raw SAR data (or 52 minutes of raw SAR signal data per day), GR 30 × 30m. For further details, see NASA Alaska SAR Facility Prelaunch Science Working Team (1989).

Data availability

ERS-1 delivers 2 types of data:

  i) Low Bit Rate (LBR): 1Mbps (from AMI-Wave and SCATT Modes, ATSR-M and ALT). Global coverage (onboard recording, capacity 100 minutes per orbit). Simultaneous transmission of recorded and real-time data when within range of Kiruna (Sweden), Maspalomas (Canary Islands) and Prince Albert (Canada).

 ii) High Bit Rate (HBR): 105 Mbps (SAR Imaging Mode). Too great for onboard recording. Only operated within line-of-sight of receiving stations: Kiruna (coverage of E Arctic, including Greenland), Fucino in Italy (no polar coverage), Gatineau (coverage of much of the Canadian and Central Arctic), and West Freugh in Scotland (Greenland and Barents Seas), and the ASF, Fairbanks (W Arctic).

    Antarctic data are available from the Chilean base Bernardo O'Higgins, operated by the Alfred Wegener Institute in Bremerhaven. SAR data collected on HDDTs (one tape holds 18 minutes of data), and flown to DFVLR on a weekly basis. Special products are generated here, along the lines of the ASF products. Japan has upgraded its receiving facility at Syowa Base to receive SAR data. Negotiations are underway for building a station at McMurdo base (and possible an Australian base); if approved, may be ready for the launch of ERS-2.

ERS-1 Control and Mission Management Centres are at Darmstadt, FRG. The primary receiving station in Kiruna, which will acquire and process about 70% of all LBR data gathered worldwide. Data are assembled in Processing and Archiving Facilities (PAFs) within the framework of ESA EARTHNET. These can generate raw, geophysical, thematic and special products on request. Off-line facilities include a long-term archive of data (including all LBR data, raw data and processed data), data retrieval and the generation of historical data products, a user's interface and support function, and quality control, etc.

Important by-products of the off-line facility programme include: the development of algorithms for generating and extracting a set of products; contributions to mission and payload validation; and support to pilot projects related to ERS-1.

The major PAFs are:

i) UK Earth Observation Data Centre, at the National Remote Sensing Centre, RAE, Farnborough, GU14 6TD, UK, telephone (0252)–24461, telex 858134 NRSC G. The UK EODC is the primary UK point of contact for distributing FD data products generated by ESA at their Kiruna facility. Raw data tapes delivered by ESA within one week of data collection. Higher level products (ie gridded or thematic) generated at specialist centres supplied by UK EODC. Latter is primary ESA processing, archiving and dissemination facility for SAR image, ATSR-M and ALT ice/land products; also back-up responsibility for processing data from the AMI Wave Mode and archiving Wind SCATT data. Archive includes relevant complementary polar data from other sources, eg PIPOR. *Product Support Team Newsletters* and other reports are available from UK EODC.

The UK is considering the generation of UK-FD products over the North Atlantic area within range of the West Freugh receiving station, eg a faster turnaround ranked wind vector product, an ice margin product, a more sophisticated wind/wave product from the ALT, and SST, cloud and water vapour products from the ATSR-M (from which no ESA-FD products are generated).

ii) France: based on two data centres (CERSAT and AVISO) and ALGOS group, to develop models and algorithms for the data centres. IFREMER in Brest is the PAF.

    a) CERSAT: off-line processing and archiving of LBR ERS-1 data up to geophysical level. Primary archive for ALT (non-ice), Wind and Wave SCATT data (including FD products), and ancillary data. The secondary archive of ATSR-M data.

    b) AVISO: processing and archiving of thematic products (from level 2 upwards), derived from historical satellite data over the oceans, ERS-1 and future missions (eg TOPEX/Poseidon). Enquiries to B. Nutten, Earth Sciences and Applications Division, Centre National d'Etudes Spatiales, 2 Place Maurice Quentin, F-75001 Paris, France.

iii) Germany: plans to process SAR data from Kiruna and from Bernardo O'Higgins base (Antarctica) to standard and precision level at DFVLR, Oberpfaffenhofen. Special products generated along the lines of the ASF products. Japanese data (from Syowa) may become available under international agreement.

iv) Italy: at Matera. SAR and LBR data, but coverage limited to Mediterranean.

v) An EARTHNET Central User Service (CUS) at ESRIN allows customers to scan available archives of data remotely and submit requests and schedules for ERS-1 products and data acquisition. Any enquiries may be addressed to: ESRIN EARTHNET Programme Office, Via Galileo Galilei, 00044, Frascati, Italy, telephone (06)–94011, telex 610637 ESRIN I; or National Points of Contact, eg Mr M. Hammond, National Remote Sensing Centre, Space Department, RAE, Farnborough, Hampshire GU14 6TD, telephone 0252 24461, telex 858442 PE MOD G. For further details, consult: *ERS-1 Ground Stations Products Specification, Volumes I and II*, November 1988, ER-15-EPO-GS-0201 15/RE. Fact sheets and *ESA Bulletins* available from ESRIN.

vi) In addition to the PAFs and CUS, many research groups are involved in processing and validating ERS-1 data. An agreement signed by ESA and NASA allows the acquisition of ERS-1 SAR data at ASF, and allows NASA/NOAA to acquire onboard recorded LBR data at Fairbanks as well as Kiruna and Gatineau. An on-line catalogue and inventory allows users to select interactively and order data products; accessible via the Space Physics Analysis Network (SPAN) and NASA Science Network. Users are issued with a password on the Archive Catalog System computer, which is a menu-driven system; scenes can be ordered while still on-line. For details of data availability from the ASF, contact the Director, Alaska SAR Facility, Geophysical Institute, University of Alaska-Fairbanks, AK 99775–0800, USA.

ERS-1 SAR data ingested into the AES Ice Data Integration and Analysis System (IDIAS) in Ottawa, Canada, along with ice data from other sources; geocoded and available at 50×50km pixel dimensions. Japan receives ERS-1 data at Hatoyama and Kumamoto receiving stations, offering coverage of the Sea of Okhotsk; for relevant addresses, please see JERS-1.

References and important further reading

Anderson, A. and others. 1988. Arctic geodynamics: a satellite altimeter experiment for the European Space agency Earth remote Sensing Satellite. *EOS 69*, 39: 873–81.

Askne, J. and Johansson, R. 1988. Ice ridge observations by means of SAR. *Proceedings of IGARSS '88*, Edinburgh, 13–16 September 1988. *ESA SP-284 (IEEE 88CH2497–6)*, Volume 2: 801–3.

Drinkwater, M. 1987. Radar altimetric studies of polar ice. Unpublished PhD Thesis, University of Cambridge, UK.

ESA. 1989. ERS-1: a new tool for global environmental monitoring in the 1990s. *ESA BR-36*. European Space Agency, Paris.

Gorman, M. and others. 1986. SAR data and SAR image processing over land ice and sea ice. *UK ERS-1 product support team document DC-TN-PST-SA-0002*.

Liu, A., Holt, B. and Vachon, P. 1989. Wave evolution in the marginal ice zone: model predictions and comparisons with on-site and remote data. *Proceedings of IGARSS '89*, Vancouver BC, 10–14 July 1989: 1,520–3.

McNutt, L. and Mullane, T. 1986. An overview of operational SAR data collection and dissemination plans for ERS-1 ice data in Canada. *Proceedings of IGARSS '86*, Zurich. *ESA SP-254*: 147–52.

NASA Alaska SAR Facility Prelaunch Science Working Team. 1989. Science plan for the Alaska SAR Facility Program. Phase 1: data from the first European Remote Sensing Satellite, ERS-1. JPL Publication 89–14. NASA Jet Propulsion Laboratory, Pasadena, California.

Partington, K., Ridley, J., Rapley, C. and Zwally, J. 1989. Observations of the surface properties of the ice sheets by satellite radar altimetry. *Annals of Glaciology*: 267–75.

Prata, A., Cechet, R., Barton, I. and Llewellyn-Jones, D. 1990. The Along Track Scanning Radiometer for ERS-1 – scan geometry and data simulation. *IEEE Transactions of Geoscience and Remote Sensing* 28, 1: 3–13.

Shuchman, R., Sutherland, L., Johannessen, O. and Leavitt, E. 1988. Geophysical information on the winter marginal ice zone obtained from SAR. *Proceedings of IGARSS '88*, Edinburgh, 13–16 September 1988. *ESA SP-284 (IEEE 88CH2497–6)*, 2: 1,111–14.

Ulander, L. 1988. Observations of ice types in satellite altimeter data. *Proceedings of IGARSS '88*, Edinburgh, 13–16 September 1988. *ESA SP-284 (IEEE 88CH2497–6)*, 2: 643–6.

Widmar, P. and Joy, M. 1989. The ERS-1 central user service – a user's perspective. *Proceedings of IGARSS '89*, Vancouver BC, 10–14 July 1989: 1,801–4.

27. TOPEX/Poseidon

Operational lifespan and orbital characteristics

Dates: Scheduled for a launch in mid-1992. Planned as a 3-year mission with a possible 2-year extension.

Orbital height: 1,335km.

Inclination: 66.02°

Notes: A joint US/French oceangraphic research satellite (Figure II.21). 10-day repeat cycle.

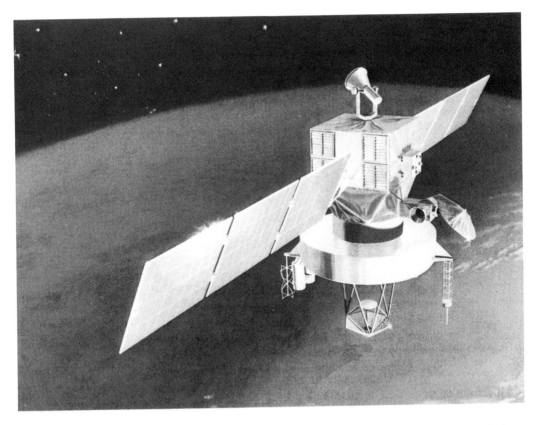

Figure II.21 An artist's conception of the joint US/French TOPEX/Poseidon satellite. From NASA Earth System Sciences Committee (1986).

Sensor characteristics

Altimeter (ALT): Dual frequency active microwave, 2 separate instruments sharing the same antenna:

i) 5.3GHz (C-band), provided by NASA. Bandwidth 320MHz. Antenna beamwidth 1.1–2.8°. Pulsewidth 102.4µs (uncompressed), 3.125ns (compressed). Repetition frequency 4000/1000Hz. Footprint size 20 × 2–10km (depending on nature of surface). Altitude precision <13cm rms. SWH accuracy ±10% in the range of 1–20m. Wind speed accuracy <2ms⁻¹.

ii) The Poseidon ALT, provided by the French CNES: 13.65GHz frequency (K_u-band); bandwidth 320MHz; pulse duration 100µsec; PRF 1,700Hz; altitude precision ±10cm (without correction for ionospheric delay).

TOPEX Microwave Radiometer (TMR):
  i)   Channel 1: 18.00GHz, NET 0.11K, GR 50.86km.
  ii) Channel 2: 21.00GHz, NET 0.11K, GR 39.76km.
  iii) Channel 3: 37.00GHz, NET 0.15K, GR 27.37km.

Tracking Systems:
    Tranet II (ephemeris rms error ±13cm). DORIS (ephemeris rms error ±10–20cm). Global Positioning System (ephemeris rms error ±3-6cm). LASER (data will be used for orbit determination on an as-available basis).

Polar applications

Information on broad-scale ocean circulation, precise sea surface topography and geoid measurement. SWH, surface wind speed, currents and eddies in the open ocean adjacent to the ice edge. Ice edge boundary delineation (up to a maximum latitude of 63°) and coverage of S Greenland Ice Sheet. The radiometer is designed to provide atmospheric correction data for the ALTs, although channels 1 and 3 are similar to Nimbus-7 SMMR and DMSP SSM/I frequencies; as such, they may provide supplementary sea ice data. For further comments of potential applications of ALT data, see Skylab, GEOS-3, Seasat, GEOSAT and ERS-1.

Limitations

No coverage poleward of latitude 63°, which excludes the Arctic Ocean, Greenland Sea and much of the Southern Ocean and the great ice sheets. Both ALTs cannot operate simultaneously. See also Skylab, GEOS-3, Seasat, GEOSAT, and ERS-1.

Data format

CCTs only. Format to be determined.

Data availability and archiving dates

The mission will begin to produce GDRs using verified algorithms approximately six months after launch. Data will be archived at and available from a NODS centre, most probably the Jet Propulsion Laboratory, or NOAA/NESDIS. Both operational and research data. The former will be supplied to US Naval Fleet Numerical Oceanography Center (FNOC) within four hours of reception. Primary research data – height data corrected, and SWHs, wind speed, precise position of spacecraft with geoidal and tidal height at nadir point. Also ionospheric content and columnar water vapour content will be derived from the radiometer. The French archive may be either at the Toulouse Space Laboratory of CNES or at the Centre Océanologique de Bretagne at Brest. A primary sensor of the World Ocean Circulation Experiment.

References and important further reading

Abadie, J., Lamboley, M., Raizonville, P. and Dumont, J. 1987. Poseidon radar altimeter description and signal processing. *Proceedings of IGARSS '87*, Ann Arbor, 18–21 May 1987, Volume 1: 153–60.

Born, G., Stewart, R. and Yamarone, C. 1985. TOPEX – a space-borne ocean observing system. In Chelton, D. (ed.), 1988, WOCE/NASA Altimeter Algorithm Workshop. *US WOCE Technical Report No. 2*. US Planning Office for WOCE, College Station, Texas.

Melbourne, W. 1984. GPS-based tracking system for TOPEX orbit determination. *Proceedings of SPIE, the International Society for Optical Engineering, Recent Advances in Civil Space Remote Sensing, Oceanography*, Arlington, Virginia, 3–4 May 1984, SPIE volume 481: 181–92.

NASA Earth System Sciences Committee, 1986. *Earth system science overview*. NASA, Washington DC.

Stewart, R., Fu, L. and Lebebvre, M. 1986. Science opportunities from the TOPEX/Poseidon Mission. *JPL Publication 86–18*, NASA Jet Propulsion Laboratory, Pasadena, CA 91109, USA.

TOPEX Science Working Group. 1981. Satellite altimetric measurements of the ocean. *JPL Document 400–III*, NASA Jet Propulsion Laboratory, Pasadena, CA 91109, USA.

28. Japanese Earth Resources Satellite-1 (JERS-1)

Operational lifespan and orbital characteristics

Dates: Scheduled launch February 1992. 2-year life expectancy.
Orbital height: 568km.
Inclination: 98.5°.
Notes: Sun-synchronous orbit. Repeat coverage cycle 41 days. Descending node 10:30 LST.

Sensor characteristics

Synthetic Aperture Radar (SAR):
   Active microwave, frequency 1.275GHz (L-band), wavelength 23.5cm. HH (linear) polarisation. SNR 7dB. RF bandwidth 12MHz. Pulse width 35μs. PRF 1550–1690pps. Off-nadir incidence angle 35° to the right of the spacecraft. GR 18 × 30m, 4 looks. Swath width 75km. Data rate 60Mpbs (including VNIR). Onboard storage capacity 10^{11} bits. Coverage between 79°S and 84°N.
Visible and Near Infrared Radiometer (VNIR):

Spatial resolution 18.3m × 24.2m, swath width 150km (all bands).
  i) Band 1: VIS (wavelength 0.52–0.60μm).
 ii) Band 2: VIS (0.63–0.69μm).
iii) Band 3: NrIR (0.76–0.86μm), nadir-looking.
 iv) Band 4: NrIR (0.76–0.86μm), forward-looking. The forward-nadir combination in bands 3 and 4 is for stereo viewing with an angle of 15°. The following bands may also be included: Band 5: IR (1.6–1.71μm); Band 6: IR (2.1–2.12μm); Band 7: IR (2.13–2.25μm); Band 8: IR (2.27–2.4μm).

Polar applications

SAR:
   for applications of L-band SAR, see Seasat and SIR-B. See also Radarsat, SIR-B/X-SAR and ERS-1. Complementary VIS and IR data available (see below).
VNIR:
   a high resolution radiometer, optimised in the IR spectral region. Comparison with SAR data. Its potential use in polar regions is similar to that of SPOT HRV, Landsat MSS and MOS-1 and -1b MESSR. Band 5 allows the automatic discrimination between snow/ice and cloud – snow appears dark at this wavelength, and clouds brighter.

Limitations

Polar coverage limited to a maximum latitude of 84.0°. Limited power availability, onboard recording capacity (20 minutes per orbit) and number of suitably equipped receiving stations.
SAR:
   for the limitations of L-band SAR, see Seasat and SIR-B. The JERS-1 SAR is a fixed-look angle instrument. The repeat period is suboptimal for sea ice dynamics studies in particular. Few direct L-band scatterometer or calibrated SAR measurements have been taken over ice by which scattering theories, models and algorithms can be tested. Doubts exist about the operation of L-band SAR through the ionospheric irregularities which are especially frequent and strongly developed near the auroral ovals. See also Radarsat, SIR-B/X-SAR and ERS-1.
VNIR:
   limited by cloud and darkness. Very narrow swath. The long repeat period is suboptimal for many polar applications, including the study of sea ice floe dynamics. For further discussion of limitations in using these (and similar) spectral bands, see Landsat MSS, SPOT HRV and MOS-1 MESSR.

Data format

SAR data digitised and recorded onboard for subsequent playback to suitable receiving stations. SAR and VNIR data formats to be determined.

Data availability

Tracking will be shared by NASDA stations and US TDRS satellites. Data will be received at the Earth Observation Centre, National Space Development Agency of Japan (NASDA),

1401 Ohashi, Hatoyama-machi, Hiki-gun, Saitama-ken 350–03, Japan. Data will be available from:
  i) The Remote Sensing Centre of Japan (RESTEC), University-Roppongi Building, 7–15–17, Roppongi, Minato Ku, Tokyo 106, Japan, telephone 03 403 1761, FAX (81)–3–403–1766, telex 02426780 RESTEC J; and
 ii) The Earth Resources Satellite Data Analysis Centre (ERSDAC), No. 39 Mori Building, 2–4–5 Azabudai, Minato-Ku, Tokyo 106, Japan.
Direct readout of SAR and optical data will also take place at ASF in Fairbanks, and NASA access to the global JERS-1 dataset allowed under an international agreement. Future ESA/EARTHNET access to data will also be allowed under a memorandum of understanding signed by ESA and NASDA regarding an exchange of data between JERS-1 (and MOS-1 and 1b) and ERS-1. JERS-1 data will be received at Syowa station in Antarctica (and possibly the German station at the Chilean base Bernardo O'Higgins and a US station at McMurdo base). New technical developments are being carried out to ensure full compatability of the receiving stations of JERS-1, ERS-1 and Radarsat. See Radarsat and ERS-1.

References and important further reading

Horikawa, Y. 1985. Studies on the Japanese Earth Resources Satellite 1. In Schnapf, A. (ed.), *Monitoring Earth's ocean, land and atmosphere from space – sensors, systems and applications*, American Institute of Aeronautics and Astronautics, New York: 606–20.
Igarishi, T. 1984. MOS-1 and ERS-1. *Proceedings of the 18th International Conference on Remote Sensing*: 167–86. ERIM, Ann Arbor, Michigan.
Quegan, S. and Lamont, J. 1986. Ionospheric and tropospheric effects on synthetic aperture radar performance. *International Journal of Remote Sensing* 7: 525–39.

29. Sea Wide Field Sensor (SeaWiFS)

Operational lifespan and orbital characteristics

Dates: 1993. Design mission lifetime >5 years.
Orbital height: 705km.
Notes: A dedicated ocean colour mission. Sun-synchronous orbit. Equator crossing time noon LST, ascending.

Sensor characteristics

Sea Wide Field Sensor (SeeWiFS):
  8 spectral bands, all VIS: 0.402–0.422, 0.433–0.453, 0.48–0.5, 0.51–0.53, 0.555–0.575, 0.655–0.675, 0.745–0.785 (notched between 0.76–0.77) and 0.845–0.885μm. GR 1km LAC and 4km GAC at nadir. Swath width 2,800km (58.3° either side of nadir). Radiometric accuracy <5% absolute each band. Relative precision <1% linearity of

signal output to radiance input. Between band precision <5% relative band to band over 50–90% of saturation. Polarisation sensitivity <2% worst case. Dynamic sensitivity 10 bits quantisation; 4 gains individually adjustable onboard. Scan plane tilt +20.0 and −20°. Location accuracy <1km, all bands co-registered to 1/3 pixel.

Polar applications

Similar to Nimbus-7 CZCS. Improved estimates of the mean rate of primary oceanic productivity and its variability. Quantification of the oceanic portion of the global cycle of carbon and other elements and trace gases. Quantification of how the polar oceans affect, and respond to, changes in global biogeochemical and climatic processes. Sensitivity intermediate between those of CZCS and EOS MODIS-T. See also JERS-1 and MOS-2 OCTS.

Limitations

As for Nimbus-7 CZCS. See also JERS-1 and MOS-2 OCTS.

Data format

Data products and format to be determined.

Data availability

Data to be calibrated and validated at NASA Goddard Space Flight Center, Greenbelt, MD 20771, USA. Contact Dr Bob Kirk, Code 970.2, for further details.
The following ground stations may receive SeaWiFS data:
  i) University of Dundee, Scotland: coverage of E Greenland to S Svalbard.
 ii) University of Alaska: Bering and Beaufort Seas, Canadian Arctic Islands up to Ellesmere.
iii) Prince Albert, Canada: much of Hudson Bay and Labrador.
iv) The two NSF AVHRR stations at Palmer and McMurdo Sound bases in Antarctica.

References and important further reading

SeaWiFS Working Group. 1987. *SeaWiFS NASA-EOSAT Joint Working Group Report*. NASA GSFC, Greenbelt, MD 20771.

30. Radarsat

Operational lifespan and orbital characteristics

Dates: Approved for a 1995 launch. Planned 5-year mission (originally extended to 10 years with in-flight servicing by the Space Shuttle).
Orbital height: 792km.
Inclination: 98.6°.
Period: 100.7 minutes. Exactly repeating orbit with repeat period 16 days, subcycle 3 days.
Notes: Sun-synchronous orbit. An ascending node of 18:00 and a descending node of 06:00 LST. A dedicated, operational SAR mission.

Sensor parameters

Synthetic Aperture Radar (SAR):
Active microwave, frequency 5.3GHz (C-band), wavelength 5.7cm. HH polarisation. Dynamic range ⩾30dB. Choice of 3 transmit pulses, bandwidths 11.6, 17.3 and 30.0MHz. Sampling rate 12.9, 18.5 and 32.3MHz. Transmit pulse length 42s. PRF 1,270–1,390Hz. Broadside-looking to the north, with incidence angles 20–59° (although the satellite will be rotated 180° about its yaw axis twice during the mission for a period of 2 weeks). Average data rate 73.9–100Mbps. Operation of the SAR for a maximum of 28 minutes per orbit. Two high speed (10-minute capacity) onboard tape recorders; capacity 10^{11} bits.

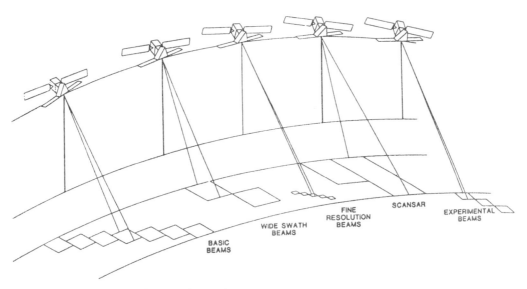

Figure II.22   The Radarsat SAR imaging modes.

Has the following coverage parameters (Figure II.22):

i) standard 'operational' mode. Swath width of ⩾100km. GR 28m × 30m (4-looks). Incidence angles of 20–49°.

ii) 'High resolution' SAR mode. Swath width 55km. GR 8 × 8m (1-look). Incidence angles of 37–49°.

iii) These 'operational' and 'high resolution' swaths can be selected from within an 'accessibility swath' of 500km. Incidence angles of 20–49°.

iv) Two wide swath beams (150km each). GR 28 × 35m (4-looks). Incidence angles of 20–40°.

v) The SCANSAR mode. Offers a choice of 2 very wide swaths: a) 300km wide, GR 100 × 50m, (8-look); and b) 500km wide, GR 100 × 100m (6-looks). Incidence angles 20–49°. This mode is intended to provide survey imagery of the entire 500km accessibility swath at any one time.

vi) Experimental mode. Swath width 75–800km in the range of 500–800km from subsatellite track (incidence angles 49° and 59°). GR of 28 × 30m.

Polar applications

SAR:

all-season, all-weather capability. HH polarisation at C-band offers greater contrast between FY and MY sea ice than VV polarisation (eg ERS-1) (see Figure 4.31). Beam-switching capability very useful for ice reconnaissance. Low incidence angles useful for ice type discrimination, and the far beams for the detection and monitoring of pressure ridges and icebergs. The near beams should distinguish undeformed ice in summer (with water covered or saturated surfaces) from smooth open water. In the far beams, pattern recognition of surface drainage features will be more important in the separation of ice from smooth open water.

The experimental swaths are available for routine use, and imagery from these will be useful for point target detection and ice feature/iceberg location. Experimental swath imagery is however inferior in quality to that obtained from operational swaths. The two wide swaths will be used primarily for synoptic ice reconnaissance. The capability of the satellite to swing the SAR beam from the right side to the left side of the flight path by a yaw manoeuvre of the spacecraft will provide complete coverage of the central Antarctic land mass. This coverage will be obtained once during maximum, and once during minimum, sea ice extent. Wave spectra near ice edge – surface features should be visible in all swaths for wind speeds of ⩾2ms⁻¹, except perhaps in the far region of the experimental swath.

Due to its steering capability, the SAR will be considerably more flexible and suitable for a wider range of applications than the fixed swath systems of Seasat, JERS-1 and ERS-1. The SCANSAR mode will provide the unique ability to cover nearly the entire Arctic, from 70° latitude to the North Pole, every day. The cumulative coverage over a 3-day period will enable total access to anywhere on the globe N of 48°N or N of 48°S. The configuration of orbits is optimised to collect data for the operational support of hydrocarbon exploration, operation and shipping routes, as well as for geophysical research purposes. A dawn–dusk orbit will provide more power for the satellite, hence more data collection will be possible, particularly in the Arctic; the sensor can be turned on at any time. Coverage of the North West Passage will be maximised. Radarsat is designed to be more sensitive to weak signal variations than ERS-1, and should be more reliable in distinguishing smooth open water from FY ice.

This SAR has a number of unique features:

i)   the elevation beam steering and shaping capability;

ii) the very high processor capacity which produces priority imagery at one-quarter real time rates, and potentially achieves a zero backlog of unprocessed data over each day; and

iii) the flexible use of three radar bandwidths to achieve high and nominally uniform ground range resolution across the swath or to match its operation to special requirements.

iv) solar panels will provide all of the satellite's power, as they will continually be facing the sun. This eradicates the need for heavy batteries.

For a more detailed discussion of the polar applications and potential of C-band SAR, see SIR-C/X-SAR and ERS-1 SAR Imaging Mode. See also Seasat SAR, SIR-B and JERS-1 SAR (all L-band).

Limitations

SAR:

effective coverage limited by narrow swath, power constraints, the high data rate, and the cost and complexity of data processing. A maximum on-time of 20 minutes in sunlight and 8 minutes in eclipse may be possible on each orbit. Coverage of Antarctica normally limited by the storage capacity of the onboard recorder and to the masks of receiving stations. When the spacecraft is rotated in its Antarctic mode, both the thermal design and power generating capabilities may limit maximum SAR operation to 15 minutes per orbit – this is adequate to provide coverage of the Antarctic. Only one subswath of data (within the 500km accessibility swath) collected per orbit. For more detailed discussions of the limitations of C-band SAR in polar regions, see SIR-C/X-SAR and ERS-1 AMI SAR Imaging Mode. See also SEASAT SAR, SIR-B and JERS-1.

Data format

SAR:

products expected from Canadian facilities are:

i)   a georeferenced low resolution digital product, full swath width times continuous length. $100 \times 100m$ GR with 64 looks, 1–3 hour turnaround time. An operational regional survey product, used to generate product (ii).

ii) Geocoded (internally) high resolution imagery. Subswath, width squared. $25 \times 28m$ resolution, with a nominal 4 looks, 6–8 hour turnaround time. A regional analysis product, QL, spatially corrected within the image, used for 'nowcast' and forecast.

iii) Geocoded high resolution imagery. Subswath, width squared. $25 \times 28m$ resolution, with a nominal 4 looks, 7-day turnaround time. A regional analysis, research product corrected within the image, and located accurately from satellite data.

iv) Precision geocoded, high resolution imagery. $50 \times 50km$ coverage. $25 \times 25m$ resolution, with a nominal 4 looks, turnaround time 3 weeks, or request data after collection. Specialised, sight specific product, spatially corrected to a UTM base map reference. For research, especially with integrated data sets.

v) Stereoscopic data sets can be made by combining 2 co-located images taken with different incidence angles.

vi) On-line and off-line products, similar to the practices of SPOT and Landsat. Prices to be determined.

Data availability

Radarsat is part of the new Canadian Space Agency (CSA), officially formed on 1 March 1989. Ground receiving stations (Prince Albert and Gatineau) and radar applications remain the responsibility of the Canada Centre for Remote Sensing. The Canadian ERS-1 high-speed processing facility (developed McDonald, Dettwiler and Assoc. Ltd, Richmond, BC) will be upgraded to handle the increased throughput of Radarsat. Satellite control, data acquisition, processing and distribution will be located at Canada Centre for Remote Sensing (CCRS), 2464 Sheffield Rd., Ottawa, Canada K1A 0E4, telephone 613 9939900, telex 0533777. NASA will provide the launch services and will operate the ASF in Fairbanks in exchange for data. The ANIK communication satellites will be used for the high data-rate haul from the ground stations to Ottawa. Emphasis placed upon the real-time throughput of data. The processing speed is projected to be 1/46 real-time rate for the production of either 6-look ERS-1 images or 4-look Radarsat images. For Seasat one-third swath 4-look images, the rate will be slower. A private sector company, Radarsat International (RSI), has been incorporated to distribute the data. NOAA will facilitate the participation of the US private sector in the distribution of data.

The Ice Centre of AES in Ottawa is developing an Ice Data Integration and Analysis System (IDIAS) to synthesise ice information from Radarsat and various other sources, all corrected to one geographical coordinate system (satellite data, aircraft observations, ship-/shore reports and met data). This system is being designed for compatibility with ERS-1 data. For further information, contact Ice Centre, Atmospheric Environment Service, 365 Laurier Avenue West, Ottawa, Ontario K1A 0H3, Canada.

Under international agreements, data will be received not only at Ottawa and the ASF (see ERS-1), but also by the ESA receiving station at Kiruna, Sweden – will enable an intercomparison with near-simultaneous ERS-1 C-band data and JERS-1 L-band data, and extensive coverage of the Arctic. Antarctic data will also be available from a German receiving station at the Chilean base Bernardo O'Higgins. SAR data will be collected at latter on HDDTs (each holding 18 minutes of data), and flown to DFVLR in Germany on a weekly basis. Special products will be generated here, along the lines of the ASF products. Japan has upgraded its receiving facility at Syowa Base to receive SAR data. Australia is also looking into the possibility of constructing a receiving station at either Casey or Davis Bases (or Hobart). Negotiations are underway for building a station at McMurdo base. These stations, if approved, may be ready for the launch.

For further information, contact: Radarsat Project Office, 110 O'Connor Street, Ottawa, Ontario K1A 0Y7, Canada. They publish an *Ice Community Newsletter*.

References and important further reading

Ahmed, S., Gray, R., Warren, H. and Fearn, D. 1989. The new RADARSAT: an all-weather, multi-purpose Earth observation spacecraft. *Space Technology* 9, 3: 267–79.
Freeman, N., and McNutt, L. 1985. Oceans and ice measurements from Canada's RADARSAT. *Marine Technology Society Journal* 20, 2: 87–100.

Hirose, T., Paterson, S. and McNutt, L. 1989. A study of textural and tonal information for classifying sea ice from SAR imagery. *Proceedings of IGARSS '89*, Vancouver BC, 10–14 July 1989: 747–50.

Ramsay, B., Henderson, D. and Carson, L. 1988. Real-time processing of digital image data in support of the Canadian sea ice analysis and prediction programme. *Proceedings of IGARSS '88*, Edinburgh, 13–16 September 1988. *ESA SP-284 (IEEE 88CH2497–6)*, Volume 3: 1,707–11.

Sack, M. and Princz, G. 1986. The ERS-1/RADARSAT SAR Canada ground segment. *Proceedings of the 10th Canadian Symposium on Remote Sensing*, 5–8 May 1986, Edmonton, Alberta, Volume 2: 957–69.

31. Advanced Earth Observation Satellite (ADEOS)

Operational lifespan and orbital characteristics

Dates: 1995
Orbital height: 796.6km
Inclination: 98.6°
Period: 100.92 minutes. 41-day exact repeat (subcycle 4 days).
Notes: An operational mission. Sun-synchronous orbit. Equator crossing time 10:30 LST. NSCAT originally destined for launch on the now defunct US NROSS satellite. Design life 3 years.

Sensor characteristics

Radar Scatterometer (NSCAT):
   Active microwave. Frequency 13.995Ghz (K_u-band). 6 dual-polarisation fan beam antennae. Transmitted pulse length 5.0μs. Receiver noise 4.0dB. Transmission duty cycle 31%. Resolution cell size for wind vectors will be 50km, and the radar backscattering cross-section will be processed at 25km. Swath width 2 × 600km, with a 300km gap centred below the satellite track. Wind speed accuracy of ±1.3ms^{-1} over range of 4–26ms^{-1}. Wind direction accuracy of ±16° over the range of 10–360°. Data rate 5 Kbps. Contributed by NASA JPL.

Advanced Visible and Near-Infrared Radiometer (AVNIR):
   pushbroom scanner. 4 bands, all VIS: centred on 0.45, 0.55, 0.67 and 0.87μm. GR 16m at nadir. Swath width 80km. A pointing mechanism will compensate for the long repeat cycle (due to the narrow swath width).

Ocean Colour and Temperature Scanner (OCTS):
   8 VIS bands, centred on 0.44, 0.49, 0.515, 0.56, 0.62, 0.66, 0.77 and 0.88μm; 1 IR band, centred on 3.7μm; and 3 TIR bands centred on 8.5, 11.0 and 12.0μm. GR 0.7km. Swath width 1,400km.

TOMS:

see Nimbus-7. Contributed by NASA GSFC.

Polar applications

NSCAT:

primarily designed to give near-surface wind speed and direction over open oceans. It will offer a high spatial resolution and global coverage. Accurate data on surface wind stress and the curl of the wind stress: these parameters are important in the study of global ocean circulation. NSCAT is an upgraded design based heavily upon the Seasat SASS, but with important improvements, eg wind directional ambiguity problems will be alleviated by the use of six antennae instead of four, and Earth rotation ambiguities will be alleviated by the use of digital Doppler filtering. A primary sensor for the World Ocean Circulation Experiment (WOCE). See also Seasat SASS and ERS-1 AMI Wind SCATT Mode.

AVNIR:

applications similar to MOS-1 and -1b MESSR and JERS-1 VNIR.

OCTS:

similar to Nimbus-7 CZCS and SeaWiFS. See also MOS-2 OCTS.

TOMS:

primary mission to monitor trends in the ozone holes of Arctic and Antarctic.

Limitations

NSCAT:

see Seasat SASS, ERS-1 (AMI Wind SCATT Mode), Skylab and Radarsat. Comments as for the SASS, except that design improvements will reduce wind direction retrieval ambiguities in comparison to the SASS. Even so, unresolved problems remain in determining wind stress direction and in undersampling of the global wind field with respect to time. Wind (and wave) measurements from a single satellite are seriously subsampled in the cross-track direction because of the large longitude separation between consecutive orbits, although this problem recedes towards the poles.

AVNIR:

narrow swath, long repeat cycle. Operational under daylight conditions only. Cloud affected. See MOS-1 and -1b MESSR and JERS-1 VNIR.

OCTS:

similar to Nimbus-7 CZCS and SeaWiFS. See also MOS-2 OCTS.

TOMS:

not suitable for monitoring surface features.

Data format

Data products and format to be determined.

Data availability

Probable main archive: Remote Sensing Technology Center (RESTEC), Uni-Roppongi Building, 7-15-17, Roppongi, Minati-Ku, Tokyo 106, Japan, telephone 03 403 1761, FAX (81)-3-403-1766, telex 2426780 RESTEC J. Data will also be archived at JPL/NODS, NASA Jet Propulsion Laboratory, Pasadena, California, USA.

References and important further reading

Hara, N., Homma, M., Igarashi, T., Tsuiki, A. and Ohta, K. 1988. Results of ADEOS conceptual study. *Proceedings of IGARSS '88*, Edinburgh, 13–16 September 1988. *ESA SP-284 (IEEE 88CH2497–6)*, Volume 1: 179–82.

NASA. 1985. NSCAT, the NASA Scatterometer. *JPL Fact Sheet*. Jet Propulsion Laboratory, Pasadena, CA 91109.

Index